FORECASTING
ECONOMIC TIME SERIES

This is a volume in
ECONOMIC THEORY, ECONOMETRICS, AND MATHEMATICAL
  ECONOMICS

A Series of Monographs and Textbooks

Consulting Editor: KARL SHELL

A complete list of titles in this series appears at the end of this volume.

FORECASTING ECONOMIC TIME SERIES

C. W. J. GRANGER

Department of Economics
University of California, San Diego
La Jolla, California

PAUL NEWBOLD

Department of Mathematics
University of Nottingham
Nottingham, England

ACADEMIC PRESS, INC.

(Harcourt Brace Jovanovich, Publishers)

Orlando   San Diego   New York   London
Toronto   Montreal   Sydney   Tokyo

ACADEMIC PRESS, INC.
Orlando, Florida 32887

United Kingdom Edition published by
ACADEMIC PRESS, INC. (LONDON) LTD.
24/28 Oval Road, London NW1 7DX

Library of Congress Cataloging in Publication Data

Granger, Clive William John.
    Forecasting economic time series.

    (Economic theory and mathematical economics series)
    Bibliography: p.
    1.   Economic forecasting.   2.   Time-series analysis.
I.   Newbold, Paul, joint author.   II.   Title.
HB3730.G67   1976   338.5'442   76-9156
ISBN 0-12-295150-6

To Alice and Pat

CONTENTS

CHAPTER THREE
BUILDING LINEAR TIME SERIES MODELS

CHAPTER FOUR
THE THEORY OF FORECASTING

CHAPTER FIVE
PRACTICAL METHODS FOR UNIVARIATE TIME SERIES FORECASTING

CHAPTER SIX
FORECASTING FROM REGRESSION MODELS

CHAPTER SEVEN
MULTIPLE SERIES MODELING AND FORECASTING

CHAPTER EIGHT
THE COMBINATION AND EVALUATION OF FORECASTS

CHAPTER NINE
NONLINEARITY, NONSTATIONARITY, AND OTHER TOPICS

PREFACE

The academic literature on forecasting is very extensive, but in reality it is not a single body of literature, being rather two virtually nonoverlapping sets concerned with the theoretical aspects of forecasting and the applied aspects. A typical member of one set is very unlikely to mention any member of the other set. It was this realization that motivated the sequence of research projects that eventually resulted in this book. One of the few exceptions to the above statement about nonoverlapping sets is the well-known book by Box and Jenkins, and our own approach owes a lot to their work. However, we have tried to take the state of the art further by introducing new multivariate techniques, by considering questions such as forecast evaluation, and by examining a wider range of forecasting methods, particularly those which have been applied to economic data, on which this present book concentrates. It is one of our aims to further bridge the gap between the theoretical and applied aspects of forecasting.

The analysis of economic data has also been approached from two different philosophies, that proposed by time series analysts and the more classical econometric approach. Although we favor the former, it is quite clear that both approaches have a great deal to contribute and that they need to be brought together to a much greater extent. In the past couple of years, a number of steps have been taken in this direction, and we have tried to encourage the merger movement in this book by showing how a combined approach to data analysis can lead to potential benefits.

We have many individuals and organizations to warmly thank for their help and encouragement in the preparation of this book and in the research projects that led up to it. Gareth Janacek, John Payne, and Harold Nelson gave considerable help with various aspects of the research, Rick Ashley and Allan Andersen have read and corrected large sections of the manuscript, as have many of our graduate students who had parts of the text inflicted on them for course reading; Alice Newbold prepared all the diagrams for us; Elizabeth Burford and Linda Sykes prepared the final version of the manuscript with great ability and patience; and Robert Young proofread it for us. The Social Science Research Council of the United Kingdom provided the funds to start our research on forecasting in 1970 at the University of Nottingham, and the National Science Foundation of the United States gave us a grant to finally complete the task at the University of California, San Diego. Both universities provided us with excellent facilities, as well as delightful surroundings. Finally, we would like to thank Mike Godfrey, Herman Karreman, and Marc Nerlove for permission to use parts of their own work for illustrations in ours. Of course, we shall have to assume the usual responsibility for those errors that undoubtedly still lurk somewhere in the book.

CHAPTER ONE

INTRODUCTION TO THE THEORY OF TIME SERIES

If we could first know where we are and whither we are tending, we could better judge what to do and how to do it.

A. LINCOLN

1.1 Introducing Time Series

The majority of statistical procedures are designed to be used with data originating from a series of independent experiments or survey interviews. The resulting data, or sample, x_i, $i = 1, \ldots, n$, are taken to be representative of some population. The statistical analysis that follows is largely concerned with making inferences about the properties of the population from the sample. With this type of data, the *order* in which the sample is presented to the statistician is irrelevant. With time series data this is by no means the case. A time series is a sequence of values or readings ordered by a time parameter, such as hourly temperature readings. Since the order of the data is now of considerable importance, most of the classical statistical techniques are no longer relevant and so new techniques have to be devised. Time series are found in many fields, such as economics (e.g., monthly employment figures), sociology (crime figures), meteorology (rainfall, temperature, wind speed), medicine (electrocardiograms and electroencephalograms), vibrating physical systems (such as the ride of a car traveling over a rough surface), seismology, oceanography, and geomorphology. They also occur in astronomy (star brightness, solar activity), as outputs of certain electronic devices, and in industrial processes, for example, the thickness of steel plate from a continuous rolling mill. The methods devised to deal with such data can also frequently be applied to data not gathered through time but ordered along a line, for example, height above sea level along a line of latitude. Although the methods of time series analysis work perfectly well in such situations, interpretation of the results is more difficult when time is not involved.

1

For some series it is possible to take measurements at every moment of time, so that a trace results. Such data, which may be denoted by $x(t)$, are said to form a *continuous* time series. However, most available series, particularly in the social sciences, consist of readings taken at predetermined, equal-interval time points, so that one might get hourly, daily, monthly, or quarterly readings. Such data form a *discrete* time series, denoted by x_t. In this work we shall deal exclusively with discrete equal-interval time series since so much actual data are of this form and because a continuous series can always be well approximated by a discrete series by suitable choice of the sampling interval.

A further classification of series is occasionally needed. A discrete series is said to be *instantaneously recorded* if it *could* have been measured at every moment of time even though it is in fact only recorded at the sampling points, examples being temperature, prices, and interest rates. Some series cannot be measured at every moment of time because they are accumulations of values. Examples of *accumulated* series are rainfall, production figures, and volume of sales. (These two types of series are called respectively stock and flow variables by economists.) For most purposes it is not necessary to distinguish between these two types since the methods and theory to be introduced will usually apply equally well to either, but there is the occasional circumstance where this is not so.

An actual observed series x_t, $t = 1, \ldots, n$, may be considered as a realization of some theoretical process which will be called the stochastic process.[1] In classical statistics one has the vital concepts of population and sample, and the equivalent concepts with time series are the (theoretical) stochastic process and the realization or observed series. The initial objective of time series analysis is to make inferences about the properties or basic features of the stochastic process from the information contained in the observed series. The first step in the analysis is usually to form certain summary statistics, but the eventual aim is to construct a *model* from the data, a model that it is hoped has similar properties to those of the generating mechanism of the stochastic process. A simple example of a stochastic process would be a sequence of random variables generated by the iterative scheme $X_t = 0.5X_{t-1} + \epsilon_t$ where ϵ_t is a sequence of purely independent and identically distributed random variables. The process is seen to be the output of a generating mechanism. Many other examples are possible and two of the essential stages in model building are to determine the class of models that seem appropriate and then to estimate the parameter values of the model.

Once a model has been obtained it can be used either to test some hypothesis or theory about the generating mechanism of the process, it can be used to forecast future values of the series, and it may be used to decide on a system to control future values. This last use, although very important, will

---

[1] *Stochastic* simply means *random*. For simplicity, the word *stochastic* may sometimes be dropped in subsequent exposition.

not be discussed in this book. We shall concentrate only on forecasting problems and will introduce those parts of time series theory and analysis that will be needed for the subsequent chapters dealing with forecasting.

In this chapter only univariate time series theory will be discussed. The generalization to multivariate series will be considered in Chapter 7.

1.2 Covariances and Stationarity

Consider a process X_t, defined for all integer values of t. In general, the process will be generated by some scheme involving random inputs, and so X_t will be a random variable for each t and $(X_{t_1}, X_{t_2}, \ldots, X_{t_N})'$ will be an $N \times 1$ vector random variable. To fully characterize such random variables, one needs to specify distribution functions; but, for reasons that will soon become apparent, it will usually be too ambitious to attempt to specify fully or to estimate these functions. Nevertheless, theoretically such distribution functions will exist; and so one can use the usual expectation notation without ambiguity. The mean of X_t will be defined by

$$\mu_t = E[X_t] \tag{1.2.1}$$

and the covariance between X_t and X_s will be

$$\lambda_{t,s} = \text{cov}(X_t, X_s) = E[(X_t - \mu_t)(X_s - \mu_s)] \tag{1.2.2}$$

so that $\lambda_{t,t}$ is the variance of X_t. The *linear* properties of the process can be described in terms of just these quantities. If it is assumed that the process is Gaussian, by which is meant that $(X_{t_1}, X_{t_2}, \ldots, X_{t_N})$ is an N-dimensional normal distribution for every set $t_1, \ldots, t_N$ and every finite integer N, then the values of $\mu_t, \lambda_{t,s}$ will be sufficient for a complete characterization of the distributional properties of the process. If normality is not assumed but if the generating process is taken to be linear, in the sense that X_t is generated by a linear combination of previous X_t's and past and present values of other processes, then once more the major properties of the process are captured in the means and covariances. Throughout most of this work only linear processes will be considered, largely because an adequate and usable theory is available only for such processes. A brief discussion of some nonlinear processes is given in Chapter 9.

It is instructive to ask the question, How would one estimate μ_t? For some processes it is possible to get a number of realizations. An example would be the thickness of steel wire made on a continuous extraction machine. One wire would constitute a single realization, but it is possible for the process to be stopped, the machine to be serviced, and then the process to be started once more. The new wire could be taken to be another realization from the same stochastic process. If the realizations are denoted by x_{jt}, $t = 1, \ldots, n$, $j = 1, 2, \ldots, k$, then a possible estimate for μ_t would be

$$\hat{\mu}_t = \frac{1}{k} \sum_{j=1}^{k} x_{jt} \tag{1.2.3}$$

However, for very many situations, it is not possible to obtain more than a single realization. One cannot, for example, stop the economy, go back to some starting point, and let it go once more to see if a different pattern would emerge. With a single realization, it is clearly quite impossible to estimate with any precision μ_t for every t if the μ_t sequence is allowed to take any set of values. It is even more ridiculous to try to estimate $\lambda_{t,t}$, the variance of X_t, at every t if only a single observed series is available.

To overcome such problems, the time series analyst is forced to adopt some restrictive assumptions about the way in which means and covariances can change over time. A restrictive but usable assumption is that of *stationarity*, which for our purposes may be defined as follows: a process X_t will be said to be stationary if

$$\text{mean of } X_t = \mu, \qquad \text{variance of } X_t = \sigma_x^2 < \infty \qquad (1.2.4)$$

$$\text{covariance } X_t, X_s = \lambda_{t-s}$$

so that $\sigma_x^2 = \lambda_0$, and the notation usually used will be

$$\text{cov}(X_t, X_{t-\tau}) = \lambda_\tau \qquad (1.2.5)$$

Thus, a stationary process will have mean and variance that do not change through time, and the covariance between values of the process at two time points will depend only on the distance between these time points and not on time itself. Essentially, a stationarity assumption is equivalent to saying that the generating mechanism of the process is itself time-invariant, so that neither the form nor the parameter values of the generation procedure change through time. It certainly cannot be claimed that an assumption of stationarity is generally realistic, but it does enable one to formulate some basic theory and it will be possible to relax the assumption somewhat in later sections.

A stronger form of stationarity than that just introduced is often defined. Denote the distribution function of X_{t+j}, $j = 1, 2, \ldots, N$, by $F(X_{t+1}, X_{t+2}, \ldots, X_{t+N})$. Then the stochastic process is said to be stationary in the stronger sense if, for any finite positive integer N, F does not depend on t. It follows immediately that strong stationarity implies the weaker stationarity defined earlier. Further, if the process is Gaussian, the two definitions are equivalent. In practical applications strong stationarity is virtually impossible to test for, and it is usual to work with the weaker form.

A further requirement is that the process be *ergodic*. Since this is a yet more difficult concept, which cannot be adequately explained without the introduction of considerable mathematical apparatus not required elsewhere, we shall provide only a heuristic explanation (for a rigorous account, see Hannan [1970, p. 201]). What is required is that values of the process sufficiently far apart in time are almost uncorrelated, so that by averaging a series through time one is continually adding new and useful information to the average. Thus, the time average

$$\bar{x}_n = \frac{1}{n} \sum_{t=1}^{n} x_t \qquad (1.2.6)$$

is an unbiased and consistent estimate of the population mean μ, so that $\text{var}(\bar{x}_n) \downarrow 0$ as $n \to \infty$ and $E[\bar{x}_n] = \mu$, all n. Similarly, estimates of λ_τ, to be introduced later, will also be consistent. Thus, given stationarity and ergodicity, one can form good estimates of the quantities of immediate interest by averaging through time rather than being forced to depend on the ensemble averages across realizations considered earlier. Unfortunately, it is not possible to test for ergodicity using a single realization, but one would not expect the data to include strictly cyclical components. A necessary condition for ergodicity, but by no means a sufficient one, is that $\lambda_\tau \to 0$ at a sufficiently fast rate as τ increases. Ergodicity will be assumed to hold in all situations considered in later sections.

The covariances λ_τ will be called *autocovariances* and the quantities

$$\rho_\tau = \lambda_\tau / \lambda_0 \qquad (1.2.7)$$

will be called the *autocorrelations* of a process. The sequence ρ_τ, $\tau = 0, 1, \ldots$, indicates the extent to which one value of the process is correlated with previous values and so can, to some extent, be used to measure the length and strength of the "memory" of the process, that is, the extent to which the value taken at time t depends on that at time $t - \tau$. From the definition (1.2.7), one has

$$\rho_0 = 1, \qquad \rho_{-\tau} = \rho_\tau \qquad (1.2.8)$$

The plot of ρ_τ against τ for $\tau = 0, 1, 2, \ldots$ is called the theoretical correlogram and the values comprising this diagram will be the major quantities that will be used to characterize the (linear) properties of the generating mechanism of the process. However, it is by no means easy to look at a theoretical correlogram and immediately decide on these properties. What is needed are some plausible models that provide correlograms of recognizable shapes. The simplest possible model is that of a sequence of independent (actually, uncorrelated since only linear features are being considered) and identically distributed random variables. The notation to be used in this book whenever possible for such a sequence is ϵ_t. For a sequence of this kind, which henceforth will be called *white noise*,[1] the autocorrelation sequence is

$$\rho_0 = 1, \qquad \rho_\tau = 0, \quad \tau \neq 0 \qquad (1.2.9)$$

so the correlogram takes a very specific and easily recognized shape. One would not expect such a simple model to represent many actual series well, although it will be seen later that one of the objectives of model building will be to transform a given process to a white noise process.

1.3 Some Mathematical Tools

Before more sophisticated models are considered, three mathematical concepts need to be introduced: the backward operator, generating functions, and difference equations.

---

[1] This is a useful and easily remembered phrase, whose origin cannot be explained until Chapter 2.

The *backward operator B*, which is frequently employed for notational convenience, is an operator on a time sequence with the property[1] $BX_t = X_{t-1}$. Thus, on reapplication

$$B^k X_t = X_{t-k} \tag{1.3.1}$$

This operator will often be used in polynomial form, so that

$$d_0 X_t + d_1 X_{t-1} + d_2 X_{t-2} + \cdots + d_p X_{t-p}$$

can be summarized as $d(B)X_t$ where

$$d(B) = d_0 + d_1 B + d_2 B^2 + \cdots + d_p B^p \tag{1.3.2}$$

A *generating function* is a compact and convenient way of recording the information held in some sequence. Consider the sequence $a_0, a_1, a_2, \ldots, a_j, \ldots$, which may be of infinite length; the generating function of this sequence is

$$a(z) = \sum_j a_j z^j \tag{1.3.3}$$

For example, if $a_j = (\lambda^j/j!)e^{-\lambda}$, then $a(z) = \exp(\lambda(z-1))$. The function $a(z)$ often can be manipulated in simpler ways than can the whole sequence a_j. The quantity z does not necessarily have any interpretation and should be considered as simply the carrier of the information in the sequence, although on occasions it can be given a specific and useful interpretation. Frequently z will be taken to be $z = e^{i\theta}$ since then one can appeal directly to Fourier theory for the existence of the generating function in many circumstances. In what follows, problems of existence will not be explicitly considered. Generating functions occur frequently in probability theory and in mathematical statistics. For example, suppose X is a discrete random variable taking only nonnegative integer values, with $\text{Prob}(X = j) = p_j$; then $p(z) = \sum p_j z^j$ is called the probability generating function, with the obvious property that $p(1) = 1$. However $p(e^{it})$ is also the characteristic function of X and so expanding it as a power series in t will give the noncentral moments; i.e.,

$$p(e^{it}) = \sum_j \mu_j \frac{(it)^j}{j!} \qquad \text{where} \qquad \mu_j = E[X^j]$$

Similarly $p(e^t)$ is the moment generating function, so that

$$p(e^t) = \sum_j \mu_j \frac{(t)^j}{j!}$$

Generating functions have the following additive property: if a_j, b_j, $j = 0, 1, \ldots$, are two sequences, and if $c_j = a_j + b_j$, then, with obvious notation

$$c(z) = a(z) + b(z) \tag{1.3.4}$$

---

[1] Strictly, the operator lags the whole sequence, so that $B(\ldots, X_{t-1}, X_t, X_{t+1}, \ldots) = (\ldots, X_{t-2}, X_{t-1}, X_t, \ldots)$.

A more useful property is that of *convolution*. Given two sequences a_j, b_j, $j = 0, 1, \ldots$, define another sequence c_j, known as their convolution, by

$$c_j = a_0b_j + a_1b_{j-1} + a_2b_{j-2} + \cdots + a_jb_0 = \sum_{k=0}^{j} a_k b_{j-k} \qquad (1.3.5)$$

then, the generating functions of the three sequences are related by the equation

$$c(z) = a(z)b(z) \qquad (1.3.6)$$

This can be seen by just multiplying out the right-hand side of (1.3.6).

Three particular generating functions will be used frequently:

(i)   given a process X_t, its generating function is

$$X(z) = \sum_{\text{all } t} X_t z^t$$

and may alternatively be called the z-transform of the process;

(ii)   given an observed series x_t, $t = 1, \ldots, n$, its generating function is

$$x(z) = \sum_{t=1}^{n} x_t z^t$$

(iii)   given a sequence of autocovariances λ_τ, the autocovariance generating function will be

$$\lambda(z) = \sum_{\text{all } \tau} \lambda_\tau z^\tau \qquad (1.3.7)$$

with corresponding autocorrelation generating function

$$\rho(z) = \sum_{\text{all } \tau} \rho_\tau z^\tau = \frac{\lambda(z)}{\lambda_0} \qquad (1.3.8)$$

Here "all τ" means that τ runs through all integers from $-\infty$ to ∞.

Consider a stationary process X_t, with $E(X_t) = 0$, and write

$$X_n(z) = \sum_{t=1}^{n} X_t z^t \qquad (1.3.9)$$

Then $X_n(z)X_n(z^{-1}) = \sum_{t, s} X_t X_s z^{t-s}$ so that

$$E[X_n(z)X_n(z^{-1})] = \sum_{\tau = -n}^{n} (n - |\tau|)\lambda_\tau z^\tau$$

It follows that in (1.3.7)

$$\lambda(z) = \lim_{n \to \infty} \frac{1}{n} E[X_n(z)X_n(z^{-1})] \qquad (1.3.10)$$

In Section 2.1 the function $f(\omega) = (2\pi)^{-1}\lambda(e^{-i\omega})$, called the power spectral function, will be discussed and given a specific and useful interpretation.

A *linear difference equation* of order p is an iterative equation of the form

$$X_t = \sum_{j=1}^{p} a_j X_{t-j} + Y_t \qquad (1.3.11)$$

which determines the sequence X_t from the values of the given time sequence Y_t together with some starting up values. Using the backward operator B, the equation may be written in the form

$$a(B)X_t = Y_t \quad \text{where} \quad a(B) = 1 - \sum_{j=1}^{p} a_j B^j \quad (1.3.12)$$

Suppose the process starts at time $t = -N$, so that $Y_t = 0$, $t < -N$, and that starting values $X_t = \overline{X}_t$, $t = -N - j$, $j = 1, \ldots, p$, are provided.

The *general solution* of the difference equation (1.3.11) takes the form

$$X_t = X_{1t} + X_{2t}$$

where X_{1t} is the solution of the homogeneous equation $a(B)X_t = 0$ and X_{2t} is given by

$$X_{2t} = b(B)Y_t,$$

where $b(z) = 1/a(z)$, and is called a *particular solution*.

Consider the homogeneous, first-order equation

$$X_t = aX_{t-1} \quad (1.3.13)$$

Then by continual substitution

$$X_t = a^2 X_{t-2} = a^3 X_{t-3} = \cdots = a^{t+N} X_{-N} \quad (1.3.14)$$

so that

$$X_t = Aa^t \quad (1.3.15)$$

where A is a suitable constant, dependent on N.

Now consider the homogeneous second-order equation

$$X_t = a_1 X_{t-1} + a_2 X_{t-2} \quad (1.3.16)$$

The previous result suggests that a possible solution is of the form

$$X_t = A_1 \theta_1^t + A_2 \theta_2^t \quad (1.3.17)$$

Substituting this into Eq. (1.3.16) and rearranging gives

$$A_1 \theta_1^t [1 - a_1 \theta_1^{-1} - a_2 \theta_1^{-2}] + A_2 \theta_2^t [1 - a_1 \theta_2^{-1} - a_2 \theta_2^{-2}] = 0$$

and so the solution is correct if θ_1^{-1} and θ_2^{-1} are chosen to be the roots of the equation $a(z) = 1 - a_1 z - a_2 z^2 = 0$, with the constants A_1 and A_2 chosen so that the initial starting values are accounted for.

The more general pth order homogeneous difference equation

$$X_t = \sum_{j=1}^{p} a_j X_{t-j} \quad (1.3.18)$$

will have the solution

$$X_t = \sum_{k=1}^{p} A_k \theta_k^t \quad (1.3.19)$$

where θ_k^{-1}, $k = 1, \ldots, p$, are the roots of the equation $a(z) = 0$, assuming no multiple roots, and where again the constants A_k are determined by the initial values.

Provided the a_j are real, Eq. (1.3.19) must produce real values for the sequence X_t, so any complex value of θ_k^{-1} must be matched by its complex conjugate. The corresponding terms will pair to become a term of the form $B|\theta_k|^t \cos(kt + \phi_k)$. The solution to Eq. (1.3.18) will thus either be a mixture of exponentials through time or a mixture of exponentials plus amplitude changing oscillatory terms. For large t the form of the solution will be dominated by $A_0|\theta_0|^t$, where $|\theta_0| = \max(|\theta_k|, k = 1, \ldots, p)$. If $|\theta_0| < 1$, the solution will be said to be stationary; and if $|\theta_0| > 1$, the solution is said to be explosive. From the definition of θ_k, it is seen that a necessary and sufficient condition for the solution to be stationary is that all the roots of the equation $a(z) = 0$ lie outside the unit circle $|z| = 1$. This will be called the *stationarity condition*.

The particular solution to Eq. (1.3.11), X_{2t}, may be written in the form

$$X_{2t} = \sum_{j=0}^{t+N} b_j Y_{t-j} \tag{1.3.20}$$

It is easily seen that the coefficients b_j will increase in magnitude as j increases unless the stationarity condition holds. A simple example is provided by the first-order equation

$$X_t = aX_{t-1} + Y_t \tag{1.3.21}$$

which may be written $(1 - aB)X_t = Y_t$ so that

$$X_{2t} = \frac{1}{1 - aB} Y_t = \sum_{j=0}^{t+N} a^j Y_{t-j} \tag{1.3.22}$$

The stationarity condition for this equation is just $|a| < 1$ since $a(z) = 1 - az$.

In the general case, provided the stationarity condition holds and if one supposes that the process began in the indefinite past, then since[1] $X_{1t} = O(|\theta_0|^{t+N}) \to 0$ as N becomes large, the complete solution is just

$$X_t = b(B)Y_t \qquad \text{for all } t \tag{1.3.23}$$

It is worth noting that multiplying X_t by z^t and summing over t gives $X(z) = b(z)Y(z)$ since the right-hand sides of Eqs. (1.3.20) and (1.3.23) are convolutions.

1.4 The Linear Cyclic Model

If one plots a white noise series through time, a rather jagged and uneven graph results. (Figure 1.1b, p. 14, is a plot of such a series.) Many actual series have quite a different appearance, being much "smoother" than a white

---

[1] A function of N, $g(N)$, is said to be $O(N)$ if $g(N)/N$ tends to a nonzero constant c as N tends to infinity, and to be $o(N)$ if $g(N)/N \to 0$ as $N \to \infty$.

noise series. This smoothness can be explained by ρ_1, the autocorrelation between adjacent values of the series, being positive. The nearer ρ_1 is to unity, the smoother the appearance of the series. There are a number of models that can be introduced to explain this smoothness. Historically, the first to be considered was a process that contained cycles or strictly periodic components. A possible reason for this is that many of the time series first analyzed did appear to contain regular cyclical components when plotted. These series included brightness of rotating twin stars, solar surface activity (sunspot series), and agricultural prices containing a clear seasonal component. A simple model to explain such fluctuations is of the form

$$X_t = d \cos(\omega t + \theta) + \epsilon_t \qquad (1.4.1)$$

so that the process is the sum of a cosine wave with amplitude d, frequency ω (and hence period of $2\pi/\omega$), and phase θ plus a white noise component. However, there is no reason to expect a cyclical component, such as an annual cycle, to be representable by a single cosine term. A periodic function $P(t)$ of period p, defined to have the property

$$P(t + kp) = P(t) \qquad \text{any integer } k$$

may always be represented by the sum of cosines,

$$P(t) = \sum_{j=1}^{p} d_j \cos(j\omega t + \theta_j) \qquad (1.4.2)$$

where $\omega = 2\pi/p$, provided p is an integer multiple of the sampling period. This is the Fourier series representation, ω being known as the *fundamental frequency*, and the frequencies $j\omega$ are the *harmonics* of ω. The somewhat more general model now becomes

$$X_t = P(t) + \epsilon_t = \sum_{j=1}^{p} d_j \cos(j\omega t + \theta_j) + \epsilon_t \qquad (1.4.3)$$

If the period p is known, the parameters d_j, θ_j can be estimated by various techniques including regression.

Suppose now that such a periodic component is fitted to the data and the residuals $X_t - \hat{P}(t)$ are plotted. If again they are too smooth to be represented by a white noise series, it is not unnatural to suppose that this residual smoothness is caused by the presence of further, undetected cycles. The model that would result from such reasoning is of the form

$$X_t = \sum_{k=1}^{m} P_k(t) + \epsilon_t \qquad (1.4.4)$$

where $P_k(t)$ is a periodic function with period p_k and ϵ_t is white noise. Using the Fourier expansion for each periodic function, the model becomes a complicated sum of cosines plus the white noise residual; that is,

$$X_t = \sum_{j,k} d_{jk} \cos(j\omega_k t + \theta_{jk}) + \epsilon_t \qquad (1.4.5)$$

where $\omega_k = 2\pi/p_k$. This will be called the *linear cyclical model*. If the periods p_k, $k = 1, \ldots, m$, are known, the parameters d_{jk}, θ_{jk} can be found by regression methods. If the periods of the components are not known, the classical method of finding them is to use a procedure known as *periodogram analysis*. There are a number of periodogram techniques, but the best known and most used is that due to Schuster [1898]. Given an observed series x_t, $t = 1, \ldots, n$, form the quantity

$$I_n(\lambda) = \frac{1}{n^2} \left[\left(\sum_t x_t \cos t\lambda \right)^2 + \left(\sum_t x_t \sin t\lambda \right)^2 \right] \qquad (1.4.6)$$

which may be called the intensity[1] at frequency λ. The plot of $I_n(\lambda)$ against $2\pi/\lambda$ is called the Schuster periodogram, but it is often more convenient to plot $I_n(\lambda)$ against λ, and this plot has been called the *spectrogram*. Both of these diagrams should have clear peaks at points corresponding to the "hidden periods" p_k of the linear cyclic model. This may be seen by considering the very simple case where

$$x_t = a \cos \omega t \qquad (1.4.7)$$

Then

$$\frac{1}{n} \sum_{t=1}^{n} x_t \cos \lambda t = \frac{a}{2n} \sum_{t=1}^{n} (\cos(\omega - \lambda)t + \cos(\omega + \lambda)t) \qquad (1.4.8)$$

Since

$$\sum_{t=1}^{n} \cos tx = \frac{\cos nx \sin((n+1)/2)x}{\sin(x/2)} - 1 \qquad (1.4.9)$$

it follows that

$$\frac{1}{n} \sum_{t=1}^{n} x_t \cos \lambda t = O(n^{-1}), \qquad \lambda \neq \omega$$

but, since $\lim_{x \to 0}(\sin Nx/\sin x) = N$, one has

$$\frac{1}{n} \sum_t x_t \cos \omega t = \frac{a}{2} + O(n^{-1}) \qquad (1.4.10)$$

Similarly

$$\frac{1}{n} \sum_{t=1}^{n} x_t \sin \lambda t = O(n^{-1}) \qquad (1.4.11)$$

so, for this simple model,

$$I(\lambda) = O(n^{-2}), \qquad \lambda \neq \omega$$

$$= (a/2)^2, \qquad \lambda = \omega \qquad (1.4.12)$$

and, if n is sufficiently large, the peak at $\lambda = \omega$ should be apparent. The

[1] It is worth noting that the intensity can be written in the form $I_n(\lambda) = |n^{-1}x(z)|^2$, with $z = e^{-i\lambda}$.

theoretical shape for the linear cyclical model can be derived in a similar fashion, provided that σ_ϵ^2, the variance of the white noise, is small compared to the portion of the variance of x_t attributable to the cyclical components. There are many problems involved with using the periodogram in practice, but these are of no immediate concern.

Since the linear cyclical process contains deterministic time functions, it is difficult at first sight to see how such a process can be stationary. It is, however, possible to make the underlying stochastic process stationary even though any particular realization will not seem to be so. Consider the simple process

$$X_t = a \cos(\omega t + \theta) \tag{1.4.13}$$

where θ is a random variable rectangularly distributed on the interval $[-\pi, \pi]$. A realization is formed by choosing a value of θ from this population at a time before the starting value of the realization. The process will have a zero mean, by noting that

$$E[X_t] = a \cos \omega t E[\cos \theta] - a \sin \omega t E[\sin \theta] = 0$$

since $E[\cos \theta] = E[\sin \theta] = 0$ if θ is rectangularly distributed on $[-\pi, \pi]$. Since $E[X_t] = 0$, the autocovariances for this simple cyclical process are

$$\lambda_\tau = E[X_t X_{t-\tau}] = a^2 E[\cos(\omega t + \theta) \cos(\omega(t - \tau) + \theta)] \tag{1.4.14}$$

From the formula

$$\cos A \cos B = \tfrac{1}{2}[\cos(A + B) + \cos(A - B)] \tag{1.4.15}$$

it is seen that

$$\lambda_\tau = (a^2/2)E[\cos(\omega(2t - \tau) + 2\theta) + \cos \tau\omega] = (a^2/2) \cos \tau\omega \tag{1.4.16}$$

since the first term vanishes. So λ_τ is seen to be independent of time.

These arguments can be extended to cover the general linear cyclic model (1.4.5) provided the phases θ_{jk} are all independent rectangular random variables. It is found that

$$\mu = E[X_t] = E[\epsilon_t] \tag{1.4.17}$$

and

$$\lambda_\tau = \operatorname{cov}(X_t, X_{t-\tau}) = \tfrac{1}{2} \sum_{j,\,k} d_{jk}^2 \cos(\omega_{jk}\tau) + \delta(\tau)\sigma_\epsilon^2 \tag{1.4.18}$$

where

$$\delta(\tau) = 1 \qquad \text{if} \quad \tau = 0$$

$$= 0 \qquad \text{if} \quad \tau \neq 0$$

Although stationarity has been achieved by the use of what may seem to be an artificial device, it is quite justified to consider realizations from the linear cyclic process as though they are stationary, even if they are essentially only marginally stationary.

1.5 The Autoregressive Model

A process generated by the equation

$$X_t = aX_{t-1} + \epsilon_t \qquad (1.5.1)$$

where ϵ_t is a zero-mean white noise is called a first-order autoregressive process and is a special case of a much analyzed class of stochastic processes known as Markov processes. Taking expectations of (1.5.1) and denoting $\mu_t = E[X_t]$, one finds that μ_t obeys the simple difference equation

$$\mu_t = a\mu_{t-1} \qquad (1.5.2)$$

The solution of this equation, according to Eq. (1.3.14), is

$$\mu_t = a^{t+N}\mu_{-N}$$

if the process starts at time $-N$. Assuming that the process began in the infinite past and that $|a| < 1$, it follows that $\mu_t = 0$ for all t. The model (1.5.1) can easily be generalized to represent stationary processes with nonzero mean μ for, writing $X_t - \mu = a(X_{t-1} - \mu) + \epsilon_t$, it follows from an argument exactly analogous to that just given that X_t has mean μ for all t. Thus, in considering covariance properties, it can be assumed that the process is zero-mean, a state of affairs that can, if necessary, be achieved by subtracting a fixed mean at each time point.

Squaring both sides of (1.5.1) and taking expectations yields the difference equation

$$\operatorname{var}(X_t) = a^2 \operatorname{var}(X_{t-1}) + \sigma_\epsilon^2 \qquad \text{where} \quad \sigma_\epsilon^2 \equiv \operatorname{var}(\epsilon_t)$$

This follows since ϵ_t and X_{t-1} are uncorrelated, for the white noise ϵ_t is by definition uncorrelated with values occurring before time t. (This is readily verified by noting that X_{t-1} can be expressed from (1.5.1) as a linear function of $\epsilon_{t-j}, j = 1, 2, \ldots$, plus a distant "starting up" value.) If $|a| < 1$ and the process started in the infinite past, then the solution to the above difference equation is simply the particular solution

$$\lambda_0 = \operatorname{var}(X_t) = \sigma_\epsilon^2/(1 - a^2) \qquad (1.5.3)$$

so that variance is constant over time.

The first autocovariance of the process (1.5.1) is given by

$$\lambda_1 = E[X_t X_{t-1}] = aE[X_{t-1}^2] + E[\epsilon_t X_{t-1}]$$

and, since the second term in this expression vanishes, one can write from (1.5.3)

$$\lambda_1 = a\lambda_0 \qquad (1.5.4)$$

Thus the autocovariance λ_1, and hence the first autocorrelation, take the same sign as a, the consequence being that for positive a the series X_t will be smoother than a white noise process (smoothness increasing with the magnitude of a), but if a is negative, the process will be less smooth. This is illustrated in Fig. 1.1 which shows plots of generated series from first-order autoregressive processes with $a = -0.5$, 0 (white noise), 0.3, 0.8.

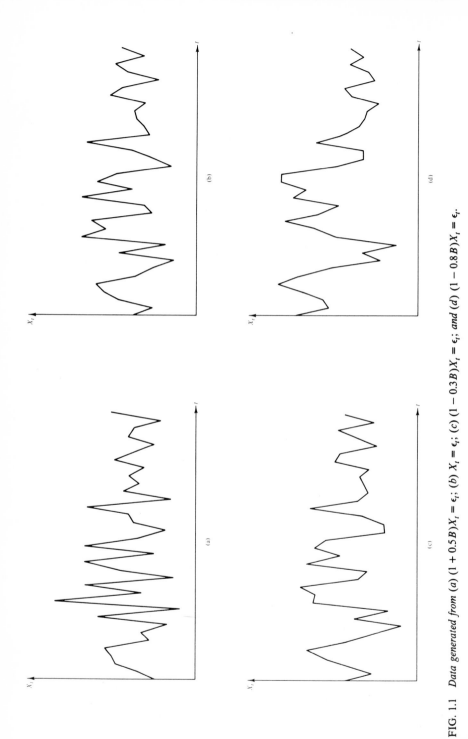

FIG. 1.1 *Data generated from (a)* $(1 + 0.5B)X_t = \epsilon_t$; *(b)* $X_t = \epsilon_t$; *(c)* $(1 - 0.3B)X_t = \epsilon_t$; *and (d)* $(1 - 0.8B)X_t = \epsilon_t$.

Multiplying both sides of (1.5.1) by $X_{t-\tau}$, $\tau > 0$, and taking expectations gives $\lambda_\tau = a\lambda_{\tau-1}$, so

$$\lambda_\tau = a^\tau \lambda_0, \qquad \tau \geqslant 0 \qquad (1.5.5)$$

Hence the autocorrelation function is simply $\rho_\tau = a^\tau$, $\tau \geqslant 0$. Autocorrelation functions for the processes generating the data of Fig. 1.1 are shown in Fig. 1.2, the typical exponential decay of such functions as τ increases being clear from these plots. It is seen that X_t is stationary provided $|a| < 1$ since the first and second moments of the process are all time invariant.

Autoregressive processes were first introduced by Yule [1927] and it is not difficult to persuade oneself that they could well arise in practice. For example, the total number of unemployed in one month might be thought to consist of a fixed proportion a of those unemployed in the previous month, the others having obtained jobs, plus a new group of workers seeking jobs. If the new additions are considered to form a white noise series with positive mean $\mu(1 - a)$, then the unemployment series is first-order autoregressive for if X_t denotes numbers unemployed at time t

$$X_t = aX_{t-1} + (1 - a)\mu + \epsilon_t$$

where ϵ_t is zero-mean white noise. Hence

$$X_t - \mu = a(X_{t-1} - \mu) + \epsilon_t$$

A second example can be constructed as follows. A tank of some chemical holds 10 liters, of which 1 liter is used each working week. The liter removed is replaced at the end of each week from a bottle of the chemical purchased from a manufacturer. The chemical is a mixture of two components; let X_t be the proportion of the first component in the tank in week t. If the bottles of chemical have concentrations of the first component varying randomly over time, with the bottle used at the start of week t having concentration ϵ_t, it follows that

$$X_t = \tfrac{9}{10} X_{t-1} + \tfrac{1}{10} \epsilon_t$$

so a first-order autoregressive series again arises.

Some economic series may be thought to be generated by a mechanism

value at time $t + 1$ = expectation made at time t of value at time

$t + 1$, plus error term

If the error term is white noise and the expectation is just a proportion of current value, a first-order autoregressive process results. However, if expectations are formed from a weighted sum of past and present values of the series, a higher order autoregressive process is obtained.

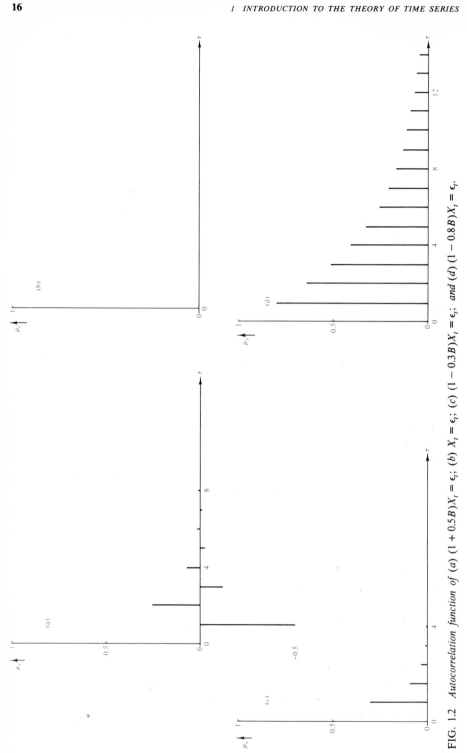

FIG. 1.2   Autocorrelation function of (a) $(1 + 0.5B)X_t = \epsilon_t$; (b) $X_t = \epsilon_t$; (c) $(1 - 0.3B)X_t = \epsilon_t$; and (d) $(1 - 0.8B)X_t = \epsilon_t$.

The general zero-mean autoregressive process is one generated by

$$X_t = \sum_{j=1}^{p} a_j X_{t-j} + \epsilon_t \tag{1.5.6}$$

For processes with nonzero mean μ, the extension is

$$X_t - \mu = \sum_{j=1}^{p} a_j (X_{t-j} - \mu) + \epsilon_t$$

It will again be assumed, without loss of generality, that the process of interest has zero mean. If a process X_t is generated by an equation of the form (1.5.6), it will be called a pth-order autoregressive process and denoted $X_t \sim \mathrm{AR}(p)$. For the process to be stationary, the difference equation will need to obey the stationarity condition introduced in Section 1.3. Thus, the roots of $a(z) = 0$ must all lie outside the unit circle $|z| = 1$, where

$$a(z) = 1 - \sum a_j z^j \tag{1.5.7}$$

In terms of the backward operator, (1.5.6) may be written

$$a(B)X_t = \epsilon_t \tag{1.5.8}$$

and the general solution to this difference equation is just the particular solution

$$X_t = \frac{1}{a(B)} \epsilon_t \tag{1.5.9}$$

provided the stationarity condition holds. Thus, when this condition holds, X_t may be written in the form

$$X_t = \sum_{j=0}^{\infty} b_j \epsilon_{t-j}, \qquad b_0 = 1 \tag{1.5.10}$$

where $b(B) = \sum b_j B^j = 1/a(B)$. In particular, if $X_t \sim \mathrm{AR}(1)$, as in (1.5.1), then $a(z) = 1 - az$ and so

$$X_t = \sum_{j=0}^{\infty} a^j \epsilon_{t-j} \tag{1.5.11}$$

Taking expectations, it is immediately seen that $\mu = E[X_t] = 0$, provided $E[\epsilon_t] = 0$. Squaring both sides of (1.5.11), taking expectations, and noting that $E[\epsilon_t \epsilon_s] = 0$, $s \neq t$, it is immediately seen that

$$\mathrm{var}\, X_t = \sigma_\epsilon^2 \sum_{j=0}^{\infty} a^{2j} = \frac{\sigma_\epsilon^2}{1 - a^2}$$

thus confirming (1.5.3). This proof very clearly illustrates the explosive nature of the solution for X_t if $|a| \geqslant 1$.

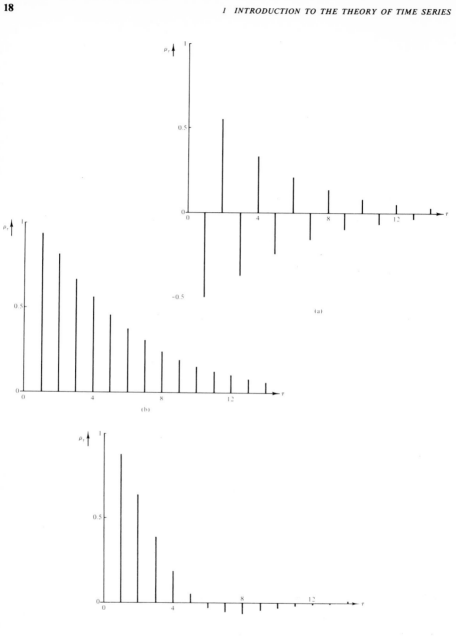

FIG. 1.3   *Autocorrelation function of (a)* $(1 + 0.3B - 0.4B^2)X_t = \epsilon_t$; *(b)* $(1 - 1.3B + 0.4B^2)X_t$
$= \epsilon_t$; *and (c)* $(1 - 1.3B + 0.5B^2)X_t = \epsilon_t$.

Multiplying both sides of (1.5.10) by ϵ_s, $s \geqslant t$ and taking expectations gives

$$E[\epsilon_s X_t] = 0, \qquad s > t$$
$$= \sigma_\epsilon^2, \qquad s = t \tag{1.5.12}$$

This result is needed to find the autocovariance sequence for the AR(p) process. If both sides of (1.5.6) are multiplied by $X_{t-\tau}$, $\tau \geqslant 0$, taking expectations gives

$$\lambda_\tau = \sum_{j=1}^{p} a_j \lambda_{\tau-j}, \qquad \tau > 0 \tag{1.5.13}$$

and

$$\lambda_0 = \sum_{j=1}^{p} a_j \lambda_j + \sigma_\epsilon^2 \tag{1.5.14}$$

If $a(z) = \prod_{j=1}^{p}(1 - \theta_j z)$ and $\theta_j \neq \theta_k$, $j \neq k$, with $|\theta_j| < 1$ all j, then the results of Section 1.3 show that the solutions to the difference equations (1.5.13), (1.5.14) are of the form

$$\lambda_\tau = \sum_{k=1}^{p} A_k \theta_k^\tau + \delta(\tau)\sigma_\epsilon^2 \tag{1.5.15}$$

where

$$\delta(\tau) = 0, \qquad \tau \neq 0$$
$$= 1, \qquad \tau = 0$$

If

$$\frac{1}{a(z)} = \sum_{j=1}^{p} \frac{\phi_j}{1 - \theta_j z} \tag{1.5.16}$$

then it may be shown that

$$A_k = \sigma_\epsilon^2 \frac{\phi_k}{a(\theta_k)} \tag{1.5.17}$$

If any θ_j is complex, then some other θ will be its complex conjugate.

Denoting $\max(|\theta_k|, k = 1, \ldots, p)$ by $|\theta_m|$, it follows from (1.5.15) that for τ large $|\lambda_\tau| \approx A_m |\theta_m|^\tau$, so $|\lambda_\tau|$ will decline exponentially to zero as τ becomes large.

It will be shown in Section 1.6 that the autocovariance generating function for an AR(p) process is given by $\lambda(z) = \sigma_\epsilon^2 / a(z)a(z^{-1})$.

Dividing (1.5.13) by the variance λ_0 yields a system of equations relating autocorrelations of a pth-order autoregressive process

$$\rho_\tau = \sum_{j=1}^{p} a_j \rho_{\tau-j}, \qquad \tau > 0 \tag{1.5.18}$$

These are known as the Yule–Walker equations, from Yule [1927] and Walker [1931]. Note that the set of equations (1.5.18) with $\tau = 1, 2, \ldots, p$ can be solved for the coefficients a_j in terms of the first p autocorrelations.

The behavior of the autocorrelation function of an autoregressive process is that of a mixture of damped exponentials and/or sine waves. To illustrate, consider in more detail the second-order process

$$X_t - a_1 X_{t-1} - a_2 X_{t-2} = \epsilon_t$$

From the first two equations of (1.5.18), since $\rho_{-\tau} = \rho_\tau$ and $\rho_0 = 1$,

$$\rho_1 = a_1 + a_2\rho_1, \qquad \rho_2 = a_1\rho_1 + a_2$$

Hence

$$\rho_1 = \frac{a_1}{(1 - a_2)} \; , \qquad \rho_2 = \frac{a_1^2}{(1 - a_2)} + a_2$$

and ρ_τ, $\tau = 3, 4, \ldots$, can be obtained directly from (1.5.18). The stationarity requirement is that the roots of

$$1 - a_1 z - a_2 z^2 = 0 \tag{1.5.19}$$

lie outside the unit circle $|z| = 1$. If the roots of this equation are real, the autocorrelations die out exponentially, as shown in Figs. 1.3a and 1.3b for the processes

$$(1 - 0.5B)(1 + 0.8B)X_t = (1 + 0.3B - 0.4B^2)X_t = \epsilon_t$$

and

$$(1 - 0.5B)(1 - 0.8B)X_t = (1 - 1.3B + 0.4B^2)X_t = \epsilon_t$$

If the roots of (1.5.19) are complex, then the autocorrelation function exhibits sinusoidal decay. This is illustrated in Fig. 1.3c for the process

$$(1 - 1.3B + 0.5B^2)X_t = \epsilon_t$$

1.6 The Moving Average Model

Suppose that ϵ_t is a zero-mean white noise, then the series

$$X_t = \epsilon_t + \epsilon_{t-1} \tag{1.6.1}$$

will be smoother than the original white noise series, as will be clear either by simple experimentation or by noting that the first autocorrelation is 0.5. This is a very simple example of a moving average, the more general form for the process being

$$X_t = \sum_{j=0}^{q} b_j \epsilon_{t-j}, \qquad b_0 = 1 \tag{1.6.2}$$

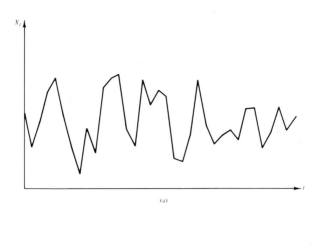

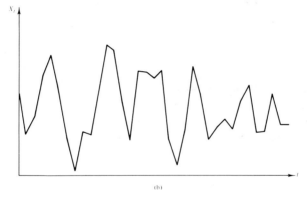

FIG. 1.4 *Data generated from (a) $X_t = (1 + 0.4B)\epsilon_t$; and (b) $X_t = (1 + 0.8B)\epsilon_t$.*

If a process is generated by such an equation, it is said to be a moving average of order q and denoted $X_t \sim MA(q)$. If $q = 0$, X_t will be just white noise. The first-order moving average process is, then, $X_t = \epsilon_t + b\epsilon_{t-1}$. For positive b the process will be smoother than white noise, smoothness increasing with the magnitude of b. Figure 1.4 shows generated data from first-order moving average processes with $b = 0.4$ and 0.8.

The mean of the process (1.6.2) is clearly zero. Obviously the model can be extended to deal with the nonzero mean case by writing

$$X_t - \mu = \sum_{j=0}^{q} b_j \epsilon_{t-j}, \qquad b_0 = 1$$

However, it will again be assumed that, if necessary, the mean has been removed by transformation and subsequent analysis will consider (1.6.2). The form of the autocovariances can be seen by writing

$$X_t = \epsilon_t + b_1\epsilon_{t-1} + \cdots + b_{\tau-1}\epsilon_{t-\tau+1} + b_\tau\epsilon_{t-\tau} + b_{\tau+1}\epsilon_{t-\tau-1} + \cdots + b_q\epsilon_{t-q}$$

$$X_{t-\tau} = \qquad\qquad\qquad\qquad \epsilon_{t-\tau} + b_1\epsilon_{t-\tau-1} + \cdots + b_{q-\tau}\epsilon_{t-q}$$

$$+ \cdots + b_q\epsilon_{t-q-\tau}$$

Remembering that $E[\epsilon_t\epsilon_s] = 0$, $t \neq s$, it follows immediately that

$$\lambda_\tau \equiv E[X_t X_{t-\tau}] = \sigma_\epsilon^2[b_\tau + b_1b_{\tau+1} + b_2b_{\tau+2} + \cdots + b_{q-\tau}b_q],$$

$$\text{for} \quad |\tau| \leqslant q \tag{1.6.3}$$

and

$$\lambda_\tau = 0 \qquad \text{for} \quad |\tau| > q \tag{1.6.4}$$

Thus, λ_τ vanishes for sufficiently large τ.

Taking $\tau = 0$ gives

$$\text{var } X_t \equiv \lambda_0 = \sigma_\epsilon^2 \sum_{j=0}^{q} b_j^2 \tag{1.6.5}$$

It is immediately seen that an MA(q) process, with $q < \infty$, is always stationary. It follows from (1.6.4) that, for a moving average process of order q, the autocorrelations ρ_τ are all zero for $\tau > q$. Thus, the autocorrelation function takes a simple and easily recognized form.

Denoting

$$b(z) = \sum_{j=0}^{q} b_j z^j \tag{1.6.6}$$

then, in terms of the backward operator, the MA(q) model (1.6.2) may be written

$$X_t = b(B)\epsilon_t \tag{1.6.7}$$

and, since the right-hand side of (1.6.2) is a convolution, in terms of generating functions, the model may be written

$$X(z) = b(z)\epsilon(z) \tag{1.6.8}$$

Noting that

$$b(z)b(z^{-1}) = \sum_{j,\,k=0}^{q} b_j b_k z^{j-k} = \sum_{s=-q}^{q} z^s \sum_{j=0}^{q} b_j b_{j+s} \tag{1.6.9}$$

by putting $j - k = s$ and taking $b_j = 0$, $j > q$, it follows immediately from (1.6.3) and (1.6.4) that the autocovariance generating function is

$$\lambda(z) = \sigma_\epsilon^2 b(z)b(z^{-1}) \tag{1.6.10}$$

It is of some interest to ask if a set of numbers $c_j, j = 0, 1, \ldots,$ with $c_j = 0$, $j > q$, can be the autocovariances of an MA(q) scheme. To show that just any set of numbers cannot be used, consider the MA(1) process $X_t = \epsilon_t + b\epsilon_{t-1}$ which has first autocorrelation

$$\rho_1 = b/(1 + b^2) \tag{1.6.11}$$

and clearly $|\rho_1| \leqslant 0.5$. It can easily be proved that the largest possible first autocorrelation ρ_1 achievable from an MA(q) process is

$$\rho_1(\text{max}) = \cos[\pi/(q + 2)] \tag{1.6.12}$$

It follows, for example, that if $\rho_1 = 0.8$, $\rho_j = 0$ for $j > 1$, then there is no corresponding MA(1) process with such autocorrelations. A necessary and sufficient condition that there exists an MA(q) process corresponding to a set of "covariances" $c_j, j = 0, 1, \ldots, q$, has been given by Wold [1954a, p. 154] and it is easily shown that the condition is equivalent to $f(\omega) \geqslant 0$, $-\pi \leqslant \omega \leqslant \pi$ where

$$f(\omega) = \lambda(z) = \sum_{j=-q}^{q} c_j z^j \tag{1.6.13}$$

and $z = e^{-i\omega}$. In terms of the power spectral function, to be more fully introduced in Section 2.1, the condition merely states that the spectrum must be nonnegative. The result (1.6.12) has been extended by Davies *et al.* [1974], who show that for a moving average process of order q, the highest value that can be taken by the τth autocorrelation is

$$\rho_\tau(\text{max}) = \cos[\pi/(N + 1)] \qquad \text{if} \quad \tau \text{ divides } q + 1$$

$$= \cos[\pi/(N + 2)] \qquad \text{if} \quad \tau \text{ does not divide } q + 1$$

where N is the largest integer not exceeding $(q + 1)/\tau$.

Moving averages were introduced by Yule [1926] and studied in more detail in 1938 by Wold (see Wold [1954a]). If some economic variable is in equilibrium but is moved from the equilibrium position by a series of buffeting effects from unpredictable events either from within the economy, such as strikes, or from outside, such as periods of exceptional weather, and if the system is such that the effects of such events are not immediately assimilated, then a moving average model will arise. An example might be a small commodity market that receives a series of news items about the state of crops in producing countries. A particular news item will have both an immediate effect on prices and also a weighted, or discounted, effect over the next few days as the market assimilates the importance and relevance of the news. Let X_t denote price change at time t. The discounting of shocks entering the system may be such that a particular item exerts an influence on price change up to q days after its occurrence. In this case, an appropriate model would be the moving average process

$$X_t = \epsilon_t + b_1\epsilon_{t-1} + \cdots + b_q\epsilon_{t-q}$$

where ϵ_{t-j} is the initial value of the shock that occurs at time $t - j$. If this discounting takes an exponential form, with weight $b_j = a^j$, $0 < a < 1$, so that a news item j days earlier has effect proportional to a^j, then the moving average that results takes the form

$$X_t = \sum_{j=0}^{\infty} a^j\epsilon_{t-j}$$

However, Eq. (1.5.11) indicates that X_t may be represented by

$$X_t = aX_{t-1} + \epsilon_t$$

so in this case an MA(∞) process is found to be equivalent to an AR(1) process.

It was shown in Section 1.5 that a stationary AR(p) process could always be written in an MA(∞) form. It is natural to ask if an MA(q) process is equivalent to an AR(∞) process. Considering (1.6.2) as a difference equation for the ϵ_t given the sequence X_t, the answer is seen immediately to be yes, provided the condition previously called the stationarity condition is obeyed, that is, the roots of $b(z) = 0$ all lie outside the unit circle $|z| = 1$. In this context, this condition will be called the *invertibility condition*. An example where the invertibility condition does not hold is the MA(1) process $X_t = \epsilon_t + b\epsilon_{t-1}$ where $|b| > 1$. Invertibility will be discussed further in Sections 3.4 and 4.9.

If $X_t \sim$ AR(p), then the corresponding equations using backward operators will be

$$a(B)X_t = \epsilon_t \quad \Rightarrow \quad X_t = b(B)\epsilon_t$$

and in generating function form

$$a(z)X(z) = \epsilon(z) \quad \Rightarrow \quad X(z) = b(z)\epsilon(z) \qquad \text{where} \quad a(z)b(z) = 1$$

It follows from this and (1.6.10) that the autocovariance generating function for an AR(p) process is

$$\lambda(z) = \sigma_\epsilon^2 / a(z)a(z^{-1})$$

1.7 The Mixed Autoregressive–Moving Average Model

An obvious generalization of the MA and AR models that includes them as special cases is the mixed model in which X_t is generated by

$$X_t = \sum_{j=1}^{p} a_j X_{t-j} + \sum_{j=0}^{q} b_j \epsilon_{t-j} \qquad (1.7.1)$$

or if X_t has mean μ

$$X_t - \mu = \sum_{j=1}^{p} a_j(X_{t-j} - \mu) + \sum_{j=0}^{q} b_j \epsilon_{t-j}$$

where ϵ_t is a zero-mean, white noise and $b_0 = 1$. If X_t is generated in this fashion, it is called a mixed ARMA process and denoted $X_t \sim \text{ARMA}(p, q)$. Using the operator B, the model is

$$a(B)X_t = b(B)\epsilon_t \tag{1.7.2}$$

so that the corresponding generating function form is

$$a(z)X(z) = b(z)\epsilon(z) \tag{1.7.3}$$

where

$$a(z) = 1 - \sum_{j=1}^{p} a_j z^j \tag{1.7.4}$$

and

$$b(z) = \sum_{j=0}^{q} b_j z^j \tag{1.7.5}$$

Mixed processes were first studied by Wold in 1938 (see Wold [1954a]) and Bartlett [1946].

From the considerations of the previous two sections it is clear that:

(i)   the process is stationary if the roots of $a(z) = 0$ all lie outside the unit circle $|z| = 1$;

(ii)   if the process is stationary, then there is an equivalent MA(∞) process

$$X_t = \sum_{j=0}^{\infty} c_j \epsilon_{t-j}, \qquad c_0 = 1 \tag{1.7.6}$$

where

$$c(z) = \sum c_j z^j = \frac{b(z)}{a(z)} \tag{1.7.7}$$

(iii)   there is an equivalent AR(∞) process

$$X_t = \sum_{j=1}^{\infty} d_j X_{t-j} + \epsilon_t \tag{1.7.8}$$

where

$$d(z) = 1 - \sum_{j=1}^{\infty} d_j z^j = \frac{a(z)}{b(z)} \tag{1.7.9}$$

provided the roots of $b(z) = 0$ all lie outside the unit circle $|z| = 1$, that is, provided the invertibility condition holds.

It thus follows that a stationary ARMA process can always be well approximated by a high-order MA process and that if the process obeys the invertibility condition, it can also be well approximated by a high-order AR process.

A specific form for the autocovariance sequence λ_τ is rather more difficult to find than it was for the AR and MA models. If the model is put into the form

$$\sum_{j=0}^{p} \alpha_j X_{t-j} = \sum_{j=0}^{q} b_j \epsilon_{t-j} \qquad (1.7.10)$$

where $\alpha_0 = 1$, $\alpha_j = -a_j$, $j = 1, \ldots, p$, then multiplying both sides of (1.7.10) by $X_{t-\tau}$ and taking expectations gives

$$\sum_{j=0}^{p} \alpha_j \lambda_{\tau-j} = g_\tau \qquad (1.7.11)$$

where

$$g_\tau = \sum_{j=0}^{q} b_j \theta_{j-\tau} \qquad (1.7.12)$$

and

$$\theta_{j-k} = E[X_{t-k}\epsilon_{t-j}] \qquad (1.7.13)$$

From (1.7.6) it is seen that

$$\theta_{j-k} = 0 \qquad \text{for} \quad j < k \qquad (1.7.14)$$

Multiplying both sides of (1.7.10) by ϵ_{t-k} and taking expectations gives

$$\sum_{j=0}^{p} \alpha_j \theta_{k-j} = \sigma_\epsilon^2 b_k \qquad (1.7.15)$$

In principle an expression for λ_τ can be found by solving the difference equation (1.7.15) for the θ's, substituting into (1.7.12) to find the g's and then solving the difference equation (1.7.11) for the λ's. Since, in general, the results will be rather complicated, the eventual solution is not presented. However, for $\tau > q$, g_τ will be zero by (1.7.14) and so the λ's obey the homogeneous difference equation $\sum_{j=0}^{p}\alpha_j\lambda_{\tau-j} = 0$ so that for $\tau > q$, the λ's will be similar in form to those of the AR(p) process $a(B)X_t = \epsilon_t$, as given in Eq. (1.5.15). It follows that for large enough τ, $|\lambda_\tau|$ will take the exponential form found for AR(p) processes. Dividing by the variance λ_0, it follows that for an ARMA(p, q) process the autocorrelations obey

$$\rho_\tau = \sum_{j=1}^{p} a_j \rho_{\tau-j}, \qquad \tau > q$$

Thus the autocorrelation function for the mixed process eventually, after the setting of q starting values determined by the moving average operator, takes the same shape as that of the autoregressive process $a(B)X_t = \epsilon_t$.

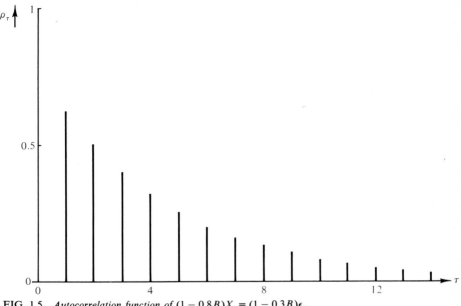

FIG. 1.5   *Autocorrelation function of* $(1 - 0.8B)X_t = (1 - 0.3B)\epsilon_t$.

Since the right-hand side of (1.7.12) is a convolution between the b sequence and the reversed θ sequence, it follows that the generating function form of (1.7.11) is

$$a(z)\lambda(z) = b(z)\theta(z^{-1}) \tag{1.7.16}$$

and the generating function form of (1.7.15) is

$$a(z)\theta(z) = \sigma_\epsilon^2 b(z) \tag{1.7.17}$$

Combining these equations gives

$$\lambda(z) = \sigma_\epsilon^2 \, \frac{b(z)b(z^{-1})}{a(z)a(z^{-1})} \tag{1.7.18}$$

In theory at least, the value of λ_τ can be found as the coefficient of z^τ in the power series expansion of the autocovariance generating function $\lambda(z)$.

For the ARMA(1, 1) process $X_t - aX_{t-1} = \epsilon_t + b\epsilon_{t-1}$ it may be verified from the above equations that

$$\lambda_0 = \frac{(1 + 2ab + b^2)}{1 - a^2} \, \sigma_\epsilon^2, \qquad \lambda_1 = \frac{(1 + ab)(a + b)}{1 - a^2} \, \sigma_\epsilon^2, \qquad \lambda_\tau = a\lambda_{\tau - 1}, \quad \tau \geqslant 2$$

Similarly, the autocorrelations obey $\rho_\tau = a\rho_{\tau - 1}$, $\tau \geqslant 2$. However, the distinction between this process and the first-order autoregressive process is that the relation $\rho_1 = a\rho_0 = a$ now no longer holds unless $b = 0$. Figure 1.5 shows the autocorrelation function of the ARMA (1, 1) process $X_t - 0.8X_{t-1} = \epsilon_t - 0.3\epsilon_{t-1}$.

1.8 Interpreting the Mixed Model

In Sections 1.5 and 1.6 fairly realistic ways in which autoregressive processes and moving average processes could be caused have been suggested, in terms of expectations and by unexpected shocks on the economic system. It would be possible to combine these causes and thereby suggest that a mixed model should arise, but it is difficult to make such explanations completely convincing. There are, however, a number of convincing reasons the mixed model is likely to arise in practice. To illustrate two of these reasons, it is necessary to prove a theorem concerning the form of a process that is the sum of two independent processes.

Suppose X_t and Y_t are two independent, stationary processes with zero means and let

$$Z_t = X_t + Y_t \tag{1.8.1}$$

$$\operatorname{cov}(X_t, X_{t-\tau}) = \lambda_{X,\tau} \tag{1.8.2}$$

and similarly for Y_t, Z_t. Then it follows immediately that

$$\lambda_{Z,\tau} = \lambda_{X,\tau} + \lambda_{Y,\tau} \tag{1.8.3}$$

It then follows, from the condition stated just before Eq. (1.6.13), that if $X_t \sim \mathrm{MA}(m)$, $Y_t \sim \mathrm{MA}(n)$ then

$$Z_t \sim \mathrm{MA}(r) \qquad \text{where} \quad r \leqslant \max(m, n)$$

for both $\lambda_{X,\tau}$ and $\lambda_{Y,\tau}$ are zero for all $\tau > \max(m, n)$

THEOREM If $X_t \sim \mathrm{ARMA}(p, m)$, $Y_t \sim \mathrm{ARMA}(q, n)$, X_t and Y_t are independent, $Z_t = X_t + Y_t$; then $Z_t \sim \mathrm{ARMA}(x, y)$ where

$$x \leqslant p + q, \qquad y \leqslant \max(p + n, q + m) \tag{1.8.4}$$

Proof.　　Let

$$a_1(B)X_t = b_1(B)\epsilon_t \qquad \text{and} \qquad a_2(B)Y_t = b_2(B)\eta_t$$

where a_1, a_2, b_1, b_2 are polynomials in B of order p, q, m, n respectively and ϵ_t, η_t are independent, zero-mean white noise processes.

Since $Z_t = X_t + Y_t$ it follows that

$$a_1(B)a_2(B)Z_t = a_2(B)a_1(B)X_t + a_1(B)a_2(B)Y_t$$

$$= a_2(B)b_1(B)\epsilon_t + a_1(B)b_2(B)\eta_t$$

The first term on the right-hand side is $\mathrm{MA}(q + m)$, and the second term is $\mathrm{MA}(p + n)$ so the whole of the expression on the right-hand side is $\mathrm{MA}(y)$ where $y \leqslant \max(p + n, q + m)$. The order of the polynomial $a_1(B)a_2(B)$ is not more than $p + q$, and so the theorem is established.

The theorem also holds if X and Y have nonzero means μ_X and μ_Y. It has just been shown that

$$a(B)(X_t + Y_t - \mu_X - \mu_Y) = b(B)e_t$$

where

$$a(B) = 1 - a_1 B - \cdots - a_{p+q} B^{p+q} = a_1(B)a_2(B)$$

$$b(B)e_t = 1 + b_1 e_{t-1} + \cdots + b_y e_{t-y} = a_2(B)b_1(B)\epsilon_t + a_1(B)b_2(B)\eta_t$$

and e_t is white noise. Hence, it follows that, if $Z_t = X_t + Y_t$,

$$a(B)(Z_t - \mu_Z) = b(B)e_t \qquad \text{where} \quad \mu_Z = \mu_X + \mu_Y.$$

The need for the inequalities in the expressions for any x and y in the above theorem partly arises from the fact that the polynomials $a_1(B)$ and $a_2(B)$ may contain common roots and so part of the operator need not be applied twice. For example, if

$$(1 - \alpha B)X_t = \epsilon_t \qquad \text{i.e.,} \quad X_t \sim \text{AR}(1)$$

$$(1 - \alpha B)(1 - \beta B)Y_t = \eta_t \qquad \text{i.e.,} \quad Y_t \sim \text{AR}(2)$$

then with $Z_t = X_t + Y_t$ one has

$$(1 - \alpha B)(1 - \beta B)Z_t = (1 - \beta B)\epsilon_t + \eta_t$$

that is,

$$Z_t \sim \text{ARMA}(2, 1)$$

In general, if the polynomials $a_1(B)$, $a_2(B)$ have just k roots in common, the inequalities (1.8.4) become

$$x \leqslant p + q - k, \qquad y \leqslant \max(p + n - k, q + m - k) \qquad (1.8.5)$$

The inequalities in this expression are still necessary since the possibility remains of cancellation on both sides of the equation. For example, in the simple case just considered write

$$(1 - \beta B)\epsilon_t + \eta_t = (1 + bB)e_t$$

Equating variances and first autocovariances on both sides of this expression produces

$$(1 + \beta^2)\sigma_\epsilon^2 + \sigma_\eta^2 = (1 + b^2)\sigma_e^2, \qquad -\beta\sigma_\epsilon^2 = b\sigma_e^2$$

Hence

$$\frac{b}{1 + b^2} = \frac{-\beta}{(1 + \beta^2) + \sigma_\eta^2/\sigma_\epsilon^2}$$

and if either $\beta < \alpha < 0$ or $\beta > \alpha > 0$ it is possible to find a $\sigma_\eta^2/\sigma_\epsilon^2$ such that $b = -\alpha$, in which case Z_t follows the AR(1) process $(1 - \beta B)Z_t = e_t$.

That the inequality for y in (1.8.5) is still required is further demonstrated by the following example. Suppose

$$(1 - \alpha B)X_t = \epsilon_t \qquad \text{i.e.,} \quad X_t \sim \text{AR}(1)$$

$$(1 + \alpha B)Y_t = \eta_t \qquad \text{i.e.,} \quad Y_t \sim \text{AR}(1)$$

and also the variance of ϵ = variance of $\eta = \sigma^2$.

If $Z_t = X_t + Y_t$, then $(1 - \alpha B)(1 + \alpha B)Z_t = (1 + \alpha B)\epsilon_t + (1 - \alpha B)\eta_t$. Denote the right-hand side by $Q_t = \epsilon_t + \alpha\epsilon_{t-1} + \eta_t - \alpha\eta_{t-1}$. Then

$$\text{var } Q_t = 2(1 + \alpha^2)\sigma^2 \qquad \text{and} \qquad E\{Q_t Q_{t-k}\} = 0, \quad \text{all } k > 0$$

so Q_t is a white noise process and $Z_t \sim \text{AR}(2)$ rather than ARMA(2, 1), which would generally occur when two independent AR(1) processes are added together. Those situations in which a simpler model arises than might generally be expected will be called "coincidental situations."

Two situations where series are added are of particular interpretational importance. The first is where series are aggregated to form some total, and the second is where the observed series is the sum of the true process plus observational error, corresponding to the classical "signal plus noise" situation. Most macroeconomic series, such as GNP, employment, or exports, are aggregates and there is no particular reason to suppose that all of the component series will obey exactly the same model. Virtually any macroeconomic series, other than certain prices or interest rates, contains important observation errors. It would be highly coincidental if the "true" series and the observational error series obeyed models having common roots, apart possibly from the root unity, or that the parameters of these models should be such that the cancelling of terms produces a value of y in (1.8.4) less than the maximum possible.

It follows from the theorem that if the series being analyzed is the sum of two independent components each of which is AR(1), then the series will be ARMA(2, 1), for example. If the observed series is the sum of a "true" series that is AR(p) plus a white noise observation error, then an ARMA(p, p) series results. It may be thought to be unrealistic to suppose that the components of an aggregate series are independent. However, if the components can be written in the form

$$X_t = c_1 F_t + W_{1t}, \qquad Y_t = c_2 F_t + W_{2t}$$

where F_t is some comon factor explaining the relatedness of X_t and Y_t, W_{1t} and W_{2t} are independent AR processes and F_t is itself an AR process, then

$$Z_t = (c_1 + c_2)F_2 + W_{1t} + W_{2t}$$

and applying the basic theorem twice, the sum of the three components will in general be an ARMA process.

The situations considered in this section have been analyzed in more detail by Granger and Morris [1976], where it is also pointed out that the mixed model may also arise in various other ways, for example, from an AR process with lags that are not integer multiples of the sampling period or from a feedback mechanism between two or more series. Thus, on purely theoretical grounds, it is seen that the mixed model is very likely to arise in practice, and that arguments as to the difficulty of its interpretation (for example, Chatfield and Prothero [1973]) can be overcome in a number of ways.

There is also a sound statistical reason for considering ARMA models. It is often the case that one has the choice of either fitting an ARMA(p, q) model or an AR(p') model to some data. Experience suggests that the mixed model may achieve as good a fit as the AR model but using fewer parameters, i.e., $p + q < p'$. Since the actual amount of data available is usually very limited, the statistician prefers to fit a model involving as few parameters as possible. Box and Jenkins [1970] have called this the *principle of parsimony* and the application of this principle will often lead one to a mixed model.

1.9 Filters

If a series Y_t is formed by a linear combination of terms of a series X_t, so that

$$Y_t = \sum_{j=-s}^{m} c_j X_{t-j} \tag{1.9.1}$$

then Y_t is called a *filtered* version of X_t. If only past and present terms of X_t are involved, so that

$$Y_t = \sum_{j=0}^{m} c_j X_{t-j} \tag{1.9.2}$$

then Y_t might be called a one-sided or backward-looking filter. In terms of the backward operator B, this type of filter may be written as

$$Y_t = c(B)X_t \tag{1.9.3}$$

where

$$c(z) = \sum_{j=0}^{m} c_j z^j \tag{1.9.4}$$

Provided $\sum_j c_j^2 < \infty$, m need not be finite.

It is seen immediately that if Y_t is an MA process, then it is a backward-looking filter applied to a white noise process. If X_t is an AR(p) process, it may be described in the following way: an appropriate finite backward-looking filter applied to X_t will produce a white noise series. Similarly, the ARMA process can be said to be one for which a finite, backward-looking filter applied to the process produces a backward filter applied to a white noise process.

In terms of generating functions, (1.9.1) may be written $Y(z) = c(z)X(z)$. It then follows from (1.3.10) that the autocovariance functions of X_t and Y_t, denoted $\lambda_x(z)$ and $\lambda_y(z)$, are related by the equation

$$\lambda_y(z) = c(z)c(z^{-1})\lambda_x(z) \tag{1.9.5}$$

Filters are frequently used to remove, or reduce in importance, components of series that are troublesome. However, it is inconvenient to try to determine the effects of filters using the tools introduced in this chapter. These effects can be very simply described by using the spectral theory to be introduced in Chapter 2.

Filters are also associated with the idea of a "black box" relationship between two series. Suppose a "box" is observed to have an input series X_t and output series Y_t, with the box having a memory property, so that Y_t can be a function of current and past X_t. Diagramatically, this can be shown as

$$X_t \longrightarrow \boxed{\begin{array}{c} \text{BLACK} \\[4pt] \text{BOX} \end{array}} \longrightarrow Y_t$$

The analyses associated with this type of construct involve either deriving the properties of the output series given those of the input series and the box, or characterization of the properties of the box given the input and output series. The first problem is simply that of filtering discussed above, while the second involves estimation of the "transfer function" $c(z)$, defined in (1.9.4), of the box. A method for doing this is briefly mentioned in Section 2.5, and black box constructs are discussed further in Chapter 7.

If Y_t is a backward-looking filter, with finite m, applied to an ARMA(p, q) process, then $Y_t \sim$ ARMA($p, q + m$). This is seen by taking $a(B)X_t = b(B)\epsilon_t$ where a, b are polynomials of order p, q. Write $Y_t = c(B)X_t$ and hence $a(B)Y_t = c(B)a(B)X_t = c(B)b(B)\epsilon_t$, and $c(B)b(B)$ will be a polynomial in B of order $q + m$.

1.10 Deterministic Components

A time sequence M_t will be called *deterministic* if there exists a function of past and present values $g_t = g(M_{t-j}, j = 0, 1, \ldots)$ such that

$$E\left[(M_{t+1} - g_t)^2\right] = 0 \tag{1.10.1}$$

If the function g_t is a linear function of $M_{t-j}, j \geqslant 0$, then M_t will be called *linear deterministic*. Consider the sequence $M_t = ae^{bt}$ so that $M_{t+1} = e^b M_t$. If b is known, then the sequence is linear deterministic. If b is not known, it can be perfectly estimated from the past of the series, but the estimate will be a nonlinear function of the sequence; for instance

$$b = \tfrac{1}{2}\left[\log M_t^2 - \log M_{t-1}^2\right]$$

allowing for the fact that a may be negative.

Other functions that are linearly deterministic are:

(i) the periodic sequence

$$M_t = a\cos(\omega t + \theta) \tag{1.10.2}$$

provided ω is known and there exists an integer k such that $2\pi k/\omega$ is an integer. For example, if $M_t = a\cos(2\pi t/12 + \theta)$, then

$$M_{t+1} = M_{t-11} \qquad (1.10.3)$$

so that M_t is clearly linear deterministic. If ω has to be estimated from past data, the sequence becomes nonlinear deterministic.

(ii)
$$M_t = \sum_{j=0}^{m} d_j t^j \qquad (1.10.4)$$

so that the sequence is a polynomial in t. To show that such polynomials are linear deterministic, consider the simple sequence $M_t = a + bt$. Then $M_t - M_{t-1} = b$ so, if b is known, it follows that

$$M_{t+1} = b + M_t \qquad (1.10.5)$$

The difference operator is defined by

$$\Delta M_t = M_t - M_{t-1} \qquad (1.10.6)$$

It can be shown that if M_t is a polynomial or order m, then $\Delta^m M_t = m!\, d_m$ and

$$\Delta^{m+1} M_t = 0 \qquad (1.10.7)$$

It is thus seen that the M_t sequence obeys a homogeneous linear difference equation, and so is linear deterministic. However, one needs to know the value of m, or at least a value m' such that $m' > m$, to use this procedure.

These types of functions are of considerable importance in practice since the plots of many time series, particularly economic series, appear to contain trends and seasonal components. A classical model involving these components is to assume an economic variable to be represented by

$$X_t = T(t) + S(t) + Y_t \qquad (1.10.8)$$

where $T(t)$ is a deterministic component representing the trend, the seasonal $S(t)$ is also deterministic and is a periodic component with period 12 months and Y_t is a stationary process with no deterministic components.

If the variable measured some macroeconomic quantity, it was once usual to add further periodic components, with periods greater than 12 months, to represent business cycles. It is now generally accepted that this is not a useful way to represent the business cycle component, since these "cycles" are by no means strictly periodic. In fact, there is virtually no evidence that modern macroeconomic series contain periodic components other than seasonal ones.

Given a length of series representable as (1.10.8), the estimated variance can be approximately decomposed:

$$\widehat{\mathrm{var}}(X_t) = \widehat{\mathrm{var}}(T(t)) + \widehat{\mathrm{var}}(S(t)) + \widehat{\mathrm{var}}(Y_t) \qquad (1.10.9)$$

In practice, the first two components dominate in the sense that they contribute a great deal more to the overall variance of the series than does Y_t. However, for testing hypotheses or investigating relationships between variables, it is the term Y_t that is often of greatest interest. For this reason, as well as others to be explained later, it is usually thought desirable to implement techniques that either remove the trend and seasonal components or, at least, greatly reduce their importance.

The trend term is very difficult to define, given only a finite series to analyze. It is usually taken to be some monotonically increasing or monotonically decreasing function of time. For an observed series it is often possible to approximate the trend by a polynomial in t or an exponential function of time. To extrapolate this trend outside of the observed period can lead to disastrous results. Suppose, for example, one measured outdoor temperatures at some location every minute from 4 A.M. to 11:30 A.M. The data would almost certainly appear to contain a clear-cut upward trend, but when this segment of data is considered in the context of temperature readings taken over several days, the apparent trend is seen to be just a segment of a daily cycle in temperature.

There are two basic methods of estimating a trend term. One is to assume that $T(t)$ can be well approximated by some time function, such as a polynomial in t, an exponential in t, or some combination of these, and then to estimate the parameters of the function by a regression procedure. The alternative is to view trend as the current mean of the series, i.e., $T(t) = E[X_t]$, where clearly X_t is now nonstationary. One might estimate this current mean by either

$$\hat{T}(t) = \frac{1}{2m+1} \sum_{j=-m}^{m} x_{t-j} \qquad (1.10.10)$$

or

$$\hat{T}(t) = \alpha \hat{T}(t-1) + (1-\alpha)x_t \qquad (1.10.11)$$

where α is near one. This second estimate may be written as

$$\hat{T}(t) = (1-\alpha) \sum_{j=0}^{\infty} \alpha^j x_{t-j} \qquad (1.10.12)$$

so that the most recent value of the series is given greatest weight, but if $E[X_t] = \mu$, a constant, then $E[\hat{T}(t-1)] = \mu$. Estimates of the form (1.10.11) form the basis of a forecasting technique known as "exponential smoothing" which will be discussed in more detail in Chapter 5. Both of these estimates involve filters that greatly smooth the input series x, and this underlying smooth component is then equated with the trend. Once trend has been "estimated," trend removal (strictly trend reduction) consists of forming a new series

$$x'_t = x_t - \hat{T}(t) \qquad (1.10.13)$$

A different approach to trend removal is considered in the following section.

If X_t is a trend-free zero-mean process but with a seasonal component, so that

$$X_t = S(t) + Y_t$$

then there is a wide variety of techniques for estimating $S(t)$. If a monthly sampling period is being used, $S(t)$ may be taken to have period 12 months and so, from (1.4.2), may be represented by

$$S(t) = \sum_{j=1}^{6} d_j \cos\left(\frac{2\pi jt}{12} + \theta_j \right) = \sum_{j=1}^{6} \left\{ d_j' \cos\left(\frac{2\pi jt}{12} \right) + d_j'' \sin\left(\frac{2\pi jt}{12} \right) \right\}$$

$$(1.10.14)$$

The quantities d_j' and d_j'' may be estimated by least-squares regression. An equivalent procedure is to use dummy variables $D_{jt}, j = 0, \ldots, 11$, such that $D_{jt} = 1$ if $(t - j - 1)/12$ is an integer and zero otherwise, so that $D_{0t} = 1$ every January and is zero in every other month, and so forth (assuming x_1 is a January figure). $S(t)$ may then be represented by

$$S(t) = \sum_{j=0}^{11} d_j D_{jt} \qquad (1.10.15)$$

with the side condition that $\sum_{j=0}^{11} d_j = 0$, and the d_j found once more by regression.

An alternative, but easier, procedure is to form

$$\bar{S}_0 = \text{average of all January } x\text{'s}$$

$$\bar{S}_1 = \text{average of all February } x\text{'s, etc.}$$

and then to take

$$\hat{S}(t) = \sum_{j=0}^{11} \bar{S}_j D_{jt} \qquad (1.10.16)$$

again assuming the series starts in January. However, one may suspect that the seasonal component is slowly changing either in amplitude or shape through time. In this case $\bar{S}_0$ may be taken to be the average of all recent January figures, or a time-changing $S_0(k)$ figure for January in year k derived from

$$\hat{S}_0(k) = \alpha \hat{S}_0(k - 1) + (1 - \alpha)[\text{most recent January figure}] \quad (1.10.17)$$

and similarly for the other months. This class of methods of estimating the seasonal components may be thought of as linear filters, and their properties can be determined from the theory to be introduced in Section 2.2. On occasions, one may fear that not only is the seasonal component changing through time, so that averages over just the recent past should be used, but that some years are very untypical, due to freak weather conditions say, and that the figures for these years should not be included in the averages. Thus, $\bar{S}_0$ may be estimated as the average over the last m January x's, excluding the largest and smallest x in this period. This is a nonlinear procedure, and its effects can be determined only by simulation. The results of such a simulation will be described in Section 2.8.

Once the seasonal component has been estimated, the *seasonally adjusted* series is $x_t' = x_t - \hat{S}(t)$. An alternative method of removing the seasonal component is considered in Section 1.14. It should be emphasized that the foregoing discussion of seasonal analysis is only a brief survey of a difficult and complicated problem.

Many early textbooks dealing with time series analysis (see, for example, Croxton and Cowden [1955]) concentrated almost exclusively on procedures for the removal of the deterministic components in (1.10.8), paying relatively little attention to the properties of Y_t. The more modern view is that, as far as possible, the trend, seasonal, and "irregular" components should be handled simultaneously in a single model aimed at depicting as faithfully as possible the behavior of a given time series. Trend is generally treated by differencing, leading to the consideration of "integrated processes" in Section 1.13. (An alternative, to be discussed in Chapter 9, is to transform the series, by taking logarithms for example, before differencing.) Seasonality can be treated, as will be seen in Section 1.14, through a generalization of the ARMA models discussed earlier.

1.11 Wold's Decomposition

The models that have been introduced so far may appear to the reader to have been selected rather arbitrarily. One might well expect that there are many other models that could have been considered. In fact, a famous theorem due to Wold [1954a] suggests otherwise. He proved that any stationary process X_t can be uniquely represented as the sum of two mutually uncorrelated processes $X_t = D_t + Y_t$ where D_t is linearly deterministic and Y_t is an MA(∞) process. The Y_t component is said to be *purely nondeterministic*.

Even though the process may be generated nonlinearly, the decomposition is a linear one and is determined entirely by the second moments of the process.

The theorem is probably more satisfying to theoreticians than to practical time series analysts. It does, however, mean that it is reasonable to hope that the true generating mechanism of a process can be adequately approximated by a mechanism that generates some simple type of linearly deterministic process plus some stationary ARMA process. It does not necessarily follow that a nonlinear model may not provide a better explanation. Wold's decomposition does suggest a wide class of stationary processes that should be considered and, given stationarity, will usually be difficult to better.

1.12 Nonstationary Processes

In series arising in many disciplines it is generally agreed that an assumption of stationarity is too restrictive and that series actually met are in some way nonstationary. The immediate problem when considering nonstationarity is that there are unlimited ways in which a process could be nonstationary.

There is naturally a strong temptation not to venture too boldly into such uncharted territory, but rather to consider models that are in some way close to those met under a stationarity assumption. One such model, in which a stationary process had added to it a trend in mean, was briefly discussed in Section 1.10. With economic data it is frequently observed that the extent of the fluctuations of a series are roughly proportional to the current level. This suggests that the model

$$X_t = T(t) \cdot Y_t \tag{1.12.1}$$

where Y_t may be a stationary series, might be appropriate. If X_t is necessarily nonnegative, for example, a price or level of production, then a logarithmic transformation of the data will produce the previous model in which trend is only found in the mean. Alternatively, a transformation such as

$$X_t' = X_t \Big/ \frac{1}{m} \sum_{j=0}^{m-1} X_{t-j} \tag{1.12.2}$$

could produce a series that is apparently stationary (see, for instance, Granger and Hughes [1971]). No completely satisfactory techniques are available for testing whether or not a series contains a trend in mean and/or variance. A number of sensible procedures can be suggested, but a decision based on the plot of the data is likely to be a reasonable one, provided the analyst is sufficiently experienced.

Even when a series appears to contain no clear-cut trend in mean or variance, there are various more subtle ways in which it can be nonstationary. An obvious class of models are ARMA processes with time-changing parameters. An example would be a process generated by

$$X_t = a(t)X_{t-1} + b(t)\epsilon_t \tag{1.12.3}$$

where ϵ_t is stationary white noise. Provided that $\prod_{j=0}^{n} a(t-j) \to 0$ as $n \to \infty$ for every t, this particular difference equation has the solution

$$X_t = \sum_{j=0}^{\infty} c_j(t)\epsilon_{t-j} \tag{1.12.4}$$

where

$$c_j(t) = \left\{ \prod_{k=0}^{j-1} a(t-k) \right\} b(t-j), \qquad j > 0 \tag{1.12.5}$$

and $c_o(t) = b(t)$. It can be shown that any ARMA process with time-changing parameters will have a solution of the form (1.12.4) with an appropriate sequence of functions $c_j(t)$ provided some kind of stability condition holds. This class of nonstationary models may be considered of real importance due to a generalization of Wold's decomposition theorem provided by Cramér [1961], which states that for *any* process X_t, there is a uniquely determined decomposition $X_t = D_t + Y_t$ where D_t and Y_t are uncorrelated, D_t is deterministic, Y_t is purely nondeterministic representable as

$$Y_t = \sum_{j=0}^{\infty} c_j(t)\epsilon_{t-j}$$

where

$$\sum_{j=0}^{\infty} \left[c_j(t)\right]^2 < \infty, \quad \text{all } t \qquad (1.12.6)$$

If the parameters of the ARMA process, or equivalently the $c_j(t)$, change too quickly with time, they clearly cannot be estimated at all satisfactorily given a single realization of the process for analysis, for the same reasons as discussed in Section 1.2. It is therefore natural to consider processes for which the parameters are only slowly changing with time. A systematic account of such processes which may be called *evolutionary processes* has been given by Priestley [1965].

It was seen in Section 1.3 that generating mechanisms with time invariant parameters could produce nonstationary outcomes. For example, if

$$X_t = aX_{t-1} + \epsilon_t \qquad (1.12.7)$$

where $a > 1$ and ϵ_t is a zero-mean white noise process, with the process starting at time $t = -N$, then the difference equation has the solution

$$X_t = Aa^{t+N} + \sum_{j=0}^{t+N} a^j\epsilon_{t-j} \qquad (1.12.8)$$

Even if initial conditions make $A = 0$, it is seen that the variance of X_t is given by

$$\text{var } X_t = \frac{a^{2(t+N+1)} - 1}{a^2 - 1} \text{ var}(\epsilon) \qquad (1.12.9)$$

which depends on time, is increasing with t, and becomes infinite as $N \to \infty$. In general, then, X_t will have a trend both in mean and variance and these effects should be noticeable from a plot of a realization of the process. Such processes may be called *explosive*. Figure 1.6 shows a plot of generated data from the process (1.12.7), with $a = 1.05$ and ϵ_t a zero-mean Gaussian white noise process with variance 16. The calculations were started by setting $X_0 = 100$.

The solution of (1.12.7) is thus explosive if $a > 1$ but is stationary if $|a| < 1$. The case $a = 1$ provides a process that is neatly balanced between an overtly nonstationary one and the stationary situation. If X_t is generated by the model

$$X_t = X_{t-1} + \epsilon_t + m \qquad (1.12.10)$$

then it is known as a *random walk*, provided ϵ_t is a zero-mean white noise process. If $m \neq 0$, X_t will be a *random walk with drift*. If the process starts at

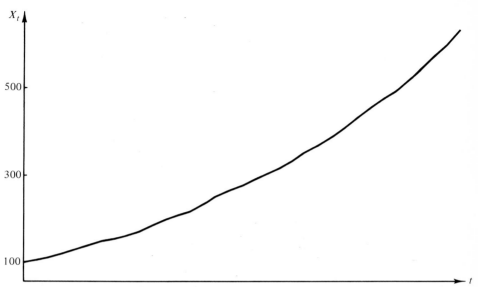

FIG. 1.6   *Data generated from* $X_t = 1.05X_{t-1} + \epsilon_t$.

time $t = -N$ with $X_{-N} = A$, then

$$X_t = A + (t + N)m + \sum_{j=0}^{t+N-1} \epsilon_{t-j}$$

so that

$$\mu_t = E[X_t] = A + (t + N)m$$

$$\lambda_{0,t} = \mathrm{var}[X_t] = (t + N)\sigma_\epsilon^2 \quad \text{and} \quad \lambda_{\tau,t} = \mathrm{cov}[X_t, X_{t-\tau}], \quad \tau \geq 0$$

$$= (t + N - \tau)\sigma_\epsilon^2$$

where σ_ϵ^2 is the variance of ϵ_t and is assumed to be finite. Thus,

$$\rho_{\tau,t} = \mathrm{corr}(X_t, X_{t-\tau}) = \frac{(t + N - \tau)}{\sqrt{(t + N)(t + N - \tau)}} = \sqrt{\frac{t + N - \tau}{t + N}}$$

Provided $t + N$ is large compared to τ, it is seen that all $\rho_{\tau,t}$ approximate unity. It follows that the sequence X_t is exceptionally smooth, but is also nonstationary since its variance is either increasing with t or, if $N = \infty$, then this variance is infinite. Figure 1.7 shows plots of the random walk (1.12.10) with $m = 0$, and the random walk with drift, obtained by setting $m = 1$ in (1.12.10). In both cases the series are generated from a zero-mean Gaussian white noise process ϵ_t with variance 16, and by setting $X_0 = 100$.

A random walk is an example of a class of nonstationary processes known as *integrated processes* that can be made stationary by the application of a time-invariant filter. If X_t is a random walk, then $\Delta X_t = X_t - X_{t-1} = m + \epsilon_t$ is

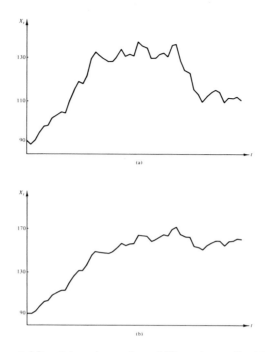

FIG. 1.7   *Data generated from (a) random walk; and (b) random walk with drift.*

a white noise process. Integrated processes are considered in more detail in the next section. Although they clearly do not represent a very wide class of nonstationary processes, they may introduce sufficient nonstationarity into the models as to produce adequate approximations to actually observed processes in many cases.

1.13 Integrated Processes

A process X_t is said to be an integrated process if it is generated by an equation of the form

$$a(B)(1 - B)^d X_t = b(B)\epsilon_t \qquad (1.13.1)$$

where ϵ_t is zero-mean white noise, $a(B)$, $b(B)$ are polynomials in B of orders p, q respectively ($a(B)$ being a stationary operator), and d is an integer. Such a process will be denoted $X_t \sim \mathrm{ARIMA}(p, d, q)$ (autoregressive integrated moving average of order p, d, q) and it will generally be assumed that the roots of $a(z) = 0$ all lie outside the unit circle. Thus the process obtained by differencing d times, $Y_t = (1 - B)^d X_t$, will be a stationary $\mathrm{ARMA}(p, q)$ process. Experience suggests that $d = 0$ or 1 will be appropriate for most observed processes, although occasionally $d = 2$ is required.

In the case $d = 1$, (1.13.1) has solution

$$X_t = \sum_{j=0}^{t+N-1} Y_{t-j}$$

if the process starts at time $t = -N$ and $X_{-N} = 0$. It follows that $\text{var}(X_t)$ $= O(N + t)$ and, noting that

$$\text{cov}(X_t, X_{t-\tau}) = \text{var}(X_{t-\tau}) + \text{cov}\left(\sum_{j=0}^{\tau-1} Y_{t-j}, \sum_{j=0}^{N+t-\tau-1} Y_{t-j-\tau}\right)$$

as the second term in this expression is $O(\tau)$, it follows that

$$\text{corr}(X_t, X_{t-\tau}) = \frac{O(N + t - \tau)}{\sqrt{O(N + t)O(N + t - \tau)}} = \sqrt{\frac{O(N + t - \tau)}{O(N + t)}}$$

so, if τ is small compared to $t + N$, $\text{corr}(X_t, X_{t-\tau}) \approx 1$. Thus, for an integrated process, the theoretical correlogram will take values near one for all nonlarge τ. For example, consider the process

$$X_t - aX_{t-1} = \epsilon_t + b\epsilon_{t-1} \tag{1.13.2}$$

The autocovariances for this process were derived in Section 1.7, from which it follows that the autocorrelations are

$$\rho_1 = \frac{(1 + ab)(a + b)}{1 + 2ab + b^2}, \qquad \rho_\tau = a\rho_{\tau-1}, \quad \tau \geqslant 2 \tag{1.13.3}$$

Now, as $a \to 1$ in (1.13.2), the resulting process is ARIMA(0, 1, 1):

$$X_t - X_{t-1} = \epsilon_t + b\epsilon_{t-1}$$

and it is seen that the autocorrelations (1.13.3) all tend to one. If d is greater than one, the same result holds. It follows that a plausible technique for determining the value of d is to form correlograms for the process differenced once, twice, etc. until a correlogram is obtained that does not display the shape found for integrated processes. This procedure is discussed further in Chapter 3.

It will be seen in subsequent chapters that integrated processes play a central role in the analysis of economic time series. This observed fact becomes crucially important when considering methods appropriate for analyzing the relationships among such series.

1.14 Models for Seasonal Time Series

In Section 1.10 the problem of dealing with time series possessing a deterministic seasonal component was discussed. There are two difficulties with such an approach. First, one would rarely, if ever, be in the happy position of *knowing* the exact functional form of the deterministic component. Further, it is extremely difficult to identify this form from actual time series of length typically available in practical applications. An analogous problem is that of fitting deterministic trend curves to a given (nonseasonal) time series. It is well known that one can very often find two or three fairly simple curves that fit the data almost equally well. However, when these curves are projected forward, the resulting forecasts can be markedly different (see Newbold [1973a]). Secondly, the seasonal component may well not be deterministic, but rather stochastic in nature.

Accordingly, it is worth investigating whether or not some extension of the nonseasonal autoregressive integrated moving average model of the previous section might provide a useful form for the representation of seasonal time series. Such a model is developed by Box *et al.* [1967] and further examined by Box and Jenkins [1970]. Suppose that X_t is a seasonal time series, with period s, so that $s = 4$ for quarterly and 12 for monthly data. One would like, in some way, to remove the effects of seasonality from this series to produce a nonseasonal series to which an ARIMA model could be built. Now, denote the resulting nonseasonal series by u_t; then it is reasonable to think in terms of the seasonal ARIMA filter

$$a_s(B^s)(1 - B^s)^D X_t = b_s(B^s) u_t \qquad (1.14.1)$$

where

$$a_s(B^s) = 1 - a_{1,s} B^s - \cdots - a_{P,s} B^{Ps}$$

$$b_s(B^s) = 1 + b_{1,s} B^s + \cdots + b_{Q,s} B^{Qs}$$

The rationale behind this is that if one were dealing with, say, monthly time series it would be expected that one January would be pretty much like the previous January, and indeed be similar to the last few previous Januarys and similarly for other months of the year. Further, if the same relationship between years held for every month of the year, (1.14.1) would be an appropriate representation, with the operator B^s indicating a relationship between points s time periods apart. It is important to note that nothing has been said about the autocorrelation properties of the series u_t. In particular, it is *not* assumed that the transformation (1.14.1) is such that

$$\text{corr}(u_t u_{t-ks}) = 0 \qquad \text{for all} \quad k > 0 \qquad (1.14.2)$$

It will not in general be possible (or, indeed, desirable) to find such a transformation. The only requirement is the weaker one that the filtered series u_t be free from seasonality. (Note that the series $u_t - a_1 u_{t-1} = \epsilon_t$, where ϵ_t is white noise, is nonseasonal without (1.14.2) holding.) Since u_t is a nonseasonal series, the possibility can be considered of approximating its behavior with the ARIMA model

$$a(B)(1 - B)^d u_t = b(B) \epsilon_t \qquad (1.14.3)$$

where

$$a(B) = 1 - a_1 B - \cdots - a_p B^p \qquad \text{and} \qquad b(B) = 1 + b_1 B + \cdots + b_q B^q$$

Combining (1.14.1) and (1.14.3) yields the multiplicative model

$$a(B) a_s(B^s)(1 - B)^d (1 - B^s)^D X_t = b(B) b_s(B^s) \epsilon_t \qquad (1.14.4)$$

The problem of fitting the model (1.14.4) to actual series will be examined in Chapter 3.

CHAPTER TWO

SPECTRAL ANALYSIS

All is flux, nothing is stationary.

HERACLEITUS 513 B.C.

2.1 Introduction

The previous chapter was largely concerned with the description of time series in terms of models—known as the time-domain approach. There is, however, an alternative approach, known as spectral or frequency-domain analysis, that often provides useful insights into the properties of a series. Since spectral results will only occasionally be used in later chapters, it will be sufficient to give just a brief survey of the basic concepts and interpretations of this theory. Considerably more detailed discussions will be found in the books listed in Section 2.9.

Let X_t be a zero-mean stationary process with autocovariances $\lambda_\tau = \text{cov}(X_t, X_{t-\tau})$ and corresponding autocorrelations $\rho_\tau = \lambda_\tau / \lambda_0$. The autocovariance generating function $\lambda(z)$ was introduced in Section 1.3 and was defined by

$$\lambda(z) = \sum_{\text{all } \tau} \lambda_\tau z^\tau \qquad (2.1.1)$$

and it will be assumed that this function exists, at least for $|z| = 1$. The function of particular interest in this chapter is defined by

$$s(\omega) = \frac{1}{2\pi} \lambda(z), \qquad z = e^{-i\omega} \qquad (2.1.2)$$

so that

$$s(\omega) = \frac{1}{2\pi} \sum_{\text{all } \tau} \lambda_\tau e^{-i\tau\omega} \qquad (2.1.3)$$

43

or

$$s(\omega) = \frac{\lambda_0}{2\pi} + \frac{1}{\pi} \sum_{\tau \geqslant 1} \lambda_\tau \cos \tau\omega \qquad (2.1.4)$$

as $\lambda_\tau = \lambda_{-\tau}$. The function $s(\omega)$ is called the power spectrum (strictly the power spectral density function) and from (1.3.10) is seen to be both real and positive. From the definition it is also seen that

$$s(\omega + 2\pi k) = s(\omega) \qquad \text{for integer } k \qquad (2.1.5)$$

so that the function $s(\omega)$ is periodic outside the interval $(-\pi, \pi)$ and thus one needs to consider the function only over this interval. From (2.1.4) it also follows that

$$s(-\omega) = s(\omega) \qquad (2.1.6)$$

and so it is usual to plot $s(\omega)$ against ω only for $0 \leqslant \omega \leqslant \pi$. One further important property follows from the fact that

$$\int_{-\pi}^{\pi} e^{i(k-\tau)\omega} \, d\omega = 0 \qquad \text{if } k \neq \tau$$
$$\qquad (2.1.7)$$
$$= 2\pi \qquad \text{if } k = \tau$$

and so by direct evaluation one finds from (2.1.3) that

$$\lambda_\tau = \int_{-\pi}^{\pi} e^{i\tau\omega} s(\omega) \, d\omega \qquad (2.1.8)$$

Thus, the sequence λ_τ and the function $s(\omega)$ comprise a Fourier transform pair, with one being uniquely determined from the other, so that the time-domain approach, which concentrates on the λ_τ sequence and the models derived from it, and the frequency-domain approach, which is based on interpretation of $s(\omega)$, are theoretically equivalent to one other. A result achieved in one domain may always be found in the other. The reason for considering the frequency domain is that on occasions results are easier to prove and interpretations are easier to make using the spectral approach.

The function $s(\omega)/\lambda_0$ has the properties of a probability density function over the range $-\pi \leqslant \omega \leqslant \pi$ since $s(\omega) \geqslant 0$ and, from (2.1.8) with $\tau = 0$,

$$\int_{-\pi}^{\pi} \frac{s(\omega)}{\lambda_0} \, d\omega = 1 \qquad (2.1.9)$$

In describing how spectral techniques are interpreted, it is convenient to define the rather more general spectral distribution function, given by

$$S(\omega) = \int_{-\pi}^{\omega} \frac{s(\omega)}{\lambda_0} \, d\omega \qquad (2.1.10)$$

so that $S(\omega)$ is monotonically nondecreasing as ω goes from $-\pi$ to π, with further properties $S(-\pi) = 0$, $S(\pi) = 1$, and $1 - S(-\omega) = S(\omega)$, from which it follows that $S(0) = \frac{1}{2}$.

Equation (2.1.8) may now be written in the alternative form

$$\rho_\tau = \frac{\lambda_\tau}{\lambda_0} = \int_{-\pi}^{\pi} e^{i\tau\omega} \, dS(\omega) \qquad (2.1.11)$$

Equation (2.1.11) is called the spectral representation of the autocorrelation sequence ρ_τ. This representation is one of the two fundamental relationships that need to be understood and interpreted for a proper appreciation of the spectral approach. The other fundamental relation, called the spectral representation of the stationary series X_t, is

$$X_t = \int_{-\pi}^{\pi} e^{it\omega} \, dz(\omega) \qquad (2.1.12)$$

where

$$E\left\{ dz(\omega) \; \overline{dz(\lambda)} \right\} = 0, \qquad \omega \neq \lambda$$
$$= \lambda_0 \, dS(\omega), \qquad \omega = \lambda \qquad (2.1.13)$$

To explain the importance and interpretation of these two representations, it is necessary to consider a sequence of models of increasing generality. First consider the simple linear cyclical process

$$X_t = a \, \cos(\omega_1 t + \theta) \qquad (2.1.14)$$

discussed in Section 1.4, where θ is a random variable rectangularly distributed on $(-\pi, \pi)$ and whose value is taken to have been determined at time $t = -\infty$ for the process to be considered stationary. It was shown in Section 1.4 that the autocovariance sequence for this process is given by $\lambda_\tau = (a^2/2) \cos \tau\omega_1$, and so (2.1.11) holds with

$$S(\omega) = \tfrac{1}{2}, \qquad 0 \leqslant \omega < \omega_1$$
$$= 1, \qquad \omega_1 \leqslant \omega \leqslant \pi$$

as may be seen by direct substitution. Thus $S(\omega)$ is the simple step function shown in Fig. 2.1.

Now consider the rather more general linear cyclical process

$$X_t = \sum_{j=1}^{m} a_j \, \cos(\omega_j t + \theta_j), \qquad 0 \leqslant \omega_j < \omega_{j+1} \leqslant \pi, \quad \text{all } j \qquad (2.1.15)$$

where θ_j, $j = 1, \ldots, m$, is a set of independent random variables, each rectangularly distributed on $(-\pi, \pi)$. By the arguments used in Section 1.4, it follows that

$$\lambda_\tau = \tfrac{1}{2} \sum a_j^2 \, \cos \tau\omega_j \qquad (2.1.16)$$

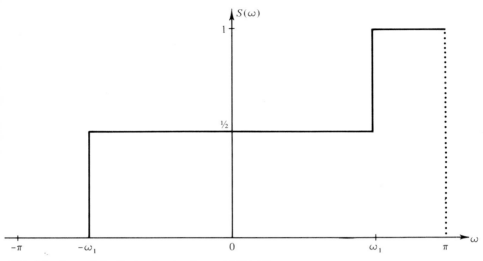

FIG. 2.1 *Spectral distribution function for model (2.1.14).*

and the power spectral distribution function corresponding to this covariance sequence is

$$S(\omega) = \tfrac{1}{2}, \qquad 0 \leqslant \omega < \omega_1$$

$$= \tfrac{1}{2}\left(\lambda_0 + \tfrac{1}{2}\sum_{j=1}^{k}a_j^2\right)/\lambda_0, \qquad \omega_k \leqslant \omega < \omega_{k+1}, \quad k = 1, \ldots, m-1$$

$$= 1, \qquad \omega_m \leqslant \omega \leqslant \pi \qquad\qquad\qquad\qquad (2.1.17)$$

where, of course, $\lambda_0 = \tfrac{1}{2}\sum_{j=1}^{m}a_j^2$ is the variance of X_t. There are a number of features of this example that are particularly noteworthy:

(i) X_t is the sum of m components, the jth component being $X_{j,t} = a_j \cos(\omega_j t + \theta_j)$, and is thus completely associated with the frequency ω_j.

(ii)   The components are uncorrelated with one another, which can be verified by noting that

$$X_{j,t} = a_j \cos\theta_j \cos\omega_j t - a_j \sin\theta_j \sin\omega_j t \quad \text{and} \quad E[\cos\theta_j] = E[\sin\theta_j] = 0$$

and then by remembering that θ_j, θ_k are independent and thus any function of θ_j will be independent of any function of θ_k.

(iii)   The variance of $X_{j,t}$ is $\tfrac{1}{2}a_j^2$ so that one can say that the contribution to the total variance of X_t is $\tfrac{1}{2}a_j^2$.

(iv) $S(\omega)$ is a monotonically nondecreasing step function, with steps at $\omega = \pm\omega_j, j = 1, \ldots, m$, and with a step of size $a_j^2/4\lambda_0$ at frequency $\omega = \omega_j$.

(v)   The probability function corresponding to this distribution function takes the form shown in Fig. 2.2, having peaks of height $a_j^2/4$ at frequencies $\omega = \pm\omega_j$.

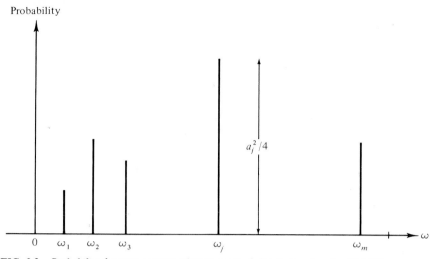

FIG. 2.2 *Probability function corresponding to spectral distribution function* (2.1.17).

A natural generalization of model (2.1.15) follows by allowing the number of components to tend to infinity, i.e., letting $m \to \infty$, giving

$$X_t = \sum_{j=1}^{\infty} a_j \cos(\omega_j t + \theta_j) \tag{2.1.18}$$

i.e.,

$$X_t = \sum_{j=1}^{\infty} (a_j \cos \theta_j \cos \omega_j t - a_j \sin \theta_j \sin \omega_j t) \tag{2.1.19}$$

with $0 \leqslant \omega_j < \omega_{j+1} \leqslant \pi$ for all j and the θ_j still being independent random variables rectangularly distributed on $(-\pi, \pi)$. Thus X_t now consists of a countably infinite number of components, but since X_t is required to have finite variance, the condition

$$\lim_{m \to \infty} \tfrac{1}{2} \sum_{j=1}^{m} a_j^2 = \mathrm{var}(X_t) < \infty$$

needs to be imposed.

The model in (2.1.18) is a fairly general one, but it is possible to go further. One could ask, Why should there be only a countably infinite number of components rather than an uncountably infinite number?[1] If one takes this further step, (2.1.19) becomes

$$X_t = \int_0^{\pi} \cos \omega t \, du(\omega) - \int_0^{\pi} \sin \omega t \, dv(\omega) \tag{2.1.20}$$

[1] If the objects in a set can each be associated with an integer, they are said to be countable, for example, the positive even integers, but if the objects in a set are too numerous for this, such as the number of points on the line (0, 1), they are said to be uncountably infinite.

where $du(\omega)$ and $dv(\omega)$ are random variables such that

$$E[du(\omega)\,du(\lambda)] = 0, \qquad \omega \neq \lambda$$

$$E[du(\omega)\,dv(\lambda)] = 0, \qquad \text{all } \omega, \lambda \qquad (2.1.21)$$

$$E[dv(\omega)\,dv(\lambda)] = 0, \qquad \omega \neq \lambda$$

Viewing the integral sign as an elongated S, denoting a sum, then (2.1.20) has X_t as a "sum" of an uncountable number of uncorrelated components of the form

$$X_t(\omega) = \cos t\omega\, du(\omega) - \sin t\omega\, dv(\omega) \qquad (2.1.22)$$

so that each component is associated with a specific frequency ω. If, further, $\text{var}(du(\omega)) = \text{var}(dv(\omega)) = 2\lambda_0\, dS(\omega)$, then

$$\text{var}(X_t(\omega)) = (\cos t\omega)^2\, \text{var}(du(\omega)) + (\sin t\omega)^2\, \text{var}(dv(\omega)) = 2\lambda_0\, dS(\omega)$$

and so

$$\lambda_0 = \text{var } X_t = \int_0^\pi \text{var}(X_t(\omega))\, d\omega = 2\int_0^\pi \lambda_0\, dS(\omega)$$

$$= 2\lambda_0(S(\pi) - S(0)) = \lambda_0$$

as required since $S(\pi) = 1$ and $S(0) = \frac{1}{2}$.

A more convenient form of (2.1.20) is achieved by defining the complex random variable

$$dz(\omega) = \frac{1}{2}\{du(\omega) + i\, dv(\omega)\} \qquad (2.1.23)$$

since then (2.1.20) becomes

$$X_t = \int_{-\pi}^\pi e^{it\omega}\, dz(\omega) \qquad (2.1.24)$$

with

$$E\left[dz(\omega)\,\overline{dz(\lambda)}\,\right] = 0, \qquad\qquad \omega \neq \lambda$$

$$= \lambda_0\, dS(\omega), \qquad \omega = \lambda \qquad (2.1.25)$$

where $\overline{X}$ denotes the complex conjugate of X. The more general model, so derived, has not been considered simply out of curiosity but because it can be shown that *any* stationary series may be represented in the form (2.1.20) or equivalently by (2.1.24) together with the condition (2.1.25). An elementary proof may be found in Chapter 17 of Wilks [1962].

There is one important way in which X_t generated by (2.1.24) is more general than X_t generated by the linear cyclical models (2.1.15) or (2.1.18). In theory at least, a series generated by such a linear cyclical model is deterministic, that is, it can be predicted without error given the infinite past of the series, but X_t generated by (2.1.24) need not have this property. (Deterministic and nondeterministic series are discussed further in Section 4.1.)

This aspect of the general model appears mathematically in the following way: there is a classical result in mathematics that states that any monotonically nondecreasing function can always be decomposed into three monotonic components, one of which is continuous and differentiable, the second is a step function, and the third has more unusual mathematical properties which cannot be interpreted in the present context and can safely be ignored. Thus the spectral distribution function $S(\omega)$ may be written

$$S(\omega) \approx S_1(\omega) + S_2(\omega) \tag{2.1.26}$$

where $S_1(\omega)$ is the continuous component and $S_2(\omega)$ the step function. This decomposition matches exactly Wold's decomposition of a stationary series, discussed in Section 1.11, with $S_1(\omega)$ being the spectral distribution function of the purely nondeterministic ($MA(\infty)$) component and $S_2(\omega)$ corresponding to the deterministic component, as in (2.1.17) for example.

Suppose now that X_t is purely nondeterministic, so that $S(\omega) = S_1(\omega)$, and let $\lambda_0 \, dS(\omega)/d\omega = s(\omega)$. Then one has

$$X_t = \int_{-\pi}^{\pi} e^{it\omega} \, dz(\omega) \tag{2.1.27}$$

with[1]

$$E\left[dz(\omega) \, \overline{dz(\lambda)} \right] = 0, \qquad \omega \neq \lambda$$
$$= s(\omega) \, d\omega, \qquad \omega = \lambda \tag{2.1.28}$$

As stated earlier, this formula together with (2.1.8) make up the two most important relationships in spectral analysis. Unfortunately it is likely to appear rather complicated and of an unfamilar form to many readers, and so some further explanation is in order. The need to use complex conjugates in the expectation term arises as follows. Let A and B be two complex random variables with zero means; then the covariance between them is defined as $\text{cov}(A, B) = E\{A \, \overline{B}\}$. This form is necessary so that the variance of A is real, remembering that variance is a measure of dispersion and thus needs to be real, i.e.,

$$\text{var}(A) \equiv \text{cov}(A, A) = E\{|A|^2\} \tag{2.1.29}$$

If $E\{dz(\omega)\} = 0$, which will be assumed to be so, then

$$\lambda_\tau \equiv \text{cov}(X_t, X_{t-\tau}) = E\left[X_t \overline{X}_{t-\tau} \right] = \int_{-\pi}^{\pi} \int_{-\pi}^{\pi} e^{it\omega} e^{-i(t-\tau)\lambda} E\left[dz(\omega) \, \overline{dz(\lambda)} \right] \tag{2.1.30}$$

The need for condition (2.1.28) is now seen as a requirement for X_t to be stationary, for otherwise λ_τ will be a function of time t, which is not possible

---

[1] The reason it is necessary to use the more awkward-looking terms $du(\omega)$ and $dz(\omega)$ rather than, say, $a(\omega) \, d\omega$, is clear from (2.1.28) where it is seen that $dz(\omega)$ is $O(\sqrt{d\omega})$ rather than $O(d\omega)$.

under stationarity. Using (2.1.28), the double integral in (2.1.30) becomes just the integral along the line $\omega = \lambda$ and so this integral becomes

$$\lambda_\tau = \int_{-\pi}^{\pi} e^{i\tau\omega} s(\omega) \, d\omega \qquad\qquad (2.1.31)$$

Thus, the spectral representation of the series X_t, alternatively called the Cramér representation, and the spectral representation of the covariance sequence λ_τ, given by (2.1.31), have been derived. The method used above to reach these relationships leads immediately to the interpretation of the spectrum $s(\omega)$ since $s(\omega) \, d\omega$ is the contribution to the variance of X_t attributable to the component $X_t(\omega)$ of X_t with frequencies in the range $(\omega, \omega + d\omega)$. Thus, if $s(\omega)$ had the shape shown in Fig. 2.3, one could say that the low-frequency (long-period) components were of little importance since they contribute little to $\text{var}(X_t)$, the components near frequency ω_1 were of greatest importance, and the high-frequency components (short periods, ω near π) are of intermediate importance. Note that, from its definition, the area under the curve $s(\omega)$ over the range $0 \leqslant \omega \leqslant \pi$ is $\frac{1}{2}$ var X_t.

To summarize this section, it has been proposed that any stationary time series can be thought of as the sum of (possibly) a noncountably infinite number of uncorrelated components, each associated with a particular frequency, and the importance of any group of components with frequencies

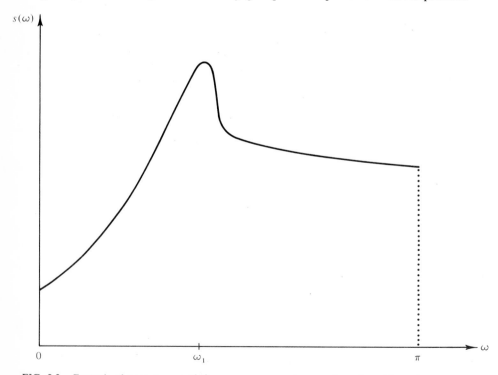

FIG. 2.3 *Example of a spectrum, with frequency components around ω_1 of most importance.*

falling into some narrow band is measured by their composite variance. This variance when plotted against frequency is the power spectral function.

As an illustration of these ideas, consider the simplest type of series, a zero-mean white noise, so that $X_t = \epsilon_t$. Then $\lambda_\tau = 0$, $\tau \neq 0$, and the corresponding spectrum is seen from (2.1.4) to be $s_\epsilon(\omega) = \sigma_\epsilon^2/2\pi$, i.e., a constant over the whole frequency range. Thus white noise is made up of components of all frequencies and all components are equally important, when measured in terms of their contribution to the variance of the series. This explains how the expression "white noise" arose to describe a purely random sequence, as if one considers visual light, a light for which one component is more important than others will appear as a color, but light in which all color components are present in equal amounts will appear white.

2.2 Filters

If X_t is a zero-mean stationary series with Cramér representation $X_t = \int_{-\pi}^{\pi} e^{it\omega} \, dz_x(\omega)$, then clearly $B^k X_t = X_{t-k} = \int_{-\pi}^{\pi} e^{i(t-k)\omega} \, dz_x(\omega)$. It then follows that if Y_t is a filtered version of X_t, given by

$$Y_t = \sum_{j=0}^{m} c_j X_{t-j} \tag{2.2.1}$$

i.e.,

$$Y_t = c(B)X_t \quad \text{where} \quad c(B) = \sum_{j=0}^{m} c_j B^j \tag{2.2.2}$$

then the Cramér representation of Y_t is just $Y_t = \int_{-\pi}^{\pi} e^{it\omega} c(e^{-i\omega}) \, dz_x(\omega)$. The spectrum of Y_t is then found by considering $\lambda_\tau^{(y)} \equiv \text{cov}(Y_t, Y_{t-\tau})$, i.e.,

$$E\left[Y_t \bar{Y}_{t-\tau} \right] = \int_{-\pi}^{\pi} \int_{-\pi}^{\pi} e^{it\omega} e^{-i(t-\tau)\lambda} c(e^{-i\omega}) c(e^{i\lambda}) E\left[dz_x(\omega) \ \overline{dz_x(\lambda)} \right]$$

which, from the orthogonality condition for $dz_x(\omega)$, becomes, as with (2.1.30),

$$\lambda_\tau^{(y)} = \int_{-\pi}^{\pi} e^{i\tau\omega} c(e^{-i\omega}) c(e^{i\omega}) s_x(\omega) \, d\omega$$

so that, comparing with (2.1.31), one obtains

$$s_y(\omega) = |c(e^{i\omega})|^2 s_x(\omega) \tag{2.2.3}$$

where $|c(e^{i\omega})|^2$ may be called the *filter function*. It is possible to interpret the effect on a series of applying a filter from this formula. Suppose, for example, that the coefficients c_j were such that the shape of $|c(e^{i\omega})|^2$ when plotted against ω is as shown in Fig. 2.4.

The spectrum of Y_t will then be the spectrum of X_t multiplied by this shape. If X_t were white noise, then the spectrum of Y_t would be shaped exactly as in Fig. 2.4. Since this shape has a peak at $\pi/2$, it follows that the component of Y_t with this frequency will be important compared to other

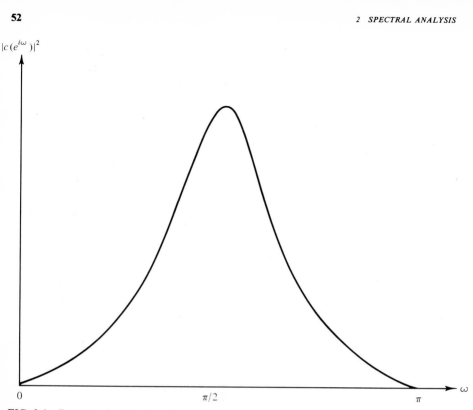

FIG. 2.4 *Example of a filter function.*

frequencies, and thus Y_t will tend to appear to contain a cycle with frequency approximately equal to $\pi/2$. Thus, in this case one outcome of filtering a white noise series has been to produce a series that appears to have a cyclical component, at least over some sample lengths. A possible set of coefficients producing such an effect is $c_0 = 1$, $c_2 = -2$, $c_4 = 1$, and all the other $c_j = 0$; then $|c(e^{i\omega})|^2 = 4(1 - \cos 2\omega)^2$ which has the shape shown in the figure. Early time series analysts were very concerned about the possibility of a filter inserting a spurious cycle into their data—the so-called Slutzky-effect. When using only time-domain analysis, it is difficult to investigate this possibility; but using spectral methods it is seen that interpretation becomes much easier.

As a second example of a filter, consider $Y_t = \sum_{j=0}^{m} X_{t-j}$, i.e., a moving average with constant weights. Then

$$|c(e^{i\omega})|^2 = [1 - \cos(m + 1)\omega]/(1 - \cos \omega)$$

which has the form shown in Fig. 2.5. If m is quite large, the first peak, around zero frequency, becomes of dominating importance, and so it follows that Y_t will consist almost exclusively of the low-frequency component of the original series X_t. Such a filter is called a *low-band pass* filter. It was suggested in Section 1.10 that such a moving average can be used to estimate the trend

$|c(e^{i\omega})|^2$

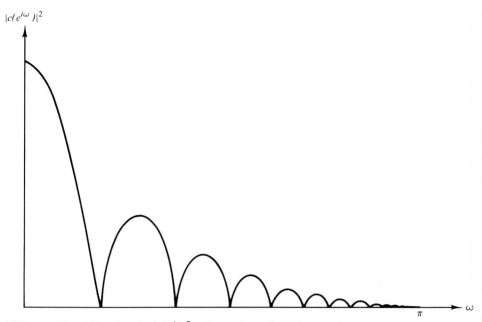

FIG. 2.5 *Plot of filter function* $|c(e^{i\omega})|^2 = [1 - \cos(m + 1)\omega]/(1 - \cos \omega)$.

component of a series. The reason for this can now be seen since a trend component will correspond to the very low (near zero) frequency component, since a trend is essentially monotonic and so does not repeat itself, meaning that trend has infinite period and hence zero frequency.

2.3 The Spectrum of Some Common Models

In the previous sections the shape of the spectrum of certain models was determined, in particular:

(i)   A white noise series has a flat spectrum over the whole range $0 \leqslant \omega \leqslant \pi$.

(ii)   A series with a cyclical component of frequency ω_0 has a spectrum with a tall, narrow peak at that frequency (theoretically, an infinitely tall, infinitely narrow, peak having finite area). In particular, a series with a seasonal component will have a spectrum with peaks at $j\omega_s$, $j = 1, 2, \ldots$, where ω_s is the principal seasonal frequency, so that $\omega_s = 2\pi/12$ for monthly data, and $\omega_s = 2\pi/4$ for quarterly data, and so forth. The secondary seasonal frequencies $j\omega_s$, $j > 1$ and $j\omega_s \leqslant \pi$, correspond to the harmonics of the seasonal component.

(iii)   A series with an important trend component will have a strong peak at the very low frequencies.

Now consider a stationary series generated by an ARMA(p, q) model, so that

$$a(B)X_t = b(B)\epsilon_t \qquad (2.3.1)$$

which has a corresponding MA(∞) representation $X_t = c(B)\epsilon_t$, where

$$c(z) = b(z)/a(z) \qquad (2.3.2)$$

The MA(∞) form indicates that X_t is merely a filtered version of the white noise series ϵ_t, and thus from (2.2.3) it follows that

$$s_x(\omega) = |c(z)|^2(\sigma_\epsilon^2/2\pi), \qquad z = e^{-i\omega} \qquad (2.3.3)$$

i.e.,

$$s_x(\omega) = |b(e^{i\omega})/a(e^{i\omega})|^2(\sigma_\epsilon^2/2\pi) \qquad (2.3.4)$$

This result can also be obtained from the definition of $s_x(\omega)$ in terms of the autocovariance generating function, obtained in Chapter 1, using Eqs. (2.1.2) and (1.7.18).

As a particular example, consider the AR(1) process $X_t = aX_{t-1} + \epsilon_t$. Then, from (2.3.4), this process has spectrum

$$s_x(\omega) = |1 - ae^{i\omega}|^{-2}(\sigma_\epsilon^2/2\pi) = \sigma_\epsilon^2/2\pi(1 + a^2 - 2a\cos\omega) \qquad (2.3.5)$$

which has the shape shown in Fig. 2.6. When the parameter a is positive, the low frequency (long-period) components are seen to be the more important, and it therefore follows that the series will be smoother than white noise. If a is negative, the high-frequency (short-period) components will dominate and the series will be less smooth, or more ragged, even than a white noise series. These interpretations agree exactly, of course, with those achieved from the time-domain considerations of Section 1.5.

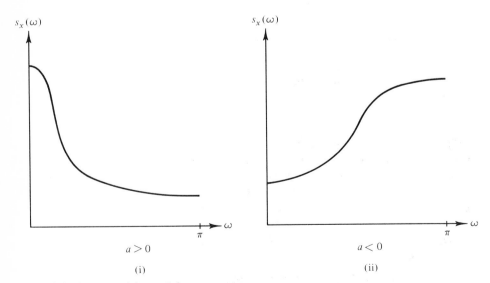

FIG. 2.6   *Spectrum of first-order autoregressive process.*

As a nears one, the low frequency peak will increase in height until, in the limit as $a \to 1$, the peak becomes infinite in height, as can be seen from the fact that

$$s_x(0) = \sigma_\epsilon^2 / 2\pi(1 - a)^2$$

Thus, in this limiting sense, the spectrum of the random walk series $X_t = X_{t-1} + \epsilon_t$ is of the form

$$s_x(\omega) = \sigma_\epsilon^2 / 4\pi(1 - \cos \omega) \tag{2.3.6}$$

Strictly speaking a random walk does not possess a spectrum since it does not have a finite variance, but apart from a possible problem at the zero frequency the function given in (2.3.6) can be taken to be the spectrum of a random walk. Similarly, the spectrum of an ARIMA(p, 1, q) process can be taken to be of the form

$$s_x(\omega) = (\sigma_\epsilon^2 / 4\pi)|b(e^{i\omega})/a(e^{i\omega})|^2(1 - \cos \omega)^{-1} \tag{2.3.7}$$

This spectrum will have basically the same shape as that of a random walk since its shape is largely determined by the term $(1 - \cos \omega)^{-1}$ in (2.3.7). This fact alone strongly illustrates the necessity of first differencing such series before further detailed analysis is possible.

If a series W_t is the sum of two zero-mean independent series X_t and Y_t, i.e., $W_t = X_t + Y_t$, then

$$s_w(\omega) = s_x(\omega) + s_y(\omega) \tag{2.3.8}$$

as can be seen from (2.1.2) and the fact that $\lambda_\tau^{(w)} = \lambda_\tau^{(x)} + \lambda_\tau^{(y)}$. Thus, if W_t is the sum of a stationary ARMA series and a seasonal component, its spectrum will have the peaks arising from the seasonal superimposed on the spectrum of the ARMA series.

It is generally not possible to decide on the exact generating model of a series by looking at its spectrum, but for the reasons just outlined the spectrum can be useful in helping to decide whether or not to first difference a series and whether to allow for a seasonal component when modeling. Further, if a series has a rather flat spectrum, this is an excellent indication that the series is a white noise, or at least very nearly so. Thus, if one believes that X_t is generated by $a(B)X_t = b(B)\epsilon_t$, then the spectrum of the filtered series $\epsilon_t = b^{-1}(B)a(B)X_t$ will be flat if one's belief is correct.

2.4 Aliasing

Economists find that the sampling interval of their data is rarely at their disposal, so that data are provided with a daily or a monthly or a quarterly sampling interval. For many series, the sampling interval could have been shorter and in some cases the series could have been recorded continuously, for example, an interest rate series or a temperature. The fact that one is sampling such a series at equally spaced intervals of time does imply certain important considerations in the interpretation of a spectrum. Suppose that the continuous series $X(t)$ contains a cyclical component of frequency $2\pi/\Delta$, where Δ is the sampling interval. Then the sampled series $\{X_t\}$ will contain no information about this component, which will seem to be a constant. As an example, if one recorded the temperature daily at noon one would have no information about daily cycle of temperature variation. However, if recordings were taken at noon and at midnight, the importance of the daily fluctuation in temperature could be estimated. In general, the highest frequency about which we have direct information is π/Δ and this is known as the "Nyquist frequency."

If this Nyquist frequency is denoted by ω_0, then its power will appear at $\omega = \pi$ in the power spectrum when defined as in the previous sections since $\lambda_\tau = E\{X_t X_{t-\tau}\}$ is strictly $\lambda_{\tau\Delta} = E\{X(t)X(t - \tau\Delta)\}$ in terms of the continuous series. If $X(t)$ contains a cycle with frequency greater than ω_0, then this component will be confused—or "aliased"—with a component having frequency less than ω_0. This follows because a high-frequency cosine wave systematically sampled at regular intervals appears the same as a low-frequency wave, as illustrated in Fig. 2.7. With Nyquist frequency ω_0, then if $\omega < \omega_0$, the frequencies ω, $2\omega_0 - \omega$, $2\omega_0 + \omega$, $4\omega_0 - \omega$, $4\omega_0 + \omega$, ... are all confounded and are aliases of one another, and the sum of all of their powers will appear at frequency ω in $s(\omega)$. As an example, suppose one's data are measured monthly, but the underlying continuous series contains a weekly cycle. Of course, there are not exactly four weeks to a month, so the weekly cycle will be seen as a peak in the spectrum—but at what frequency? Taking into account leap years, the average number of days in a month is 30.437 and so there are 4.348 weeks in the average month. Thus, the weekly cycle will induce a spike into the spectrum at frequency 0.348π since here π corresponds to the Nyquist frequency.

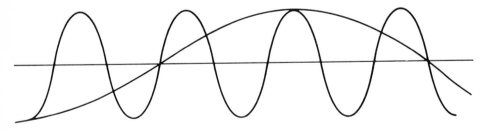

FIG. 2.7 *Illustration of aliasing.*

2.5 The Cross Spectrum

Cross-spectral analysis has nothing to do with angry ghosts, but is the generalization of the power spectrum to the two-series case and provides a sophisticated method of interpreting the relationship between a pair of series. Suppose that X_t and Y_t are both zero-mean, stationary series that are also jointly stationary so that

$$\lambda_\tau^{(x)} = E[X_t X_{t-\tau}], \quad \lambda_\tau^{(y)} = E[Y_t Y_{t-\tau}], \quad \lambda_\tau^{(xy)} = E[X_t Y_{t-\tau}] = \lambda_{-\tau}^{(yx)} \quad (2.5.1)$$

are all independent of t. This essentially means that the relationship between the two series is time invariate. Denote the cross-covariance generating function by $\lambda^{(xy)}(z)$, so that

$$\lambda^{(xy)}(z) = \sum_{\text{all } \tau} \lambda_\tau^{(xy)} z^\tau \quad (2.5.2)$$

Then the cross spectrum (strictly, cross-spectral density function) between the series X_t, Y_t is defined to be

$$s_{xy}(\omega) = \frac{\lambda^{(xy)}(z)}{2\pi}, \quad z = e^{-i\omega} \quad (2.5.3)$$

The corresponding inverse relation is

$$\lambda_\tau^{(xy)} = \int_{-\pi}^{\pi} e^{i\omega\tau} s_{xy}(\omega) \, d\omega \quad (2.5.4)$$

as is seen by multiplying both sides of (2.5.3) by $e^{i\tau\omega}$ and then integrating over $(-\pi, \pi)$. Since $\lambda_\tau^{(xy)} \neq \lambda_{-\tau}^{(xy)}$, in general, it follows that $s_{xy}(\omega)$ will usually be a complex function of ω, and so one could write

$$s_{xy}(\omega) = \text{co}(\omega) + iq(\omega)$$

where $\text{co}(\omega)$ and $q(\omega)$ are real functions, known as the cospectrum and the quadrature spectrum, respectively. These functions are all quite difficult to interpret as they stand, and so it is usual to define three further functions, that are much easier to interpret, from them. These are:

(i)
$$C(\omega) = \frac{|s_{xy}(\omega)|^2}{s_{xx}(\omega)s_{yy}(\omega)} = \frac{\text{co}^2(\omega) + q^2(\omega)}{s_{xx}(\omega)s_{yy}(\omega)} \quad (2.5.5)$$

where $s_{xx}(\omega)$, $s_{yy}(\omega)$ are respectively the spectral density functions of X_t, Y_t. $C(\omega)$ will here be called the coherence function.[1]

(ii)
$$\phi(\omega) = \tan^{-1}[q(\omega)/\text{co}(\omega)] \quad (2.5.6)$$

which is called the phase.

(iii)
$$R_{xy}(\omega) = |s_{xy}(\omega)|/s_{yy}(\omega) \quad (2.5.7)$$

known as the gain function.

All three of these functions are real and so can be conveniently plotted against ω, and due to symmetry it is only necessary to plot them over the range $0 \leqslant \omega \leqslant \pi$.

To learn how to interpret these plots, it is convenient to use the Cramér

---

[1] Some other writers call $C(\omega)$ the coherency function, whereas others use this name for $\{C(\omega)\}^{1/2}$.

representations of the series:

$$X_t = \int_{-\pi}^{\pi} e^{it\omega} \, dz_x(\omega), \qquad Y_t = \int_{-\pi}^{\pi} e^{it\omega} \, dz_y(\omega) \tag{2.5.8}$$

It was seen earlier that to ensure that X_t is stationary it is necessary that $E\{dz_x(\omega) \, \overline{dz_x}(\lambda)\} = 0$, $\omega \neq \lambda$. For X_t and Y_t to be jointly stationary requires

$$E\left\{ dz_x(\omega) \, \overline{dz_y}(\lambda) \right\} = 0, \qquad \omega \neq \lambda \tag{2.5.9}$$

Consider now $\lambda_\tau^{(xy)} = E\{X_t Y_{t-\tau}\}$, which from (2.5.8) may be written

$$\lambda_\tau^{(xy)} = \int_{-\pi}^{\pi} \int_{-\pi}^{\pi} e^{it\omega} e^{-i(t-\tau)\lambda} E\left\{ dz_x(\omega) \, \overline{dz_y}(\lambda) \right\} \tag{2.5.10}$$

Hence, from (2.5.9), one gets

$$\lambda_\tau^{(xy)} = \int_{-\pi}^{\pi} e^{i\tau\omega} E\left\{ dz_x(\omega) \, \overline{dz_y}(\omega) \right\}$$

so that

$$s_{xy}(\omega) \, d\omega = E\left\{ dz_x(\omega) \, \overline{dz_y}(\omega) \right\} \tag{2.5.11}$$

It is possible to give immediately an interpretation to the coherence $C(\omega)$ by noting that it can be written

$$C(\omega) = \frac{\left(E\left\{ dz_x(\omega) \, \overline{dz_y}(\omega) \right\} \right)^2}{E\left\{ dz_x(\omega) \, \overline{dz_x}(\omega) \right\} E\left\{ dz_y(\omega) \, \overline{dz_y}(\omega) \right\}} \tag{2.5.12}$$

Thus $C(\omega)$ is the square of the coefficient of correlation of $dz_x(\omega)$ and $dz_y(\omega)$. It follows that $0 \leqslant C(\omega) \leqslant 1$ and $C(\omega)$ should be interpreted in exactly the same way as the square of a correlation coefficient. Thus if $C(\omega)$ is near one, it means that the ω-frequency components of the two series are highly (linearly) related, but a value near zero means that these corresponding frequency components are only slightly related. The stationarity condition (2.5.9) implies that if X_t and Y_t are related, then they are related only through their corresponding frequency components. Similarly, by its construction, the gain function $R_{xy}(\omega)$ is the regression coefficient of the ω-frequency component of X_t on the corresponding ω-frequency component of Y_t, and can be interpreted accordingly.

To understand the use of the phase function $\phi(\omega)$, it is useful to consider a model of the form

$$X_t = aY_{t-k} + V_t \tag{2.5.13}$$

where Y_t and V_t are independent stationary series. Putting the terms of the model into the Cramér representation form gives

$$\int_{-\pi}^{\pi} e^{it\omega} \, dz_x(\omega) = \int_{-\pi}^{\pi} e^{it\omega} a e^{-ik\omega} \, dz_y(\omega) + \int_{-\pi}^{\pi} e^{it\omega} \, dz_v(\omega)$$

so that

$$dz_x(\omega) = ae^{-ik\omega}\, dz_y(\omega) + dz_v(\omega) \qquad (2.5.14)$$

Then multiplying by $\overline{dz_y}(\omega)$ and taking expectations, from (2.5.11), yields

$$s_{xy}(\omega)\, d\omega = ae^{-ik\omega}E\left\{ dz_y(\omega)\, \overline{dz_y}(\omega)\right\} = ae^{-ik\omega}s_{yy}(\omega)\, d\omega \qquad (2.5.15)$$

since $E\{dz_v(\omega)\, \overline{dz_y}(\omega)\} = 0$, since Y_t, V_t were taken to be independent. It follows immediately from (2.5.15) and the definition of $\phi(\omega)$ that

$$\phi(\omega) = \tan^{-1}\left| \frac{-a \sin k\omega\, s_{yy}(\omega)}{a \cos k\omega\, s_{yy}(\omega)} \right| = -k\omega$$

Thus, the phase diagram, being the plot of $\phi(\omega)$ against ω, will be useful in finding any lag relationship between the series one is investigating. If $\phi(\omega)$ is a straight line over some frequency band, then the direction of slope tells one which series is leading and the amount of the slope gives the extent of the lag. It is important to note that the lag k need not be an integer multiple of the sampling period, so that it is possible to detect a lag of say 1.5 months when using monthly data. When investigating lags using a time-domain approach, it is much more difficult to pick up such noninteger lags.

In interpreting a phase diagram it is important to realize that one has to have a causal model of the form (2.5.13) in mind since if there is a feedback relationship between the two series, then the above interpretation is no longer appropriate. If the two series X_t and Y_t are related in a more complicated manner, such as a filter form $X_t = \sum_{j=0}^{m} a_j Y_{t-j}$, then the phase diagram is no longer easy to interpret. This is seen by noting that

$$dz_x(\omega) = a(\omega)\, dz_y(\omega) \qquad \text{where} \quad a(\omega) = \sum_{j=0}^{m} a_j e^{-ij\omega}$$

so that $C(\omega) = 1$ for all ω. However,

$$\phi(\omega) = \tan^{-1}\left[\frac{-\sum a_j \sin j\omega}{\sum a_j \cos j\omega} \right]$$

which will typically be of a nonsimple form.

It should also be noted that there is a degree of ambiguity in the definition of $\phi(\omega)$ since it is a periodic function, with $\phi(\omega + 2j\pi) = \phi(\omega)$ if j is an integer. It is usual to restrict $\phi(\omega)$ to the region $-\pi < \phi(\omega) \leqslant \pi$, but the periodic nature of the function is important to remember when interpreting the diagram. An example is when $X_t = -aY_t$, since then $dz_x(\omega) = -a\, dz_y(\omega)$, which may be written $dz_x(\omega) = ae^{i\pi}\, dz_y(\omega)$ and so $\phi(\omega) = \pi$. However, the estimate of $\phi(\omega)$ will be inclined to lie about the value π; and, from the periodic nature of the definition, this means that some values will lie just under π and others just above $-\pi$, having "flipped over" by the amount 2π.

Some examples of the use of spectral and cross-spectral analysis are given in Section 2.7, after the estimation problem has been discussed in the following section.

2.6 Estimation of Spectral Functions

The estimation of a spectrum has proved to be, for the statistician, one of the more interesting and difficult estimation problems so far encountered. One aspect of this difficulty arises because one is attempting to estimate all the uncountably infinite number of points of a continuous curve $s(\omega)$, $0 \leqslant \omega \leqslant \pi$, from just a finite amount of data x_t, $t = 1, \ldots, n$. An obvious starting point in such an attempt is the definition (2.1.4)

$$s(\omega) = \frac{\lambda_0}{2\pi} + \frac{1}{\pi} \sum_{\tau=1}^{\infty} \lambda_\tau \cos \tau\omega \qquad (2.6.1)$$

Some of the λ_τ may be estimated from the data by

$$\hat{\lambda}_\tau = \frac{1}{n(\tau)} \sum_{t=1}^{n-\tau} (x_t - \overline{x})(x_{t+\tau} - \overline{x}), \qquad \tau = 0, 1, \ldots, n-1 \quad (2.6.2)$$

where $n(\tau)$ could be n or $n - |\tau|$ or some other appropriate quantity. In fact, it turns out that the choice of $n(\tau) = n$ has certain advantages, even though it means that a biased estimate of λ_τ is being used. The proposed estimate is thus

$$\hat{s}(\omega) = \frac{\hat{\lambda}_0}{2\pi} + \frac{1}{\pi} \sum_{\tau=1}^{n-1} \hat{\lambda}_\tau \cos \tau\omega = \frac{1}{2\pi} \sum_{\tau=-n+1}^{n-1} \hat{\lambda}_\tau \cos \tau\omega \qquad (2.6.3)$$

If, in (2.6.2), $n(\tau) = n$, then (2.6.3) is seen, after some algebra, to be just

$$\hat{s}(\omega) = \frac{1}{2\pi n} \left| \sum_{t=1}^{n} (x_t - \overline{x}) e^{i\omega t} \right|^2 \qquad (2.6.4)$$

so that this estimate of the spectrum is seen to be proportional to the periodogram discussed in Section 1.4. Thus, in a sense, Schuster, when introducing the periodogram at the very beginning of this century, was anticipating the development of spectral analysis. However, it has been found that in many ways the estimate (2.6.4) is an unsatisfactory estimate, and in particular it is not consistent since the variance of $\hat{s}(\omega)$ does not tend to zero as n, the sample size, tends to infinity. Further, the covariance between $\hat{s}(\omega_1)$ and $\hat{s}(\omega_2)$, that is, the covariance between estimates at two different frequencies, does tend to zero as $n \to \infty$, so that for large n, $\hat{s}(\omega)$ has a tendency to become very jagged and the possibility of finding spurious "cycles" in one's data is enhanced. This is the reason the periodogram proved to be an unsatisfactory tool of analysis and has been largely superseded by the spectrum.

To produce estimates having better properties, the following class has been considered by many writers:

$$\hat{s}_k(\omega) = \frac{1}{2\pi} \sum_{\tau=-n+1}^{n-1} k_n(\tau)\hat{\lambda}_\tau \cos \tau\omega \qquad (2.6.5)$$

where the constants $k_n(\tau)$ are derived from a function $k(\ )$ by

$$k_n(\tau) = k(\tau/M_n) \qquad (2.6.6)$$

for some number M_n depending on n. The function $k(\ )$ is known as the "lag window generator" and M_n is called the "truncation point." There has been considerable debate about how these quantities should be selected, largely because there is no single criterion that is generally accepted to be the best for use in selecting among possible estimates. A widely used function, suggested by Parzen [1961], is

$$
\begin{aligned}
k(u) &= 1 - 6u^2 + 6|u|^3, & |u| &\leqslant 0.5 \\
&= 2(1 - |u|)^3, & 0.5 &< |u| < 1.0 \qquad (2.6.7) \\
&= 0, & |u| &\geqslant 1
\end{aligned}
$$

This produces consistent estimates, provided $M_n \to \infty$ and $M_n/n \to 0$ as $n \to \infty$, and also ensures that the spectral estimate cannot take negative values. However, one pays for consistency by having a biased estimate, the bias being approximately proportional at frequency ω to the value of the actual spectrum $s(\omega)$.

If the sequence $k_n(\tau)$, for fixed n, has Fourier transform window $k_n(\omega)$, so that

$$k_n(\tau) = \int_{-\pi}^{\pi} \cos \tau\omega k_n(\omega) \, d\omega \qquad (2.6.8)$$

then it may be shown that

$$\hat{s}_k(\omega) = \int_{-\pi}^{\pi} k_n(\lambda)\hat{s}(\omega - \lambda) \, d\lambda$$

where $\hat{s}(\omega)$ is given by (2.6.4) and is proportional to the periodogram. Thus, since $k_n(\lambda)$ takes its main maximum at $\lambda = 0$, $\hat{s}_k(\omega)$ for given ω is seen to be a weighted average of periodogram-type estimates at frequencies centered on ω. This may be thought of as viewing a periodogram through a Gothic window and averaging what one sees. As one moves to an adjacent frequency value ω', $\hat{s}_k(\omega')$ will still contain part of the periodogram used in forming $\hat{s}_k(\omega)$, and this will ensure a positive covariance between $\hat{s}_k(\omega)$ and $\hat{s}_k(\omega')$, provided ω and ω' are sufficiently near, and so an estimated curve that is smoother than a periodogram will result. The value of M_n chosen will determine the "width" of the window and hence the size of this covariance. The value of M_n chosen is for the analyst to decide. As a rule of thumb, for typical sample sizes found in economics, M_n is usually not more than $n/3$, and experienced analysts find it helpful in interpreting possible cyclical peaks in the spectrum to superimpose three estimates of the spectrum using three different values of M_n, such as $n/5$, $n/4$, and $n/3$.

The cross spectrum may be estimated using (2.6.5) but with $\hat{\lambda}_\tau$ being replaced by an estimate of the corresponding cross covariance, i.e.,

$$\hat{\lambda}_\tau^{xy} = \frac{1}{n} \sum_{t=\tau+1}^{n} (x_t - \bar{x})(y_{t-\tau} - \bar{y})$$

Estimates of coherence and phase may be derived directly from the estimate of the cross spectrum.

It will be seen from (2.6.4) that, since the mean is subtracted from the data, it necessarily follows that $\hat{s}(0) = 0$ and, since $\hat{s}_k(\omega)$ is a smoothed periodogram, it then follows that spectral estimates at zero frequency are not very reliable, and similarly for the cross spectrum.

Asymptotic values for the mean, variance, and covariance of spectral estimates are available, and Neave [1970] has provided exact formulas for these quantities, although these depend on the actual autocovariance sequence of the process. Asymptotically $\hat{s}_k(\omega)$ is normally distributed, but a chi-squared approximation seems more appropriate for series of fewer than 200 terms. Some suggestions have also been made in the literature on the distribution of the coherence and phase functions, and many of these possibilities have been investigated, using simulation techniques, by Granger and Hughes [1968], although the conclusions there reached for the spectral estimate $\hat{s}_k(\omega)$ have to be somewhat modified in the light of comments made by Neave [1972]. In general terms, the simulation studies indicate that the main shape of a spectrum should be observed even when estimated from quite short series and, similarly, simple lags between series should be detected, but the estimated coherence derived from short series is less satisfactory since the estimates appear to be biased toward the value 0.5.

One important feature of the phase diagram does result from the theory and is supported by the simulations: the estimate of phase is rectangularly distributed over $(-\pi, \pi)$ if the true coherence is zero. This is intuitively sensible since one cannot hope to estimate a measure of lag between two unrelated components. It follows that if the phase diagram is extremely unstable, then this is good evidence that the two series are uncorrelated over the appropriate set of frequencies.

Considerably more detail relating to the problems of estimation and the properties of the estimates may be found in the books referenced in Section 2.9, particularly those by Parzen, Hannan, and Koopmans.

The following two sections of this chapter discuss some applications of spectral techniques.

2.7 The Typical Spectral Shape and Its Interpretation

The interpretation of the power spectrum of a single series depends upon the existence of any underlying smooth shape in the estimated spectrum plus the position and height of any possibly significant peaks. The peaks can be interpreted as indicating "cyclical" components in the series, having almost constant periods. The underlying smooth shape conceptually can be used to suggest possible models for the noncyclical component. The original uses of spectral analysis in economics concentrated on these aspects. Peaks at the seasonal frequencies were interpreted as a clear indication that the series contained a seasonal component and so looking at the spectrum of a seasonally adjusted series is helpful in evaluating the effectiveness of the adjustment process. The use of spectral techniques in dealing with problems of seasonal adjustment is considered further in the next section. The only other possibly meaningful peaks would be those corresponding to the business cycles in the economy. Although it is now generally accepted that the business cycle cannot be well represented by the sum of a few purely periodic components, the possibility of there being extra power at a low-frequency band, corresponding to the business cycle component, has been considered by a number of writers, for example, Howrey [1968]. Given sufficient data, very subtle cycles can be detected.

If the estimated spectrum appears flat, without peaks and without any clear tendency to follow a smooth curve, then this suggests that the series is white noise. Confidence bands can also be constructed to test the hypothesis that the series is white noise. This procedure has been used to test the hypothesis that prices from speculative markets follow a random walk, by forming the spectrum of price changes, in Granger and Morgenstern [1970] and Labys and Granger [1970]. If the underlying shape is not flat, this observation might be used to suggest a model for the series; but since there are usually many models that could produce almost identical shapes, the estimated spectrum is rarely found to be useful in the model identification problem to be discussed in Chapter 3.

When economic data were first analyzed by spectral methods in the early 1960s, a somewhat unexpected result was the observation that most of the estimated spectra had almost identical shapes, being very high at very low frequencies, falling steeply as frequency increased, becoming flat at high frequencies (ω near π) and with only the occasional peak at a seasonal frequency to break up the inherent smoothness. Such curves were called the "typical spectral shape" [Granger, 1966], and an example is shown in Fig. 2.8, which shows the spectrum of the Federal Reserve Board index of industrial production using 486 monthly observations, and is from Nerlove [1964]. One consequence of this shape is that it is very difficult to observe subtle features of the spectrum at low frequencies, due to an estimation problem called leakage, which means that the estimate of the spectrum near a peak is badly

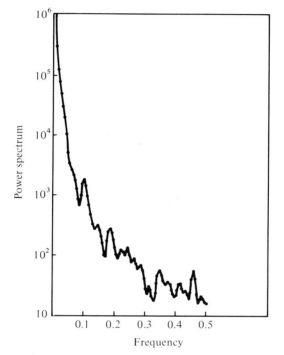

FIG. 2.8 *Spectrum of Federal Reserve Board index of industrial production.*

upward biased. The basic reason for the typical spectral shape arising is the very high correlation observed between adjacent values of the levels of most economic series when observed frequently, such as at monthly intervals. It might be said that the economy has considerable momentum and that the size of changes in economic variables is generally small compared to the current level of the variable. Since spectral analysis requires rather large samples for efficient estimation, most of the series originally analyzed were those recorded monthly or at even shorter intervals.

It was originally suggested that an appropriate model for a series having a typical spectral shape, without seasonal peaks, was a process of the form $X_t = aX_{t-1} + \epsilon_t$ where ϵ_t is white noise and a is very near one, values for a of 0.95 or greater being suggested. While such a model would produce a typical spectral shape, it is now recognized that a wider, and potentially more fruitful, class will also produce this spectral shape, namely the ARIMA class considered in Section 1.13, as can be seen from Eq. (2.3.7). In any case, the most important consequence of the observed typical spectral shape is that it will be very difficult to perform subtle analysis on the levels of most economic variables, a point which will be repeated elsewhere in this book. If the series is first filtered, to remove or at least greatly reduce the considerable peak at zero frequency, it becomes much easier to identify other features of the series. Experience suggests that an appropriate filter is the first difference, so that changes (or possibly proportional changes) should be used rather than levels. The spectra of most change series do not have the typical shape and usually are not flat, suggesting that an ARMA model is appropriate. The properties of the estimates of a spectrum that is near-flat are also much more satisfactory, this also being true for the cross-spectral estimates.

2.8 Seasonal Adjustment: An Application of the Cross Spectrum

A very casual glance at many raw economic series will indicate the presence of an important seasonal component. Obvious examples include personal consumption, production, sales and inventories, food prices, and imports and exports. Since this seasonal component is easily explained, at least superficially, and is occasionally of such overpowering importance, in terms of its contribution to the total variance of the series, that other components are difficult to observe in the plot of the data, analysts frequently want to remove the seasonal component so that they can better observe components of the series that are operationally more important, in particular the long swings. A method of removing the seasonal component from a series is called a seasonal adjustment technique. There are a number of problems that arise from the use of such techniques, and spectral analysis has proved useful in connection with several of these problems.

It is first necessary to define the seasonal component of a series, and this is best done by considering the spectrum of the series. Loosely, a possible definition of the seasonal is that component of the series that causes the spectrum to have peaks at or around seasonal frequencies, that is, frequencies $\omega_{sk} = 2\pi k / p_s$ where $k = 1, 2, 3, \ldots$ and p_s is the period of the seasonal, and thus takes the value 4 for data recorded at quarterly intervals, 12 for monthly data, and so forth. If the seasonal consists of an unchanging, strictly periodic component, then it will contribute power in the theoretical spectrum only at the seasonal frequencies and not elsewhere in the frequency band surrounding these frequencies. However, if the seasonal component is taken to change somewhat from year to year, due possibly to random elements in the causal mechanism, the seasonal component will then contribute power to the whole narrow frequency band surrounding the seasonal frequencies. There is a lot that can be said in favor of this more general definition of a seasonal component. Under either definition, the estimated spectrum will have peaks at and around the seasonal frequencies. It is important to note that the seasonal component causes there to be *extra* power at these frequencies, leading to peaks. The seasonal component cannot be defined simply as giving *some* power at the seasonal frequencies since, for example, white noise does not have zero power at these frequencies. In fact, virtually any ARMA process has a spectrum with some power at the seasonal frequencies. Although the above definition can be made both more rigorous and more satisfactory, as it stands it does provide a useful and usable criterion for whether or not a given series does contain a seasonal component. Consequently, it also provides the first step in evaluating the effectiveness of a method of seasonal adjustment. This follows from the fact that a good seasonal adjustment technique should remove the seasonal component, and no more, and thus the estimated spectrum of the seasonally adjusted series should contain no peaks at seasonal frequencies.

An example of this use of the spectrum is shown in Figs. 2.9 and 2.10 from Nerlove [1964]. The continuous lines show the spectra of total U.S. unemployed (monthly data, July 1947–December 1961) and a particular subset of the unemployed, being males aged 14–19. The broken lines show the estimated spectra of these series after being seasonally adjusted by a technique used by the Bureau of Labor Statistics in the early 1960s. The original series show strong evidence of seasonal components, that for the teenagers being extremely strong since a large cohort enters the job market at the same time each year. The adjusted series have spectra of the "typical" shape, since levels of unemployment were used, and no peaks at seasonal frequencies. Unfortunately, the peaks have often been replaced by dips, suggesting that the adjustment procedure has removed rather too much power at seasonal frequencies. Several seasonal adjustment methods are inclined to have this undesirable property. In the time domain, the presence of a seasonal component is shown by high positive correlogram values at lags 12, 24, 36, etc., if using monthly data, whereas the adjusted series may now contain negative autocorrelations at some of these lags. For integrated processes, this statement applies to the first differences of the series.

Some methods of seasonal adjustment consist of simply applying an appropriate filter to the series. An example of such a filter is the series Y_t formed from the original monthly series X_t by

$$Y_t = X_t - \frac{1}{m + s + 1} \sum_{k=-s}^{m} X_{t-12k}$$

Thus, the new January figure is the old figure minus the average January value over adjacent years. Two particular cases are important: (i) $s = m$ so that a symmetric filter is used, and (ii) $s = 0$, giving a completely one-sided filter. Symmetric filters are typically used in adjusting historical data, but to adjust a current series up to the most recent value a one-sided filter has to be used. Since the spectrum of Y_t can be determined from that of X_t and the filter function, as explained in Section 2.2, the effect of a seasonal adjustment filter can be evaluated theoretically rather than by using real data as illustrated above. To give a simple example, consider the twelfth-differencing method introduced in Section 1.14, so that $Y_t = X_t - X_{t-12}$, and, in terms of spectra,

$$s_y(\omega) = 2(1 - \cos 12\omega)s_x(\omega)$$

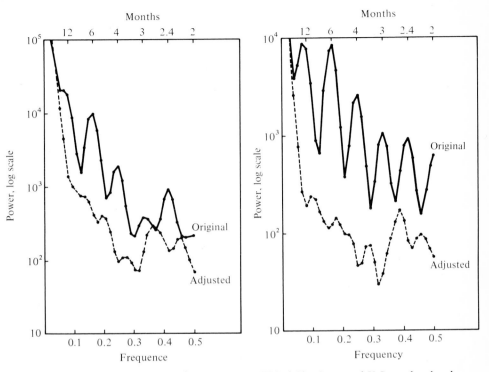

FIG. 2.9 *Spectra of total U.S. employed, before and after seasonal adjustment.*

FIG. 2.10 *Spectra of U.S. employed males, 14–19, before and after seasonal adjustment.*

If $s_x(\omega)$ is finite for all ω, it is seen that $s_y(\omega) = 0$ at $\omega = 2\pi k/12$, $k = 0, 1, \ldots$, so Y_t will have no seasonal component but will have induced zero spectral values at the zero and seasonal frequencies. It is to allow for this overadjustment that the more complicated seasonal filter considered in (1.14.1) has been suggested.

Unfortunately, the seasonal adjustment methods used in practice, particularly those of government agencies, do not correspond to simple filters and so cannot be completely evaluated by the method just proposed. The non-linearities introduced are designed to cope with values taken in exceptional years, due to events such as dock strikes, or very poor harvests, or with fairly rapidly changing seasonal patterns. A further problem is the tendency to use a symmetric filter as the current value is approached until eventually a one-sided filter is used on the most recent term of the series. This corresponds to a nonstationary filter and induces a variable lagging of the low-frequency component. Because the available theory is not always able to cope with the evaluation of actual seasonal adjustment techniques, it becomes necessary to analyze these techniques by simulation.

To illustrate this, consider the series X_t generated by

$$X_t = S(t) + X_t' \tag{2.8.1}$$

where $S(t)$ is a pure seasonal component, having spectrum with power *only* over narrow frequency bands surrounding the seasonal frequency, and X_t' contains no seasonal component. Suppose a seasonal adjustment procedure is applied to X_t, producing the series $A(X_t)$. A strong criterion for a good method of adjustment is that $A(X_t)$ should closely approximate X_t'. A first step in evaluating this approximation consists of comparing the spectra of $A(X_t)$ and X_t' to see if they have similar shapes. This corresponds to the method used by Nerlove mentioned above. The obvious second step is to estimate the cross spectrum between $A(X_t)$ and X_t'. With real data, X_t' is not observable, so that simulated data have to be used. There is little point in forming the cross spectrum between X_t and $A(X_t)$ since the presence of the important seasonal component $S(t)$ will spoil the estimates over various important frequency bands. The simulation technique has been used by Godfrey and Karreman [1967] for various simulated series and a number of adjustment methods. The kind of result they obtained is illustrated in Fig. 2.11, in which a Bureau of the Census method, known as X10, is applied to a series without a seasonal component and generated by an AR(2) model. The spectra of the original and the adjusted series suggest that the method passes the first test quite well since no "seasonal" has been removed and no dips induced in the spectrum. The method is used on historical data, not up to the last few values, and the flat phase estimate, lying around zero in part (ii) of the figure, suggests that no lags are induced in the adjusted series compared to the original nonseasonal data. However, the coherence diagram, in part (iii) of the figure, does give cause for concern about the method of adjustment. An ideal method should produce a coherence diagram with values near one at all frequencies. In practice, the low-frequency component has remained unchanged, but coherence at higher frequencies is considerably reduced. These results can be interpreted as follows: let

$$X_t = X_t^{(1)} + X_t^{(2)} \qquad \text{and} \qquad A(X_t) = A_t^{(1)} + A_t^{(2)}$$

where X_t is the original nonseasonal series and $A(X_t)$ is the adjusted series. $X_t^{(1)}$ and $A_t^{(1)}$ are the low frequency components of the series, with frequencies of say less then $2\pi/20$, and $X_t^{(2)}$, $A_t^{(2)}$ are the other frequency components. Ideally $X_t^{(1)} = A_t^{(1)}$ and $X_t^{(2)} = A_t^{(2)}$, but in practice the first equality holds but not the second, even though $X_t^{(2)}$ and $A_t^{(2)}$ have similar spectral shapes. Thus, the adjustment mechanism has removed part of the higher frequency component $X_t^{(2)}$ and replaced it with "noise" having similar spectral shape. It follows that the low-frequency component, which is the one economists are most interested in, is virtually unaffected by the adjustment method, but the higher frequency component is much affected. This certainly has important

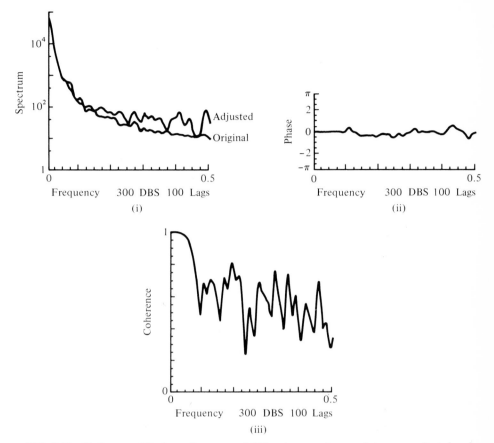

FIG. 2.11 (*i*) *Spectra,* (*ii*) *phase diagram, and* (*iii*) *coherence diagram between unadjusted and seasonally adjusted series.* (*From Godfrey and Karreman* [1967], *reprinted by permission of Princeton University Press.*)

implications for model building. Godfrey and Karreman find similar results when the input series does contain a seasonal and for various nonlinear adjustment techniques. The results might suggest that since the low frequency component is the one of greatest interest and since the higher frequency component is spoiled by the adjustment method, a simpler procedure would be to just filter out the higher frequency component of the original series, including any seasonal, leaving just that part of the series of greatest interest.

A great deal more could be said on the problem of seasonal adjustment, but to do so would take us too far from the stated objectives of this book. Nevertheless, the results shown here do suggest that time series analysts should usually prefer to work with seasonally unadjusted data and to perform their own methods of reducing the importance of any seasonal component since the consequences of the method used can usually then be determined.

2.9 Books on Time Series Analysis and Forecasting

The first two chapters of this book introduce readers to the fundamental ideas of time series analysis and the next three chapters consider single series forecasting. Of necessity, there are a number of aspects of these topics that are not covered. The following list of books is designed for those interested in acquiring a deeper understanding of both time series analysis and statistical forecasting. The books are listed chronologically under three general headings. Those books marked with a star are less concerned with mathematical rigor than the others and so are more suitable for general readers. There is a curious negative correlation between the level of mathematical rigor used and the amount of interpretation of the results attempted.

1. *Spectral Analysis.* Useful books on this topic include Granger and Hatanaka [1964]*, Jenkins and Watts [1968], Fishman [1969]*, Dhrymes [1970], and Koopmans [1974].

2. *General Time Series Analysis.* Books covering a wider range of topics in time series include Wold [1954a], Davis [1941]*, Quenouille [1957], Grenander and Rosenblatt [1957], Hannan [1960], Parzen [1967], Hannan [1970], Anderson [1971], and Kendall [1973]*.

3. *Forecasting.* Books that discuss the theory of forecasting or discuss particular time series forecasting methods include Wiener [1949], Brown [1962]*, Whittle [1963], Box and Jenkins [1970], and Nelson [1973]*.

CHAPTER THREE

BUILDING LINEAR TIME SERIES MODELS

Today is the tomorrow you were worrying about yesterday.

GRAFFITI

In this chapter, a unified approach to the fitting of linear models to a given time series is presented. Such an approach was devised by Box and Jenkins in a series of articles and a subsequent book [1970], and the material in this chapter draws heavily from Chapters 6–9 of that book. The objective, given a particular time series realization $x_1, x_2, \ldots, x_n$, is to derive a linear stochastic model that could have generated the series. As will be seen in Chapter 5, this model can then be employed to generate forecasts of future values of the series.

3.1 Model Building Philosophy

It is convenient, initially, to describe a model building procedure for nonseasonal time series. In Section 3.7 it is shown how this approach can be generalized to deal with seasonal series.

In Chapter 1, the autoregressive integrated moving average process

$$a(B)(1 - B)^d X_t = b(B)\epsilon_t \qquad (3.1.1)$$

where

$$a(B) = \left(1 - a_1 B - a_2 B^2 - \cdots - a_p B^p\right)$$

$$b(B) = \left(1 + b_1 B + b_2 B^2 + \cdots + b_q B^q\right)$$

was introduced. It is assumed that the time series under consideration can be represented by a model from the class (3.1.1) —possibly after the removal of any deterministic component, including a nonzero mean, and/or the application of some suitable transformation to the data.

A strategy for constructing autoregressive integrated moving average models can be based on a three-step iterative cycle of

(i)   model identification,
(ii) model estimation,
(iii) diagnostic checks on model adequacy.

At the identification stage one chooses a particular model from the class (3.1.1), that is, one selects values for p, d, and q. The procedures employed at this stage are, of necessity, inexact and require a good deal of judgment. However, one is not irrevocably committed to the chosen model if subsequent analysis suggests that some alternative form might provide more adequate representation of the given data. Because of the nature of the estimation techniques employed, it is also necessary at this stage to obtain initial rough estimates of the coefficients $a_1, a_2, \ldots, a_p$, $b_1, b_2, \ldots, b_q$ of the identified model.

At the estimation stage of the model building cycle, the coefficients of the identified model are estimated using efficient statistical techniques. Approximate standard errors are obtained for the estimated coefficients and, provided one is prepared to make specific distributional assumptions, tests of hypotheses and confidence intervals can be derived.

Finally, diagnostic checks are applied to determine whether or not the chosen model adequately represents the given set of data. Any inadequacies revealed may suggest an alternative model specification. If this is the case, the whole iterative cycle of identification, estimation, and diagnostic checking is repeated until a satisfactory model is obtained.

3.2 Identification

There can be little doubt that the most difficult step in the model building cycle is identification. This is so since, although a number of general principles can be laid down, there exists no surefire deterministic approach to the problem. Rather, it is necessary to exert a degree of judgment, the facility for which is greatly improved by experience. (Indeed it has been said that identification is a technique that should not be attempted for the first time.) It is worth reemphasizing that in selecting a model at this stage one is committed to no more than an assessment of its validity. The initially chosen model can always be discarded at a later stage of the analysis, should this course appear desirable. It is also possible that one may wish to carry forward from the identification stage not one, but two or more possible models.

The Tools of the Trade

The two most useful tools in any attempt at model identification are the sample autocorrelation function and the sample partial autocorrelation function. Given a time series $x_1, x_2, \ldots, x_n$, the sample autocorrelation function, which is a plot of the sample autocorrelations,

$$r_\tau = \frac{\sum_{t=\tau+1}^{n} (x_t - \bar{x})(x_{t-\tau} - \bar{x})}{\sum_{t=1}^{n} (x_t - \bar{x})^2}, \qquad \tau = 0, 1, 2, \ldots$$

against τ provides an obvious estimate of the autocorrelation function ρ_τ, defined in (1.2.7), of the underlying stochastic process.

The usefulness of the partial autocorrelation function can best be illustrated by considering the autoregressive process

$$(1 - a_1 B - a_2 B^2 - \cdots - a_p B^p) X_t = \epsilon_t$$

From (1.5.13) it is seen that the autocovariances for such a process obey

$$\lambda_\tau = \sum_{j=1}^{p} a_j \lambda_{\tau-j}, \qquad \tau > 0$$

Dividing this expression by the variance λ_0, it follows that the autocorrelations obey

$$\rho_\tau = \sum_{j=1}^{p} a_j \rho_{\tau-j}, \qquad \tau > 0 \tag{3.2.1}$$

The partial autocorrelation of order K for any stochastic process is defined as a_{KK} given by solving the set of simultaneous linear equations in a_{Kj}

$$\rho_\tau = \sum_{j=1}^{K} a_{Kj} \rho_{\tau-j}, \qquad \tau = 1, 2, \ldots, K \tag{3.2.2}$$

Thus a_{Kj}, $j = 1, 2, \ldots, K$, is substituted for a_j and K for p in (3.2.1). The first K equations of this system are then solved for a_{Kj}, and the resulting a_{KK} denoted as the partial autocorrelation. It follows from (3.2.1) that for an autoregressive process of order p, $a_{pp} = a_p$. Furthermore, for such a process, the partial autocorrelations of order greater than p will clearly all be zero. To see this, note that from (3.2.1) it follows that for any positive integer m

$$\rho_\tau = \sum_{j=1}^{p} a_j \rho_{\tau-j}, \qquad \tau = 1, 2, \ldots, p + m$$

Alternatively this system of equations can be written

$$\rho_\tau = \sum_{j=1}^{p+m} a_j^* \rho_{\tau-j}, \qquad \tau = 1, 2, \ldots, p + m$$

where

$$a_j^* = a_j, \qquad j = 1, 2, \ldots, p$$

$$= 0, \qquad j = p + 1, \ldots, p + m$$

Hence, solving the equations (3.2.2) for $K = p + m$ yields $a_{p+m, p+m} = a_{p+m}^*$ $= 0$. An obvious estimate of the partial autocorrelations is obtained by substituting the sample autocorrelations r_τ for ρ_τ in (3.2.2) and solving the resulting equations. Thus the sample partial autocorrelation of order K is $\hat{a}_{KK}$ given as the solution of the set of equations

$$r_\tau = \sum_{j=1}^{K} \hat{a}_{Kj} r_{\tau-j}, \qquad \tau = 1, 2, \ldots, K$$

A computationally efficient algorithm, which works well if $a(B) = 0$ does not have a root near the unit circle, for obtaining the sample partial autocorrelations is given by Durbin [1960].

Characterization of Stochastic Processes

It is convenient at this stage to examine the characteristic behavior of the autocorrelation and partial autocorrelation functions of the various members of the class of stochastic processes (3.1.1). First consider those processes that require differencing to induce stationarity, i.e., $d \geq 1$ in (3.1.1). Let X_t be generated by the process $(1 - B)X_t = Y_t$ where Y_t is stationary. It was shown in Section 1.13 that if this process is viewed as starting in the infinite past, then $\text{corr}(X_t X_{t-\tau}) \approx 1$ for finite τ. However, the sample autocorrelations for series generated by such processes can behave very differently. To see this, write

$$X_t = X_0 + \sum_{j=1}^{t} Y_j$$

The sample autocovariances are

$$c_\tau = \frac{1}{n} \sum_{t=\tau+1}^{n} (x_t - \bar{x})(x_{t-\tau} - \bar{x})$$

based on the sample $x_1, x_2, \ldots, x_n$, and the difference here from the population case considered earlier is that appeal cannot be made to limiting cases as t tends to infinity. To get some insight into the behavior of c_τ, note that for a stationary process Y, the population autocovariances tend to zero at high lags. (This follows, for example, from (1.7.18).) These quantities are estimated by

$$\frac{1}{n} \sum_{t=\tau+1}^{n} y_t y_{t-\tau} = \frac{1}{n} \sum_{t=\tau+1}^{n} (x_t - x_{t-1})(x_{t-\tau} - x_{t-\tau-1})$$

$$= \frac{1}{n} \sum_{t=\tau+1}^{n} [(x_t - \bar{x}) - (x_{t-1} - \bar{x})][(x_{t-\tau} - \bar{x}) - (x_{t-\tau-1} - \bar{x})]$$

Thus, apart from end terms which will make very little difference, since the above expression is close to zero, it follows that, subject to the restriction that

c_τ is bounded above by c_0,

$$2c_\tau \approx c_{\tau+1} + c_{\tau-1}$$

for sufficiently large τ. Hence the sample autocovariances and therefore the sample autocorrelations will typically behave as a very smooth function, and thus not die out rapidly, at high lags. The failure of the sample autocorrelation function to die out at high lags thus indicates that differencing is required. This is so even though the first few sample autocorrelations need not necessarily be large. The behavior of the sample autocorrelation function for the process $X_t - X_{t-1} = \epsilon_t + b\epsilon_{t-1}$ is studied in some detail by Wichern [1973] who found by simulation that, for samples of size 50, $E(r_1) = 0.62$ for $b = -0.5$ and $E(r_1) = 0.21$ for $b = -0.8$.

If differencing is found to be necessary, the sample autocorrelations and partial autocorrelations of the differenced series are far more likely than those of the original series to yield useful information about the underlying stochastic process. This is because any information contained in the latter is swamped by the behavior induced by nonstationarity, rendering further interpretation virtually impossible.

Assume now that the process X_t has been differenced a sufficient number of times as to produce the stationary process $Y_t = (1 - B)^d X_t$. Then, summarizing results derived in Chapter 1 and earlier in this chapter:

(i)   If Y_t is an autoregressive process of order p, i.e., $q = 0$, its autocorrelations will die out according to the difference equation

$$\rho_\tau = \sum_{j=1}^{p} a_j \rho_{\tau-j} \qquad \text{for all } \tau > 0$$

that is, according to a mixture of damped exponentials and/or sine waves, and its partial autocorrelations will obey

$$a_{KK} = 0 \qquad \text{for all } K > p$$

(ii)   If Y_t is a moving average process of order q, i.e., $p = 0$, its autocorrelations will obey

$$\rho_\tau = 0 \qquad \text{for all } \tau > q$$

and its partial autocorrelations will die out, though not according to any clearly recognizable pattern.

(iii)   If Y_t is a mixed autoregressive moving average process of order (p, q), with $p, q \neq 0$, its autocorrelations will die out according to

$$\rho_\tau = \sum_{j=1}^{p} a_j \rho_{\tau-j} \qquad \text{for all } \tau > q$$

and its partial autocorrelations will also die out, though again not according to any clearly recognizable pattern.

These three characteristics of members of the class of processes (3.1.1) can be employed as the basis of an attempt to identify an appropriate model for suitably differenced time series.

Model Identification in Practice

In practice, of course, one never knows the autocorrelations and partial autocorrelations of the underlying stochastic process, and must estimate them from the given time series realization. In identifying an appropriate model, then, one has to rely on the sample autocorrelation and partial autocorrelation functions imitating the behavior of the corresponding parent quantities. Clearly, the larger the sample the more likely in general this requirement is to hold. Thus, in order to have any reasonable hope of success in model identification, one requires a moderately long series of observations. It is not possible to be completely dogmatic on this point since the difficulties involved in identification are a function of the characteristics of the individual process. However, we would not be terribly confident of success with much less than 45 to 50 observations.

The first step in this identification procedure is to calculate the sample autocorrelations and partial autocorrelations of the given time series and its first one or two differences. Failure of the sample autocorrelations to die out quickly at high lags is an indication that further differencing is required. Having achieved stationarity by suitable differencing, one can then attempt to identify typical autoregressive, moving average, or mixed behavior as indicated above. As a rough guide for determining whether the parent autocorrelations are in fact zero after the qth, Bartlett [1946] shows that for a sample of size n the standard deviation of r_τ is approximately

$$n^{-1/2}\left(1 + 2\left(r_1^2 + r_2^2 + \cdots + r_q^2\right)\right)^{1/2} \qquad \text{for } \tau > q$$

Quenouille [1949] has shown that, for a pth-order autoregressive process, the standard deviations of the sample partial autocorrelations $\hat{a}_{KK}$ are approximately $n^{-1/2}$ for $K > p$. Appealing to a result of Anderson [1942] one can assume normality in moderately large samples, and hence the use of limits of plus or minus two standard deviations about zero should provide a reasonable guide in assessing whether or not the parent quantities are in fact zero. Thus, for samples of size n in the examples that follow, comparison of the sample quantities with $\pm 2n^{-1/2}$ gives a useful guide to statistical significance.

Model identification is a skill that is perhaps best acquired through practical experience. In order to illustrate some of the concepts involved, three examples will be considered.

EXAMPLE 1   First consider a series of 76 monthly observations on yields of industrial ordinary shares in the United Kingdom. Table 3.1 shows the sample autocorrelations r_K and partial autocorrelations $\hat{a}_{KK}$ for the series and its first two differences. Note that $2n^{-1/2} \approx 0.23$. The sample autocorrelations

Table 3.1 *Sample autocorrelations and partial autocorrelations for data on yields of industrial ordinary shares*

	K	1	2	3	4	5	6	7	8	9	10
	r_K	.97	.91	.86	.79	.72	.65	.57	.49	.40	.32
	$\hat{a}_{KK}$	.97	$-.29$	$-.01$	$-.21$	$-.02$	$-.05$	$-.06$	$-.15$	$-.02$	$-.06$
X_t	K	11	12	13	14	15	16	17	18	19	20
	r_K	.25	.18	.12	.07	.01	$-.04$	$-.09$	$-.16$	$-.21$	$-.26$
	$\hat{a}_{KK}$	.11	$-.00$	.03	$-.02$	$-.13$	$-.00$	$-.21$	$-.12$	.06	.00
	K	1	2	3	4	5	6	7	8	9	10
	r_K	.39	$-.00$	.17	.18	.05	.06	.09	.05	.01	$-.11$
	$\hat{a}_{KK}$	.39	$-.18$	.29	$-.03$	.04	.03	.02	.01	$-.02$	$-.16$
$(1-B)X_t$	K	11	12	13	14	15	16	17	18	19	20
	r_K	$-.15$	$-.07$	.07	.07	$-.00$	.16	.11	$-.10$	$-.07$	$-.06$
	$\hat{a}_{KK}$	$-.07$	$-.01$	.14	.04	.01	.23	$-.11$	$-.05$	$-.06$	$-.19$
	K	1	2	3	4	5	6	7	8	9	10
	r_K	$-.18$	$-.47$	.15	.11	$-.11$	$-.02$	.06	.00	.07	$-.06$
	$\hat{a}_{KK}$	$-.18$	$-.52$	$-.10$	$-.17$	$-.13$	$-.12$	$-.08$	$-.06$	.09	$-.02$
$(1-B)^2X_t$	K	11	12	13	14	15	16	17	18	19	20
	r_K	$-.10$	$-.03$	.09	.06	$-.19$	.18	.13	$-.19$	.02	.04
	$\hat{a}_{KK}$	$-.06$	$-.19$	$-.09$	$-.04$	$-.25$	.11	.03	.04	.16	.00

of the original series fail to damp out quickly at high lags, suggesting that differencing is required to produce stationarity. For the first differenced series, however, the sample autocorrelations are small after the first, leading to the tentative identification of the model

$$(1 - B)X_t = (1 + b_1 B)\epsilon_t$$

EXAMPLE 2 The second example is a series of 108 monthly values of official gold and foreign exchange holdings in Sweden. Sample autocorrelations and partial autocorrelations are given in Table 3.2 and $2n^{-1/2} \approx 0.19$. The sample autocorrelations of the undifferenced series do not die out quickly at high lags. Thus one is led to consider the first differenced series, where it is seen that the partial autocorrelations are all very small after the first, suggesting a model of the form

$$(1 - a_1 B)(1 - B)X_t = \epsilon_t$$

EXAMPLE 3 The final example is a series of 204 monthly values of the index of help wanted advertisements in the United States. Sample autocorrelations and partial autocorrelations for this series are given in Table 3.3.

Table 3.2   *Sample autocorrelations and partial autocorrelations for data on official gold and foreign exchange holdings*

	K	1	2	3	4	5	6	7	8	9	10
	r_K	.98	.95	.92	.89	.86	.83	.80	.77	.74	.71
	$\hat{a}_{KK}$	.98	−.20	−.06	−.02	−.02	.04	−.00	.00	.01	−.07
X_t	K	11	12	13	14	15	16	17	18	19	20
	r_K	.68	.65	.61	.56	.52	.48	.45	.42	.39	.36
	$\hat{a}_{KK}$	−.10	−.05	−.19	.00	.03	.09	−.02	.03	−.01	.02
	K	1	2	3	4	5	6	7	8	9	10
	r_K	.38	.19	.09	.01	−.06	−.06	−.11	−.12	−.07	.12
	$\hat{a}_{KK}$	.38	.05	.00	−.04	−.07	−.02	−.08	−.05	.02	.18
$(1-B)X_t$	K	11	12	13	14	15	16	17	18	19	20
	r_K	.02	.07	.12	.06	−.13	−.10	−.15	−.11	−.08	−.02
	$\hat{a}_{KK}$	−.09	.06	.07	−.03	−.08	−.11	−.06	.02	−.00	.02
	K	1	2	3	4	5	6	7	8	9	10
	r_K	−.39	−.04	.00	−.00	−.06	.04	−.02	−.06	−.10	.23
	$\hat{a}_{KK}$	−.39	−.23	−.13	−.09	−.07	−.07	−.07	−.15	−.28	.01
$(1-B)^2X_t$	K	11	12	13	14	15	16	17	18	19	20
	r_K	−.15	.03	.09	.01	−.00	−.02	−.07	.01	−.03	.07
	$\hat{a}_{KK}$	−.13	−.12	−.01	.03	.06	−.00	−.10	−.07	−.08	−.03

Approximate bounds for significance are ± 0.14. This series presents rather more difficulty than the two previous examples. As a first step, it is clear that differencing is required. The sample partial autocorrelations of the first differenced series are all very small after the third, suggesting that a third-order autoregressive process

$$(1 - a_1B - a_2B^2 - a_3B^3)(1 - B)X_t = \epsilon_t$$

might be appropriate. Alternatively, the sample autocorrelations after the first appear to damp out as a second-order polynomial decay, suggesting the model

$$(1 - a_1B - a_2B^2)(1 - B)X_t = (1 + b_1B)\epsilon_t$$

To see this it is necessary to note only that, for the differenced series, the relation $r_K \approx 0.5r_{K-1} + 0.3r_{K-2}$ appears to hold quite well for $K \geqslant 2$ (remembering that $r_0 = 1$). Both these possibilities will be examined in subsequent sections of this chapter in which model estimation and diagnostic checking are considered. Thus this particular example illustrates a situation in which it is desirable to carry forward to subsequent stages of the analysis two tentatively identified models.

Table 3.3 *Sample autocorrelations and partial autocorrelations for index of help wanted advertisement data*

X_t	K	1	2	3	4	5	6	7	8	9	10
	r_K	.97	.93	.88	.81	.73	.64	.55	.44	.34	.23
	$\hat{a}_{KK}$	.97	−.21	−.27	−.20	−.13	−.16	−.11	−.04	−.06	−.05
	K	11	12	13	14	15	16	17	18	19	20
	r_K	.13	.03	−.06	−.14	−.21	−.27	−.32	−.37	−.40	−.42
	$\hat{a}_{KK}$	.05	−.02	−.04	−.14	.00	−.00	−.03	−.07	.02	−.06

$(1-B)X_t$	K	1	2	3	4	5	6	7	8	9	10
	r_K	.26	.41	.29	.25	.24	.20	.12	.06	.09	−.06
	$\hat{a}_{KK}$	.26	.37	.17	.05	.06	.03	−.06	−.10	.01	−.13
	K	11	12	13	14	15	16	17	18	19	20
	r_K	−.04	−.07	−.25	−.19	−.22	−.21	−.18	−.26	−.19	−.16
	$\hat{a}_{KK}$	−.08	−.02	−.22	−.11	−.02	−.00	.04	−.07	.03	.06

$(1-B)^2X_t$	K	1	2	3	4	5	6	7	8	9	10
	r_K	−.61	.18	−.05	−.03	.03	.02	−.07	−.07	.12	−.12
	$\hat{a}_{KK}$	−.61	−.29	−.14	−.13	−.09	.00	.04	−.08	.05	−.02
	K	11	12	13	14	15	16	17	18	19	20
	r_K	.05	.09	−.15	.06	−.02	−.03	.08	−.10	.03	.00
	$\hat{a}_{KK}$	−.07	.11	−.01	−.11	−.11	−.13	−.02	−.11	−.13	−.08

3.3 Initial Estimates for Coefficients

Assume, now, that a model

$$\left(1 - a_1 B - a_2 B^2 - \cdots - a_p B^p\right)Y_t = \left(1 + b_1 B + b_2 B^2 + \cdots + b_q B^q\right)\epsilon_t$$

$$(3.3.1)$$

where $Y_t = (1 - B)^d X_t$ has been selected. The estimation procedure to be described in Section 3.5 requires initial estimates of the coefficients a_1, $a_2, \ldots, a_p, b_1, b_2, \ldots, b_q$. These can be obtained directly from the sample autocovariances employed in the identification process.

Given a series of observations $y_1, y_2, \ldots, y_n$ on the process Y_t of (3.3.1), with sample autocovariances denoted by

$$c_\tau = \frac{1}{n} \sum_{t=\tau+1}^{n} (y_t - \bar{y})(y_{t-\tau} - \bar{y}), \qquad \tau = 0, 1, 2, \ldots$$

and $c_{-\tau} = c_\tau$, it follows from (1.7.11)–(1.7.14) that the autoregressive coefficients $a_1, a_2, \ldots, a_p$ can be estimated by solving the set of equations

$$c_\tau = \sum_{j=1}^{p} \hat{a}_j c_{\tau-j}, \qquad \tau = q + 1, q + 2, \ldots, q + p$$

Assuming, now, that one can write

$$(1 - \hat{a}_1 B - \hat{a}_2 B^2 - \cdots - \hat{a}_p B^p) Y_t = \tilde{Y}_t$$

$$\approx (1 + b_1 B + b_2 B^2 + \cdots + b_q B^q) \epsilon_t$$

the coefficients $b_1, b_2, \ldots, b_q$ can be estimated using the autocovariance properties of the moving average process $\tilde{Y}_t$. Denote the sample auto-covariances of this process as $\tilde{c}_\tau$. It can then be shown that

$$\tilde{c}_\tau = \sum_{j=0}^{p} a_j^2 c_\tau + \sum_{j=1}^{p} (a_0 a_j + a_1 a_{j+1} + \cdots + a_{p-j} a_p)(c_{\tau+j} + c_{\tau-j}),$$

$$\tau = 0, 1, 2, \ldots, q$$

where $a_0 = -1$. A Newton–Raphson algorithm, due to Wilson [1969], can then be employed to generate estimates of the moving average coefficients by an iterative procedure. Define a vector

$$\boldsymbol{\beta}' = (\beta_0, \beta_1, \beta_2, \ldots, \beta_q)$$

where

$$\beta_0^2 = \sigma_\epsilon^2, \qquad \beta_j = \beta_0 b_j, \qquad j = 1, 2, \ldots, q \qquad (3.3.2)$$

Let $\boldsymbol{\beta}^{(i)}$ denote the value of $\boldsymbol{\beta}$ obtained at the ith iteration. This estimate is then updated according to

$$\boldsymbol{\beta}^{(i+1)} = \boldsymbol{\beta}^{(i)} - (\mathbf{A}^{(i)})^{-1} \mathbf{g}^{(i)}$$

where

$$\mathbf{g}' = (g_0, g_1, \ldots, g_q), \qquad g_j = \sum_{i=0}^{q-j} \beta_i \beta_{i+j} - \tilde{c}_j$$

and

$$
\mathbf{A} =
\begin{bmatrix}
\beta_0 & \beta_1 & \cdots & \beta_{q-2} & \beta_{q-1} & \beta_q \\
\beta_1 & \beta_2 & \cdots & \beta_{q-1} & \beta_q & 0 \\
\beta_2 & \beta_3 & \cdots & \beta_q & 0 & 0 \\
& & \vdots & & & \\
\beta_q & 0 & \cdots & 0 & 0 & 0
\end{bmatrix}
+
\begin{bmatrix}
\beta_0 & \beta_1 & \cdots & \beta_{q-2} & \beta_{q-1} & \beta_q \\
0 & \beta_0 & \cdots & \beta_{q-3} & \beta_{q-2} & \beta_{q-1} \\
0 & 0 & \cdots & \beta_{q-4} & \beta_{q-3} & \beta_{q-2} \\
& & \vdots & & & \\
0 & 0 & \cdots & 0 & 0 & \beta_0
\end{bmatrix}
$$

The iterative procedure is continued until satisfactory convergence is obtained, when estimates of the b_j can then be obtained from (3.3.2). The iteration process can be started off by setting $\sigma_\epsilon^2 = \tilde{c}_0$ and $b_1, b_2, \ldots, b_q = 0$.

3.4 The Autocorrelation Function as a Characteristic of Process Behavior

Up to this point it has been implicitly assumed that the behavior of the autocorrelations ρ_τ characterized the behavior of linear stationary stochastic processes of the form (3.3.1). It will now be proved that, given the requirement of invertibility introduced in Chapter 1, this is indeed so. Indeed, were it possible for two or more processes to possess the same autocorrelation structure, the identification process in Section 3.2 would be of very little value.

It is convenient initially to consider the pth-order autoregressive process

$$a(B)X_t = \epsilon_t \qquad (3.4.1)$$

whose autocorrelations are ρ_τ, and to prove that no other finite order autoregressive process can have these autocorrelations. Let

$$a^*(B)X_t = \epsilon_t^* \qquad (3.4.2)$$

denote such a process, of order p^*. It has been shown that for processes of the form (3.4.1), the autocorrelations obey

$$\rho_\tau = \sum_{j=1}^{p} a_j \rho_{\tau-j}, \qquad \tau > 0 \qquad (3.4.3)$$

Consider, then, the set of equations

$$\rho_\tau = \sum_{j=1}^{\max(p,\, p^*)} e_j \rho_{\tau-j}, \qquad \tau = 1, 2, \ldots, \max(p, p^*)$$

It follows from (3.4.3) that these equations must have the unique solution

$$
\begin{aligned}
e_j &= a_j, & j &= 1, 2, \ldots, p & &\text{if } \quad p^* \leqslant p \\
e_j &= a_j, & j &= 1, 2, \ldots, p & &\text{if } \quad p^* > p \\
&= 0, & j &= p+1, \ldots, p^* & &\text{if } \quad p^* > p
\end{aligned}
$$

Thus it follows that no finite order autoregressive process other than (3.4.1) can have autocorrelations ρ_τ, for if (3.4.2) were such a process it would be possible to write

$$\rho_\tau = \sum_{j=1}^{\max(p,\, p^*)} a_j^* \rho_{\tau-j}, \qquad \tau = 1, 2, \ldots, \max(p, p^*)$$

with $a_j^* = 0$ for $j > p^*$ if $p > p^*$, such that at least one a_j^* was different from a_j for $j = 1, 2, \ldots, p$ or different from zero for $j > p$. It has just been shown that this is impossible.

Now consider the ARMA(p, q) process

$$a(B)X_t = b(B)\epsilon_t \qquad (3.4.4)$$

whose autocorrelations are ρ_τ. Let an alternative process of order (p^*, q^*) and possessing the same autocorrelation structure be denoted as

$$a^*(B)X_t = b^*(B)\epsilon_t^* \qquad (3.4.5)$$

It follows from (1.3.8) and (1.7.18) that the autocorrelation generating function of the process (3.4.4) is given by

$$\rho(z) = \frac{\sigma_\epsilon^2}{\lambda_0} \frac{b(z)b(z^{-1})}{a(z)a(z^{-1})}$$

and that of (3.4.5) by

$$\rho(z) = \frac{\sigma_\epsilon^{*2}}{\lambda_0^*} \frac{b^*(z)b^*(z^{-1})}{a^*(z)a^*(z^{-1})}$$

where λ_0 and λ_0^* denote the variances of the two processes, and it is assumed that the operators $a(B)$ and $a^*(B)$ are stationary. If the two processes have identical autocorrelation structures, it follows that

$$\frac{\sigma_\epsilon^2}{\lambda_0} \frac{1}{a(z)b^*(z)a(z^{-1})b^*(z^{-1})} \qquad (3.4.6)$$

and

$$\frac{\sigma_\epsilon^{*2}}{\lambda_0^*} \frac{1}{a^*(z)b(z)a^*(z^{-1})b(z^{-1})} \qquad (3.4.7)$$

are equal. But (3.4.6) and (3.4.7) represent respectively the autocorrelation generating functions of the processes

$$a(B)b^*(B)Y_t = \eta_t \qquad \text{and} \qquad a^*(B)b(B)Y_t = \eta_t^*$$

provided that the roots of $b^*(B)$ and $b(B)$ all lie outside the unit circle, i.e., given invertibility (and where η_t and η_t^* are white noise processes). But it has already been shown that autoregressive operators are uniquely determined by their autocorrelation functions. Hence it follows that

$$a(B)b^*(B) = a^*(B)b(B)$$

and hence $a^*(B)/a(B) = b^*(B)/b(B)$. Thus, for a given autocorrelation structure, the representation (3.4.4) is unique in the class of invertible processes up to multiplication by a common polynomial function of B on either side of the equation. As a simple illustration, consider the process $(1 - aB)X_t = \epsilon_t$. Multiplying through by $1 - eB$ yields

$$(1 - eB)(1 - aB)X_t = (1 - eB)\epsilon_t$$

which for any $-1 < e < 1$ will clearly have the same autocorrelation structure.

The requirement of invertibility has been stressed in the above proof. It is a simple matter to demonstrate that such a requirement is necessary to ensure uniqueness of representation. Consider the two processes

$$X_t = (1 + 0.5B)\epsilon_t, \qquad X_t = (1 + 2B)\epsilon_t^* \qquad (3.4.8)$$

where ϵ_t and ϵ_t^* are white noise processes. It follows from (1.6.3) and (1.6.4) that for both processes

$$\rho_\tau = 0.4 \qquad \text{for} \quad \tau = 1$$

$$= 0 \qquad \text{for} \quad \tau > 1$$

Thus both processes have the same autocorrelation function, although of course only the first process satisfies the invertibility requirement.

The above discussion suggests two points of some practical importance in model identification.

(i) The possibility of multiple solutions arising from the multiplication of both sides of (3.4.4) by a common factor should caution against the selection of overelaborate models at the identification stage. The aim of model identification should be to choose the simplest (in terms of fewest coefficients) model compatible with the particular autocorrelation structure exhibited by the data.

(ii) Since multiple solutions do occur when the model contains moving average terms, care should be taken to ensure that the initial parameter estimates calculated are those appropriate to the (unique) invertible process.

It has been shown that, if a model of the class (3.4.4) possesses a particular autocorrelation structure, then it will be the only model in that class to do so. However, as hinted in Section 1.6, it is not true that given a particular autocorrelation structure there must be a model in the class (3.4.4) possessing such a structure. For example, consider the case

$$\rho_1 = \rho, \qquad \rho_\tau = 0, \quad \tau > 1$$

If $|\rho| \leqslant 0.5$, then there exists a first-order moving average process $X_t = (1 + bB)\epsilon_t$ with these particular autocorrelations. However, for $|\rho| > 0.5$ it is not possible to find an autoregressive moving average model with such a correlogram. If it should happen that the sample autocorrelations are such that r_1 is much greater in magnitude than 0.5, while the remaining values are close to zero, the best strategy is probably to fit a higher order moving average model.

3.5 Estimation

Having selected a particular model

$$(1 - a_1 B - a_2 B^2 - \cdots - a_p B^p)Y_t = (1 + b_1 B + b_2 B^2 + \cdots + b_q B^q)\epsilon_t$$

$$(3.5.1)$$

where $Y_t = (1 - B)^d X_t$, the next step is to obtain estimates of the coefficients $a_1, a_2, \ldots, a_p;\ b_1, b_2, \ldots, b_q$. An obvious estimating procedure in model fitting situations is, of course, the familiar least-squares approach, where one chooses the particular set of coefficients making the sum of squared errors (the ϵ's) as small as possible. The application of least squares to models of the form (3.5.1) raises two difficulties:

(i) the equations will involve unknown "starting values" Y_0, Y_{-1}, $Y_{-2}, \ldots, Y_{1-p}, \epsilon_0, \epsilon_{-1}, \epsilon_{-2}, \ldots, \epsilon_{1-q}$;

(ii) the model is, in general, nonlinear in the coefficients to be estimated.

Two approaches to the starting value problem will be described. In the first approach, the unknown "starting values" are simply replaced by some appropriate assumed values and estimation is conditional on these assumed starting values. A more efficient unconditional approach is based on estimating the starting values from the given sample data. The problem of nonlinearity can be overcome by use of familiar nonlinear regression routines, which, of course, are widely applicable in other fields.

A Conditional Approach

Equation (3.5.1) can be written as

$$\epsilon_t = Y_t - a_1 Y_{t-1} - \cdots - a_p Y_{t-p} - b_1 \epsilon_{t-1} - \cdots - b_q \epsilon_{t-q} \quad (3.5.2)$$

Thus, for a given set of data $y_1, y_2, \ldots, y_n$, the quantities ϵ_t can be calculated for this particular realization of the stochastic process for any given coefficients provided that values are specified for $Y_0, Y_{-1}, \ldots, Y_{1-p}$, $\epsilon_0, \epsilon_{-1}, \ldots, \epsilon_{1-q}$. The most obvious values to choose here are the unconditional expectations of these quantities, namely zero. This allows the calculation of ϵ_t, $t = 1, 2, \ldots, n$, for given coefficients $a_1, a_2, \ldots, a_p, b_1, b_2, \ldots, b_q$. However, if the equation $a(B) = 0$ has a root close to the unit circle, this procedure will introduce into the calculated ϵ_t series a transient that will be very slow to die out. If the identification process suggests the possibility of an autoregressive operator of this type, it is preferable to begin the calculations at time $p + 1$ so that the values computed are

$$\epsilon_t = y_t - a_1 y_{t-1} - \cdots - a_p y_{t-p} - b_1 \epsilon_{t-1} - \cdots - b_q \epsilon_{t-q}$$

$$t = p + 1, \ldots, n$$

where ϵ_{p+1-j} is set to zero for $j = 1, 2, \ldots, q$. If it is suspected that the time series to be estimated has nonzero mean μ, then for all j replace Y_j in (3.5.2) by $Y_j - \mu$, with μ treated as a further coefficient to be estimated, using the sample mean $\bar{y}$ as an initial guess.

In introducing the nonlinear regression algorithm generally employed in estimating (3.5.2) it is convenient to denote the set of coefficients to be estimated as

$$\mathbf{c}' = (c_1, c_2, \ldots, c_{p+q}) = (a_1, a_2, \ldots, a_p, b_1, b_2, \ldots, b_q)$$

and to consider ϵ_t as a function of these coefficients.

It will be recalled that, as part of the identification procedure, initial rough estimates were obtained for the coefficients of the chosen model using methods described in Section 3.3. Denote these estimates as $\mathbf{c}_0' =$

$(c_{1,0}, c_{2,0}, \ldots, c_{p+q,0})$ and consider the Taylor series expansion of ϵ_t as a function of **c**. Terminating this expansion at the first term produces

$$\epsilon_t \approx \epsilon_{t,0} - \sum_{i=1}^{p+q} (c_i - c_{i,0}) x_{t,i} \qquad (3.5.3)$$

where $\epsilon_{t,0}$ is the value of ϵ_t obtained by substituting the coefficient estimates $\mathbf{c_0}$ in (3.5.2) and $x_{t,i} = -\partial \epsilon_t / \partial c_i |_{\mathbf{c}=\mathbf{c_0}}$. In practice these partial derivatives can generally best be obtained numerically. Denote $\epsilon_t | (c_1, c_2, \ldots, c_{p+q})$ as the value obtained for ϵ_t from the coefficients $c_1, c_2, \ldots, c_{p+q}$. Then, if δ_i is chosen to be sufficiently small, $x_{t,i}$ is well approximated by

$$x_{t,i} = \delta_i^{-1} \big[\epsilon_t | (c_{1,0}, c_{2,0}, \ldots, c_{i,0}, \ldots, c_{p+q,0})$$

$$- \epsilon_t | (c_{1,0}, c_{2,0}, \ldots, c_{i,0} + \delta_i, \ldots, c_{p+q,0}) \big]$$

since the limit of this quantity as δ_i tends to zero is, by definition, minus the partial derivative of ϵ_t with respect to c_i evaluated at $\mathbf{c_0}$.

Equation (3.5.3) is now in the form of the familiar linear least-squares equation, for it can be written

$$\epsilon_{t,0} = \sum_{i=1}^{p+q} (c_i - c_{i,0}) x_{t,i} + \epsilon_t$$

The coefficients $c_i - c_{i,0}$ can then be estimated as

$$\widehat{\mathbf{c} - \mathbf{c_0}} = (\mathbf{X'X})^{-1} \mathbf{X'} \epsilon_0$$

where

$$\mathbf{X} = \begin{bmatrix} x_{1,1} & x_{1,2} & \cdots & x_{1,p+q} \\ x_{2,1} & x_{2,2} & \cdots & x_{2,p+q} \\ \vdots & \vdots & \vdots & \vdots \\ x_{n,1} & x_{n,2} & \cdots & x_{n,p+q} \end{bmatrix}, \qquad \epsilon_0 = \begin{bmatrix} \epsilon_{1,0} \\ \epsilon_{2,0} \\ \vdots \\ \epsilon_{n,0} \end{bmatrix}$$

The estimate $\widehat{\mathbf{c} - \mathbf{c_0}}$ can then be employed to update the original estimate to yield

$$\mathbf{c_0}^{(1)} = \mathbf{c_0} + \widehat{\mathbf{c} - \mathbf{c_0}}$$

The calculations are then repeated, replacing $\mathbf{c_0}$ by $\mathbf{c_0}^{(1)}$ and continuing the process of iteration until satisfactory convergence is obtained. At the final iteration stage, using a well-known result of linear least-squares theory, the variance–covariance matrix of the coefficient estimates is estimated by

$$\mathbf{V} = s^2 (\mathbf{X'X})^{-1} \qquad \text{where} \quad s^2 = \sum_{t=1}^{n} \frac{\epsilon_{t,0}^2}{n - p - q} \qquad (3.5.4)$$

and $\epsilon_{t,0}$ and **X** are here taken to refer to the values of these quantities in the final iteration. Further, if one is prepared to assume normality, approximate 95% confidence intervals for the parameters are given by $\hat{c}_i \pm 1.96 V_{ii}^{1/2}$ where $\hat{c}_i$ is the final estimate of c_i and V_{ii} is the ith diagonal element of the matrix **V** in (3.5.4).

It should be stressed that the variance estimates and corresponding interval estimates can be regarded as only approximate due to the fact that they are based on a linearization procedure which is itself, of course, an approximation.

The estimation procedure just described is by no means wholly satisfactory since the iterations may converge very slowly or not at all on occasions. Many nonlinear regression routines in current use are refinements of a procedure proposed by Marquardt [1963] which is essentially a compromise between the above linearization technique and the method of "steepest descent." A fuller account of nonlinear regression estimation is given in Chapter 10 of Draper and Smith [1966].

Assuming that the ϵ_t of (3.5.1) are independently normally distributed, the density function of the observations given the assumed starting values is of the form

$$p(\epsilon_1, \epsilon_2, \ldots, \epsilon_n \mid \text{starting values}) \propto \sigma_\epsilon^{-n} \exp\left[-\sum_{t=1}^{n} \epsilon_t^2 / 2\sigma_\epsilon^2 \right]$$

It follows, therefore, that the least squares estimates are maximum likelihood conditional on the starting values being given.

Returning now to the three series whose identification was discussed in Section 3.2; the estimates of the coefficients of the tentatively identified models are given below (together with estimated standard errors in brackets):

(i)   For the data on yields of industrial ordinary shares, the fitted model was

$$(1 - B)X_t = (1 + 0.59B)\epsilon_t$$
$$[0.09]$$

(ii)   For the data on official gold and foreign exchange holdings, the fitted model was

$$(1 - 0.38B)(1 - B)X_t = \epsilon_t$$
$$[0.09]$$

(iii)   For the series on index of help wanted advertisements it was thought desirable to proceed with two possible identifications—ARIMA(3, 1, 0) and ARIMA(2, 1, 1). The two fitted models were

$$(1 - 0.11B - 0.34B^2 - 0.16B^3)(1 - B)X_t = \epsilon_t$$
$$[0.07] \qquad [0.07] \qquad [0.07]$$

and

$$(1 - 0.45B - 0.31B^2)(1 - B)X_t = (1 - 0.36B)\epsilon_t$$
$$[0.14] \qquad [0.08] \qquad\qquad\qquad [0.15]$$

The respective error variance estimates are 14.486 and 14.483, suggesting that there is very little indeed to choose between the two models, with just a very marginal preference for the ARIMA(2, 1, 1) process. This particular series will be discussed in more detail in the following section.

An Unconditional Approach

In some situations, a more efficient treatment of the starting value problem may be required. This can be achieved by replacing unknown values by their expectations conditional on the given data set and minimizing the sum of squares

$$S = \sum_{t=-\infty}^{n} (E_c(\epsilon_t))^2 \tag{3.5.5}$$

where E_c denotes conditional expectation. That is, given an observed time series, one attempts to estimate the values of ϵ_t that generated these observations, the conditional means given the data providing an obviously sensible set of estimates. In fact, for stationary processes, these conditional expectations become very small for $t < 1 - Q$ where Q is some sufficiently large positive integer, and so one can consider minimizing the finite sum

$$S^* = \sum_{t=1-Q}^{n} (E_c(\epsilon_t))^2 \tag{3.5.6}$$

(In fact, for pure moving average processes $E_c(\epsilon_t) = 0$ for $t < 1 - q$ since the given data depend only on ϵ_t, $t \geq 1 - q$.)

That increased efficiency results from estimating the initial values from the data should be intuitively obvious. A more formal justification is provided by noting that for the process (3.5.1) it can be shown that the density function of the observations given the parameters is, under the assumption of normality,

$$p(y_1, y_2, \ldots, y_n) \propto \sigma_\epsilon^{-n} g(a_1, \ldots, a_p, b_1, \ldots, b_q) \exp\left[-\frac{\sum_{t=-\infty}^{n} (E_c(\epsilon_t))^2}{2\sigma_\epsilon^2} \right]$$

where g is a function of the coefficients but does not involve the data. It follows that for moderately large sample size minimization of S^* of (3.5.6), which is very nearly equivalent to minimizing S of (3.5.5), will yield estimates which are very nearly maximum likelihood. The likelihood functions for autoregressive, moving average, and mixed models are examined in detail in Chapter 7 of Box and Jenkins [1970]. The mixed process is further considered by Newbold [1974].

It remains now to describe for any given set of parameter values the computation of the conditional expectations required in (3.5.6). In practice, this can best be achieved by noticing that the two processes

$$a(B)Y_t = b(B)\epsilon_t, \qquad a(B^{-1})Y_t = b(B^{-1})\eta_t$$

where η_t is a white noise process having the same variance as ϵ_t, have the same autocovariance properties, and thus can equally well describe a particular stochastic process. (This can readily be verified by reference to the autocovariance generating function (1.7.18).)

Writing out these equations in full, and taking expectations conditional on the given data set yields

$$E_c(Y_t) - a_1 E_c(Y_{t-1}) - \cdots - a_p E_c(Y_{t-p}) = E_c(\epsilon_t) + b_1 E_c(\epsilon_{t-1})$$

$$+ \cdots + b_q E_c(\epsilon_{t-q}) \quad (3.5.7)$$

$$E_c(Y_t) - a_1 E_c(Y_{t+1}) - \cdots - a_p E_c(Y_{t+p}) = E_c(\eta_t) + b_1 E_c(\eta_{t+1})$$

$$+ \cdots + b_q E_c(\eta_{t+q}) \quad (3.5.8)$$

Given a time series of observations $y_1, y_2, \ldots, y_n$, then

$$E_c(Y_t) = y_t, \qquad t = 1, 2, \ldots, n \quad (3.5.9)$$

$$E_c(\epsilon_{n+j}) = 0, \qquad j = 1, 2, \ldots \quad (3.5.10)$$

since ϵ_{n+j} is an unpredictable white noise which has not entered the system at time n (clearly ϵ_{n+j} is uncorrelated with $Y_1, Y_2, \ldots, Y_n$), and similarly

$$E_c(\eta_{1-j}) = 0, \qquad j = 1, 2, \ldots \quad (3.5.11)$$

The calculations proceed iteratively and are started by setting

$$E_c(\eta_{n-p+j}) = 0, \qquad j = 1, 2, \ldots, q$$

Typically, for moderately long series, this approximation has virtually no effect. The values $E_c(\eta_t)$, $t = n - p, n - p - 1, \ldots, 1$, are then obtained from (3.5.8), using (3.5.9). Next, the values of $E_c(Y_t)$, $t = 0, -1, -2, \ldots, 1 - Q - p$, can be obtained from (3.5.8), using (3.5.11). The integer Q is chosen so that the conditional expectations will be negligibly small for $t < 1 - Q$. A reasonable rule would be to stop when three successive values of $E_c(Y_t)$ less than 1% of the standard deviation of $y_1, y_2, \ldots, y_n$ in absolute magnitude occur. (It should be reemphasized that Y_t is assumed to have zero mean or that the mean has been subtracted before the conditional expectations are calculated.) Finally, the values of $E_c(\epsilon_t)$, $t = 1 - Q, 2 - Q, \ldots, -1,$ $0, 1, \ldots, n$ are obtained from (3.5.7), setting $E_c(\epsilon_t) = 0, t < 1 - Q$.

This single iteration is generally sufficient. However, for shorter time series one can obtain from (3.5.7) values of $E_c(Y_t)$, $t = n + 1, n + 2, n + 3, \ldots,$ and hence from (3.5.8) obtain more accurate estimates of the starting up values $E_c(\eta_{n-p+j})$, $j = 1, 2, \ldots, q$. Iteration may then proceed until the sum of squares $\sum_{t=1-Q}^{n} (E_c(\epsilon_t))^2$ converges.

Although these calculations appear somewhat cumbersome, their iterative nature makes them ideally suitable for programming on an electronic computer. In fact, it is known from Barnard *et al.* [1962] that for moderately long time series very little is lost by employing the conditional approach. Just how long a series one requires for the approximation to work satisfactorily depends on the underlying model, but typically 75 observations are sufficient. However, for the seasonal models to be discussed in Section 3.7, this approximation is less adequate, and for the sample sizes generally available in practice it is preferable to work with the unconditional approach.

Finally, note that the nonlinear least-squares algorithm already described can again be employed in the minimization of the sum of squares (3.5.6). The estimation procedure is identical except that, throughout, ϵ_t and $\epsilon_{t,0}$ of (3.5.3) are replaced by $E_c(\epsilon_t)$ and $E_c(\epsilon_{t,0})$ and the range of t now runs from $1 - Q$ to n rather than from 1 to n. The appropriate error variance estimate for substitution in (3.5.4) is now

$$s^2 = \sum_{t=1-Q}^{n} \frac{\left[E_c(\epsilon_{t,0})\right]^2}{n - p - q}$$

3.6 Diagnostic Checking

It is very often the case, as the reader who attempts to follow the model building procedure just described will quickly discover, that in practice model identification is fraught with uncertainties. One might feel able to select a particular model as a "best bet," but would be unhappy about the prospect of making inference from such a model without reassurance as to its validity. Accordingly, in this section a number of diagnostic checks on the adequacy of representation of the chosen model to the given data set are described. As will be seen, any inadequacies that are revealed may well suggest an alternative model as being more appropriate.

Fitting Extra Coefficients

In identifying, possibly after suitable differencing, a particular mixed autoregressive moving average model of order (p, q), it is of course implicitly assumed that in the more general model

$$\left(1 - a_1 B - a_2 B^2 - \cdots - a_p B^p - a_{p+1} B^{p+1} - \cdots - a_{p+p^*} B^{p+p^*}\right) Y_t$$
$$= \left(1 + b_1 B + b_2 B^2 + \cdots + b_q B^q + b_{q+1} B^{q+1} + \cdots + b_{q+q^*} B^{q+q^*}\right) \epsilon_t$$

$$(3.6.1)$$

the coefficients a_{p+j}, $j = 1, 2, \ldots, p^*$, and b_{q+j}, $j = 1, 2, \ldots, q^*$, are effectively zero. To a certain extent, this assumption is testable and where it is thought desirable to do so one can extend the identified model by adding extra coefficients. The augmented model can then be estimated, as described in Section 3.5, and the standard deviations of the estimates of the added coefficients will indicate whether or not the true values differ significantly from zero.

It may well be worthwhile in this context to fit the two models

$$\left(1 - a_1 B - a_2 B^2 - \cdots - a_p B^p - a_{p+1} B^{p+1} - \cdots - a_{p+p^*} B^{p+p^*}\right) Y_t$$
$$= \left(1 + b_1 B + b_2 B^2 + \cdots + b_q B^q\right)\epsilon_t$$

and

$$\left(1 - a_1 B - a_2 B^2 - \cdots - a_p B^p\right) Y_t$$
$$= \left(1 + b_1 B + b_2 B^2 + \cdots + b_q B^q + b_{q+1} B^{q+1} + \cdots + b_{q+q^*} B^{q+q^*}\right)\epsilon_t$$

Values of $p^* = 1$ or 2 and $q^* = 1$ or 2 should prove sufficient for most general purposes.

An ARIMA(0, 1, 1) model was tentatively identified for the series on yields of industrial ordinary shares, leading to the fitted equation

$$(1 - B)X_t = (1 + 0.59 B)\epsilon_t$$

It was decided to add an autoregressive coefficient and reestimate the augmented model. This yielded the fitted equation

$$(1 + 0.13 B)(1 - B)X_t = (1 + 0.69 B)\epsilon_t$$
$$[0.19] \qquad\qquad\qquad\qquad [0.14]$$

The autoregressive coefficient is small compared with its estimated standard error, and hence this check gives no grounds on which to question the adequacy of the originally chosen model.

Perhaps a more obvious case in which the addition of extra coefficients is potentially informative is provided by the series on index of help wanted advertisements. It will be recalled that two models

$$(1 - 0.11 B - 0.34 B^2 - 0.16 B^3)(1 - B)X_t = \epsilon_t$$

and

$$(1 - 0.45 B - 0.31 B^2)(1 - B)X_t = (1 - 0.36 B)\epsilon_t$$

were both fitted to these data. A more general model, which includes both of these possibilities, is the ARIMA(3, 1, 1) process. Fitting this model to the data produced the estimated equation

$$(1 - 0.42 B - 0.29 B^2 - 0.04 B^3)(1 - B)X_t = (1 - 0.32 B)\epsilon_t$$
$$[0.31] \qquad [0.09] \qquad [0.14] \qquad\qquad\qquad [0.31]$$

The estimated standard errors are, in fact, so large as to suggest that a wide range of models could possess similar explanatory capabilities for this particular set of data. However, the actual point estimates of the coefficients (for what they are worth) are certainly closer to those of the ARIMA(2, 1, 1) model than to those of the ARIMA(3, 1, 0) model. Thus, if one were forced to choose a single model for this series, it would be the ARIMA(2, 1, 1) process. The reader might find it rather perplexing that the choice between the two models is so difficult, even though the sample of 204 observations is such a large one. In fact, the explanation for this phenomenon is that the two models are really very similar. To see this, it need only be noted that the ARIMA(2, 1, 1) model can be written in the form

$$(1 - 0.36B)^{-1}(1 - 0.45B - 0.31B^2)(1 - B)X_t = \epsilon_t$$

or

$$\left(1 + 0.36B + (0.36)^2 B^2 + (0.36)^3 B^3 + (0.36)^4 B^4 + (0.36)^5 B^5 + \cdots \right)$$

$$(1 - 0.45B - 0.31B^2)(1 - B)X_t = \epsilon_t$$

Multiplying out the polynomials in B on the left-hand side of this expression yields

$$(1 - 0.09B - 0.34B^2 - 0.12B^3 - 0.04B^4 - 0.02B^5 - \cdots)(1 - B)X_t = \epsilon_t$$

The first three terms in the autoregressive operator are very similar to those of the fitted ARIMA(3, 1, 0) model, while the remaining terms are very close to zero. Thus one would expect to find it extremely difficult to distinguish between these two models. Further, the choice is not terribly important since such similar models, when projected forward, will produce virtually identical forecasts.

It is not a good strategy to fit the more general model (3.6.1) since, as was seen in Section 3.4, the addition of extra coefficients to both sides of a correct model will lead to indeterminacy. In such situations, the point estimates of the coefficients are meaningless, and their estimated standard deviations will be exceedingly large.

Use of Autocorrelations of the Residuals

Suppose now that the fitted model is of the form

$$a(B)(1 - B)^d X_t = b(B)\epsilon_t \tag{3.6.2}$$

If this is the true model, then the residuals $\epsilon_t = b^{-1}(B)a(B)(1 - B)^d X_t$ constitute a white noise process. Given a sequence $\epsilon_1, \epsilon_2, \ldots, \epsilon_n$ from a white noise process, it has been shown by Anderson [1942] that for moderately large samples the sample autocorrelations are uncorrelated and normally distributed with standard deviations $n^{-1/2}$. Thus, if the true values of the ϵ_t sequence were known, these sample autocorrelations would give a very clear indication of any departures from white noise behavior.

Unfortunately, in practice it is necessary to estimate the coefficients of Eq. (3.6.2), and so, denoting the estimates of $a(B)$ and $b(B)$ by $\hat{a}(B)$ and $\hat{b}(B)$, what one has available are the estimated residuals

$$\hat{\epsilon}_t = \hat{b}^{-1}(B)\hat{a}(B)(1 - B)^d X_t$$

where

$$\hat{a}(B) = (1 - \hat{a}_1 B - \cdots - \hat{a}_p B^p) \quad \text{and} \quad \hat{b}(B) = \left(1 + \hat{b}_1 B + \cdots + \hat{b}_q B^q\right)$$

It seems reasonable to expect the autocorrelations

$$r_\tau(\hat{\epsilon}) = \sum_{t=\tau+1}^{n} \hat{\epsilon}_t \hat{\epsilon}_{t-\tau} / \sum_{t=1}^{n} \hat{\epsilon}_t^2$$

to yield valuable information about possible model inadequacies. The value of this information is in fact somewhat limited since, as pointed out by Durbin [1970], the standard deviation of $r_\tau(\hat{\epsilon})$ can be considerably less than $n^{-1/2}$, even asymptotically. This is especially true for small values of τ. Intuitively this is so because, in estimating a linear time series model, one is in effect choosing as parameter estimates those values that render the residuals from the fitted equation as much like white noise as possible, at least as regards their first few autocorrelations. Thus the estimation procedure itself, by its very nature, guarantees that the first few autocorrelations of the residuals will be very close to zero. In fact, asymptotic standard deviations for the autocorrelations of the estimated residuals from autoregressive–moving average models have been derived by Box and Pierce [1970], and it emerges that the standard deviations of these quantities can indeed be much less than $n^{-1/2}$ for small values of τ (less than 6 for relatively simple models), but that for larger τ the approximation remains quite good. The exact asymptotic standard deviations given by Box and Pierce are extremely difficult to calculate in practice and, moreover, depend heavily on the true underlying stochastic process for small values of τ.

Nevertheless, we feel that comparison of the $r_\tau(\hat{\epsilon})$ with bounds $\pm 2n^{-1/2}$ should provide a general indication of possible departure from white noise behavior in the ϵ_t, provided it is remembered that for small τ these bounds will underestimate the significance of any discrepancies.

To illustrate the autocorrelation checks, consider again the series on index of help wanted advertisements. (This, after all, was the series that gave most difficulty in identification. The models fitted to the other two series did in fact comfortably satisfy these checks.) The first 20 autocorrelations $r_\tau(\hat{\epsilon})$ of the errors from the fitted ARIMA(2, 1, 1) model are shown in Table 3.4.

Table 3.4   *Autocorrelations of errors from ARIMA(2, 1, 1) model fitted to series on index of help wanted advertisements*

τ:	1	2	3	4	5	6	7	8	9	10
$r_\tau(\hat{\epsilon})$:	-0.00	-0.02	0.02	-0.01	0.08	0.08	-0.03	-0.04	0.05	-0.08

τ:	11	12	13	14	15	16	17	18	19	20
$r_\tau(\hat{\epsilon})$:	0.02	0.03	-0.20	-0.10	-0.08	-0.06	-0.02	-0.14	-0.05	0.03

Since one observation is "lost" through differencing, the effective sample size is $n = 203$. Only one of the residual sample autocorrelations lies outside the range $\pm 2/\sqrt{n}$ ($= 0.14$), this being the value at lag 13, which is suggestive of little more than ill luck, and these results provide scant evidence on which to question the validity of the chosen model.

Box and Pierce [1970] also provide an overall test on the autocorrelations of the estimated residuals, which although not a very powerful tool for detecting specific departures from white noise behavior, can indicate whether these values are generally too high. The procedure is to compare

$$Q = n \sum_{\tau=1}^{M} r_\tau^2(\hat{\epsilon}) \tag{3.6.3}$$

with tabulated values of the chi-squared statistic for $M - p - q$ degrees of freedom, the hypothesis of white noise behavior in the residuals being rejected at high values of Q. The test relies for its validity on M being moderately large (generally, at least equal to 20).

Using the residual autocorrelations in Table 3.4, the Box–Pierce test statistic is found to be $Q = 203\sum_{\tau=1}^{20} r_\tau^2(\hat{\epsilon}) = 22.29$. This can be compared with the tabulated χ^2 values for 17 degrees of freedom (since $p = 1$ and $q = 2$), which are 27.59 at the 5% point and 24.77 at the 10% level. Thus the hypothesis that the true model residuals ϵ_t are white noise is not rejected at the 10% level of significance. Accordingly, it was decided to accept the ARIMA(2, 1, 1) model fitted to the index of help wanted advertising data as providing an adequate representation of the behavior of that time series.

Model Modification in Light of Residual Autocorrelation Check

Suppose that the autocorrelation function of the residuals from the fitted model (3.6.2) suggests not that the residuals are white noise but rather that they might be well described by the process

$$a^*(B)(1 - B)^{d^*}\epsilon_t = b^*(B)\eta_t \tag{3.6.4}$$

where

$$a^*(B) = 1 - a_1^* B - \cdots - a_{p^*}^* B^{p^*}, \qquad b^*(B) = 1 + b_1^* B + \cdots + b_{q^*}^* B^{q^*}$$

and η_t constitutes a white noise process.

Combining (3.6.2) and (3.6.4) then produces

$$a(B)a^*(B)(1 - B)^{d+d^*}X_t = b(B)b^*(B)\eta_t$$

This would suggest that, in order to represent the given series adequately, the appropriate degree of differencing is $d + d^*$, and an autoregressive moving average model of order $(p + p^*, q + q^*)$ should be fitted to the differenced series. A model of this nature should then be estimated, and the cycle of identification, estimation, and diagnostic checking continued until a satisfactory model is achieved.

Although we have on occasion found this modification technique to be quite useful, the procedure is not quite so simple as may at first sight appear to be the case. For example, supposing the process

$$(1 - 0.6B)(1 + 0.3B)Y_t = \eta_t \tag{3.6.5}$$

is wrongly identified as first-order autoregressive, would the residuals from the fitted equation be first-order autoregressive and hence lead simply to the correct identification? Even ignoring sampling errors this would not be the case, since for the process (3.6.5), $\rho_1 \approx 0.37$; and hence in large samples the fitted model would be close to

$$(1 - 0.37B)Y_t = \epsilon_t \tag{3.6.6}$$

Thus the residuals from the fitted model obey

$$(1 - 0.6B)(1 + 0.3B)\epsilon_t = (1 - 0.37B)\eta_t \tag{3.6.7}$$

and this process is even more complicated than the one that was not identified in the first place. The situation is rendered even more difficult by sampling errors which could render the cancellation involved in combining Eqs. (3.6.6) and (3.6.7)—that is, the occurrence of the term $1 - 0.37B$ on both sides of the amalgamated equation—very difficult to detect in practice.

3.7 Model Building for Seasonal Time Series

In Section 1.14 the model

$$a(B)a_s(B^s)(1 - B)^d(1 - B^s)^D X_t = b(B)b_s(B^s)\epsilon_t \tag{3.7.1}$$

where

$$a(B) = 1 - a_1 B - a_2 B^2 - \cdots - a_p B^p$$

$$a_s(B^s) = 1 - a_{1,s} B^s - a_{2,s} B^{2s} - \cdots - a_{P,s} B^{Ps}$$

$$b(B) = 1 + b_1 B + b_2 B^2 + \cdots + b_q B^q$$

$$b_s(B^s) = 1 + b_{1,s} B^s + b_{2,s} B^{2s} + \cdots + b_{Q,s} B^{Qs}$$

was introduced for the representation of seasonal time series of period s—that is, $s = 4$ for quarterly data and $s = 12$ for monthly data. It has been found that models of the form (3.7.1) are capable of well describing a wide range of practically occurring seasonal time series. Such models are fitted to a given set of data by employing essentially the same principles as were described for the nonseasonal case. The approach to model building again is composed of an iterative cycle of identification, estimation, and diagnostic checking. These elements of the model building cycle will be briefly described and their application to seasonal series discussed in this section.

Identification

At the identification stage, the objective is to choose suitable values for d, D, p, P, q, and Q, the degrees of the polynomial operators in (3.7.1). The identification procedure is, as before, in two steps. At the first step, the degree of differencing required to produce stationarity is determined. This is achieved by applying the two operators $1 - B$ and $1 - B^s$ until the sample autocorrelations of the differenced series die out quickly at high lags.

At the second step the sample autocorrelations and partial autocorrelations of the appropriately differenced series $Y_t = (1 - B)^d(1 - B^s)^D X_t$ are employed to suggest appropriate degrees for the four polynomial operators $a(B)$, $a_s(B^s)$, $b(B)$, and $b_s(B^s)$. In principle, the patterns followed by the autocorrelation function for various members of the class (3.7.1) can be determined from the autocovariance generating function for Y_t

$$\gamma(z) = \sigma_\epsilon^2 \frac{b(z)b_s(z^s)b(z^{-1})b_s(z^{-s})}{a(z)a_s(z^s)a(z^{-1})a_s(z^{-s})}$$

Furthermore, the results of Section 3.2, which characterized the nonseasonal process, carry over in an obvious way, so that

(i) If Y_t is a multiplicative autoregressive process of order p, P, i.e., q, $Q = 0$, its autocorrelations will die out according to the difference equation

$$a(B)a_s(B^s)\rho_\tau = 0 \qquad \text{for all } \tau > 0$$

where the operator B is on the index τ; and its partial autocorrelations will obey

$$a_{KK} = 0 \qquad \text{for all} \quad K > p + sP$$

(ii) If Y_t is a multiplicative moving average process of order q, Q, i.e., p, $P = 0$, its autocorrelations will obey

$$\rho_\tau = 0 \qquad \text{for} \qquad\qquad q < \tau < s - q$$

$$s + q < \tau < 2s - q$$

$$\vdots$$

$$(Q - 1)s + q < \tau < Qs - q$$

$$Qs + q < \tau$$

(iii) If Y_t is a multiplicative mixed autoregressive moving average process of order (p, P, q, Q), its autocorrelations will obey

$$a(B)a_s(B^s)\rho_\tau = 0 \qquad \text{for all} \quad \tau > q + sQ$$

In fact, for many members of the class (3.7.1) more comprehensive conditions than these can be derived. These are best illustrated by consideration of three fairly general cases which might be expected to include the vast majority of models that occur in practice, and which can be expanded in an obvious fashion to cover many other potentially useful models.

As a first case, consider the multiplicative moving average process

$$Y_t = (1 + b_1 B + b_2 B^2)(1 + b_{1,s} B^s + b_{2,s} B^{2s}) \epsilon_t \qquad (3.7.2)$$

As noted above, for this process

$$\rho_\tau = 0 \quad \text{for} \quad 2 < \tau < s - 2, \quad s + 2 < \tau < 2s - 2, \quad 2s + 2 < \tau \quad (3.7.3)$$

It is also straightforward to verify that

$$\rho_{s-2} = \rho_{s+2}, \qquad \rho_{s-1} = \rho_{s+1}, \qquad \rho_{2s-2} = \rho_{2s+2}, \qquad \rho_{2s-1} = \rho_{2s+1} \quad (3.7.4)$$

By setting some of the coefficients of (3.7.2) equal to zero, a number of special cases can be examined. The general conditions (3.7.4) continue to hold, while modifying (3.7.3) it is straightforward to show for the various cases:

(i)   If $b_2 = 0$: $\rho_\tau = 0$ for $1 < \tau < s - 1$, $s + 1 < \tau < 2s - 1$, $2s + 1 < \tau$.
(ii) If $b_1 = b_2 = 0$: $\rho_\tau = 0$ for all τ except $\tau = s, 2s$.
(iii) If $b_2 = b_{2,s} = 0$: $\rho_\tau = 0$ for $1 < \tau < s - 1$, $s + 1 < \tau$.
(iv) If $b_{2,s} = 0$: $\rho_\tau = 0$ for $2 < \tau < s - 2$, $s + 2 < \tau$.
(v)   If $b_1 = b_2 = b_{2,s} = 0$: $\rho_\tau = 0$ for all τ except $\tau = s$.

The second important case is the model

$$(1 - a_1 B)(1 - a_{1,s} B^s) Y_t = (1 + b_1 B)(1 + b_{1,s} B^s) \epsilon_t \qquad (3.7.5)$$

The autocorrelations of this process obey the difference equation

$$\rho_\tau = a_1 \rho_{\tau-1} + a_{1,s} \rho_{\tau-s} - a_1 a_{1,s} \rho_{\tau-s-1} \qquad \text{for all} \quad \tau > s + 1 \quad (3.7.6)$$

In practice, behavior of this kind can be recognized in the following way. Suppose that, in general, the difference equation in (3.7.6) is obeyed for all τ bigger than some number T. Then

$$\rho_\tau - a_1 \rho_{\tau-1} = a_{1,s}(\rho_{\tau-s} - a_1 \rho_{\tau-s-1}) \qquad \text{for all} \quad \tau > T \qquad (3.7.7)$$

and

$$\rho_\tau - a_{1,s} \rho_{\tau-s} = a_1(\rho_{\tau-1} - a_{1,s} \rho_{\tau-s-1}) \qquad \text{for all} \quad \tau > T \qquad (3.7.8)$$

It follows, then, that for some number k, from (3.7.8)

$$\rho_\tau - a_{1,s} \rho_{\tau-s} = k a_1^{\tau-T} \qquad \text{for all} \quad \tau > T$$

Thus, in particular,

$$\rho_{T+1} - a_{1,s} \rho_{T+1-s} = k a_1 \qquad \text{and} \qquad \rho_{T+1+K} - a_{1,s} \rho_{T+1+K-s} = k a_1^{K+1}$$

Hence it follows that

$$\rho_{T+1+K} - a_{1,s} \rho_{T+1+K-s} = a_1^K(\rho_{T+1} - a_{1,s} \rho_{T+1-s}) \qquad (3.7.9)$$

Now if K is moderately large, it follows from the stationarity requirement $|a_1| < 1$, that the right-hand side of (3.7.9) will be close to zero, and

$$\rho_{T+1+K} \approx a_{1,s} \rho_{T+1+K-s}$$

In words, this implies that for high lags the effects of the operator a_1 will become negligible. Hence behavior of the type (3.7.8) will be typified by the ratios $\rho_{T+1+K}/\rho_{T+1+K-s}$ being roughly constant for all moderately large K (say $K \geqslant 10$). Now suppose that this is so, and that the ratio is approximately equal to $\hat{a}_{1,s}$, which should be close to the value $a_{1,s}$. Then, from (3.7.8) it follows that the ratios

$$(\rho_\tau - \hat{a}_{1,s} \rho_{\tau-s})/(\rho_{\tau-1} - \hat{a}_{1,s} \rho_{\tau-1}) \qquad \text{for all} \quad \tau > T$$

should also be roughly constant, approximating a_1.

By setting some of the coefficients of (3.7.5) equal to zero, it is possible to consider a number of special cases and typify their behavior:

(i)   If $b_1 = 0$: the difference equation in (3.7.6) holds for all $\tau > s$.

(ii) If $b_{1,s} = 0$: the difference equation in (3.7.6) holds for all $\tau > 1$.

(iii) If $b_1 = b_{1,s} = 0$: the difference equation in (3.7.6) holds for all $\tau > 0$. Furthermore the partial autocorrelations obey

$$a_{KK} = 0 \qquad \text{for all} \quad K > s+1$$

(iv) If $a_{1,s} = 0$: the autocorrelations obey

$$\rho_\tau = a_1 \rho_{\tau-1} \qquad \text{for all} \quad \tau > s+1$$

(v)   If $a_{1,s} = b_1 = 0$: the autocorrelations obey

$$\rho_\tau = a_1 \rho_{\tau-1} \qquad \text{for all} \quad \tau > s$$

(vi) If $a_1 = 0$: the autocorrelations obey

$$\rho_\tau = 0 \qquad \text{for all} \quad 1 < \tau < s-1$$

and

$$\rho_\tau = a_{1,s} \rho_{\tau-s} \qquad \text{for all} \quad \tau > s+1$$

Also $\rho_{s-1} = \rho_{s+1}$.

(vii) If $a_1 = b_1 = 0$: the only nonzero autocorrelations are ρ_{js}, $j = 1, 2, 3, \ldots$, and $\rho_\tau = a_{1,s} \rho_{\tau-s}$ for all $\tau > s$.

(viii) If $a_1 = b_{1,s} = 0$: the autocorrelations obey the same conditions as in (vi) except that now

$$\rho_\tau = a_{1,s} \rho_{\tau-s} \qquad \text{for all} \quad \tau > 1$$

(ix) If $a_1 = b_1 = b_{1,s} = 0$: the only nonzero autocorrelations are ρ_{js}, $j = 1, 2, 3, \ldots$ and $\rho_\tau = a_{1,s} \rho_{\tau-s}$ for all $\tau > 0$. Further, the partial autocorrelations obey $a_{KK} = 0$ for all $K > s$.

The third case of interest concerns situations in which the multiplicativity assumed in (3.7.1) is no longer tenable. To illustrate, consider the moving average process

$$Y_t = (1 + b_1 B)(1 + b_{1,s} B^s)\epsilon_t \tag{3.7.10}$$

It has been noted that the only nonzero autocorrelations of such a process are ρ_1, ρ_{s-1}, ρ_s, and ρ_{s+1} and that

$$\rho_{s-1} = \rho_{s+1} \tag{3.7.11}$$

Consider now the alternative process

$$Y_t = (1 + b_1 B + b_s B^s + b_{s+1} B^{s+1})\epsilon_t \tag{3.7.12}$$

This differs from the process (3.7.10) if $b_{s+1} \neq b_1 b_s$. The process (3.7.12) again has only ρ_1, ρ_{s-1}, ρ_s, ρ_{s+1} as nonzero autocorrelations, but can be distinguished from (3.7.10) by noting that now the condition (3.7.11) fails in general to hold. The process

$$(1 - a_1 B)Y_t = (1 + b_1 B + b_s B^s + b_{s+1} B^{s+1})\epsilon_t$$

can be distinguished from the corresponding multiplicative process in exactly the same way. It appears to be true, in general, that for most practically occurring seasonal time series, multiplicative models provide a good representation. However, it should be remembered that the restriction of multiplicativity can always be dropped if this is thought necessary from examination of the sample autocorrelations. Some of the principles involved in the identification of seasonal models are now illustrated with some specific examples.

EXAMPLE 1   First consider a series of 108 monthly observations on employment in manufacturing in Canada. Table 3.5 shows the sample autocorrelations r_K and partial autocorrelations $\hat{a}_{KK}$ for the series and some of its differences.

The sample autocorrelations of the undifferenced series die out only slowly at high lags, suggesting nonstationary behavior. In passing, it is worth noticing that this series provides an excellent illustration of the advantages of differencing. For time series that exhibit this typical nonstationary pattern it is, as noted earlier, very often the case that, when considering the raw data alone, all other characteristics of the series will be swamped by the evidence on nonstationarity. Thus, looking only at the sample autocorrelations and partial autocorrelations of the original series, one would be hard pressed indeed to detect seasonality. The series $(1 - B)X_t$ has high autocorrelations at lags that are multiples of 12, and these appear to be dying out only very slowly. The autocorrelations of the series $(1 - B^{12})X_t$ are initially very high and die out slowly, suggesting that first differencing is required (or at least that any model fitted to this series will have an autoregressive root close to

Table 3.5 *Sample autocorrelations and partial autocorrelations for employment in manufacturing in Canada series*

	K	1	2	3	4	5	6	7	8	9	10	11	12
	r_K	.97	.94	.90	.85	.82	.79	.77	.77	.78	.77	.76	.75
	$\hat{a}_{KK}$	.97	−.13	−.07	−.13	.15	.13	.21	.06	.13	−.19	−.09	−.06
	K	13	14	15	16	17	18	19	20	21	22	23	24
X_t	r_K	.71	.67	.62	.56	.51	.48	.45	.43	.43	.41	.40	.38
	$\hat{a}_{KK}$	−.34	.00	−.07	−.09	.05	−.02	.00	.08	.04	−.09	.03	−.06
	K	25	26	27	28	29	30	31	32	33	34	35	36
	r_K	.34	.29	.24	.19	.14	.11	.09	.07	.07	.06	.05	.04
	$\hat{a}_{KK}$	−.15	.02	−.04	−.01	.05	−.04	.06	.01	−.01	.00	.04	−.02

	K	1	2	3	4	5	6	7	8	9	10	11	12
	r_K	.08	.12	.15	−.22	−.14	−.52	−.18	−.22	.12	.10	.08	.78
	$\hat{a}_{KK}$	.08	.11	.14	−.26	−.15	−.53	−.09	−.22	.44	−.09	.08	.55
	K	13	14	15	16	17	18	19	20	21	22	23	24
$(1-B)X_t$	r_K	.05	.11	.13	−.27	−.11	−.50	−.21	−.24	.10	.07	.04	.66
	$\hat{a}_{KK}$	−.10	−.12	−.02	−.14	.05	−.09	.01	−.21	.12	−.20	.01	−.04
	K	25	26	27	28	29	30	31	32	33	34	35	36
	r_K	.04	.12	.08	−.21	−.11	−.43	−.17	−.18	.06	.05	.04	.59
	$\hat{a}_{KK}$	.12	−.13	.00	.06	−.04	.03	.14	.04	−.11	−.08	−.01	.02

	K	1	2	3	4	5	6	7	8	9	10	11	12
	r_K	.96	.92	.87	.81	.74	.67	.58	.49	.40	.32	.25	.17
	$\hat{a}_{KK}$	.96	.03	−.15	−.10	−.21	−.06	−.25	−.05	.01	.07	.13	−.25
	K	13	14	15	16	17	18	19	20	21	22	23	24
$(1-B^{12})X_t$	r_K	.10	.05	.01	−.03	−.05	−.08	−.08	−.09	−.08	−.08	−.08	−.07
	$\hat{a}_{KK}$	.23	.06	.13	−.11	.06	.00	−.04	−.02	−.01	−.12	−.04	.05
	K	25	26	27	28	29	30	31	32	33	34	35	36
	r_K	−.05	−.04	−.04	−.04	−.05	−.05	−.06	−.06	−.08	−.09	−.11	−.12
	$\hat{a}_{KK}$	−.00	−.01	−.00	−.12	.01	.01	.05	−.08	.00	.07	−.04	.02

	K	1	2	3	4	5	6	7	8	9	10	11	12
	r_K	−.08	.10	.15	.27	.05	.15	.05	.06	−.09	−.02	.25	−.35
	$\hat{a}_{KK}$	−.08	.09	.17	.30	.09	.10	−.03	−.07	−.21	−.19	.24	−.29
	K	13	14	15	16	17	18	19	20	21	22	23	24
$(1-B)(1-B^{12})X_t$	r_K	−.03	−.19	.08	−.23	−.00	−.07	−.04	−.24	.19	.03	−.15	−.01
	$\hat{a}_{KK}$	−.04	−.25	.09	−.04	.09	.19	.01	−.06	.12	.06	.01	−.14
	K	25	26	27	28	29	30	31	32	33	34	35	36
	r_K	.01	.13	−.04	.07	.00	−.00	−.00	.17	−.10	−.03	−.07	−.01
	$\hat{a}_{KK}$	−.06	−.08	.05	−.04	.01	−.07	.11	−.02	−.02	−.02	−.04	−.13

unity). Accordingly one is led to consider the series $(1 - B)(1 - B^{12})X_t$. This has a high autocorrelation at lag 12, while those at lags 24 and 36 are small. (Since 13 observations are "lost" in differencing, the relevant comparison in assessing the magnitude of the sample autocorrelations is with $2/\sqrt{95} \approx 0.21$.) These considerations suggest that one might tentatively identify the model

$$(1 - B)(1 - B^{12})X_t = (1 + b_{1,s}B^{12})\epsilon_t$$

as being appropriate for this employment data.

EXAMPLE 2   The second example is a time series of 108 monthly values of unemployment in Belgium. Relevant sample autocorrelations are given in Table 3.6, and the series is graphed in Fig. 3.1.

The sample autocorrelations of the original series exhibit typical non-stationary behavior, suggesting that differencing is required. The series $(1 - B)X_t$ has high and persistent autocorrelations at lags around multiples of 12, while the series $(1 - B^{12})X_t$ has autocorrelations that suggest the need for further differencing. On the other hand, there is very little evidence to suggest that the series $(1 - B)(1 - B^{12})X_t$ is anything other than white noise. (As in the previous example, approximate 95% significance levels for sample auto-correlations under the hypothesis of white noise are given by ± 0.21.) Perhaps

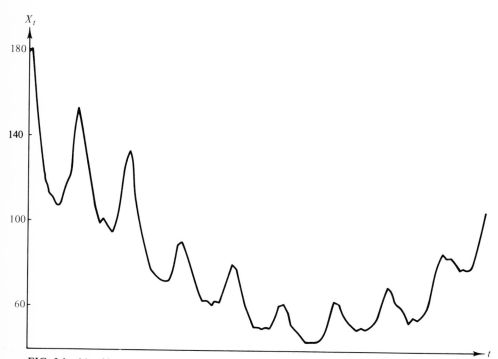

FIG. 3.1   *Monthly unemployment in Belgium.*

Table 3.6　*Sample autocorrelations and partial autocorrelations for data on unemployment in Belgium*

	K	1	2	3	4	5	6	7	8	9	10	11	12
	r_K	.92	.80	.70	.62	.58	.56	.55	.57	.61	.64	.66	.64
	$\hat{a}_{KK}$	.92	$-.27$	.07	.07	.11	.08	.11	.16	.14	.10	.03	$-.18$
	K	13	14	15	16	17	18	19	20	21	22	23	24
X_t	r_K	.56	.47	.39	.33	.28	.26	.24	.25	.28	.30	.31	.27
	$\hat{a}_{KK}$	$-.19$	$-.05$	$-.08$	$-.08$	$-.04$	$-.05$	$-.06$	.02	.07	.05	.03	$-.06$
	K	25	26	27	28	29	30	31	32	33	34	35	36
	r_K	.20	.12	.06	.00	$-.04$	$-.06$	$-.07$	$-.07$	$-.05$	$-.03$	$-.03$	$-.05$
	$\hat{a}_{KK}$	$-.09$	$-.02$	$-.02$	$-.06$	$-.04$	$-.04$	$-.06$	$-.06$	.01	.07	.03	.02

	K	1	2	3	4	5	6	7	8	9	10	11	12
	r_K	.65	.24	$-.06$	$-.26$	$-.32$	$-.33$	$-.36$	$-.34$	$-.17$	.19	.55	.68
	$\hat{a}_{KK}$	.65	$-.32$	$-.11$	$-.17$	$-.04$	$-.18$	$-.20$	$-.16$	.09	.33	.33	.17
	K	13	14	15	16	17	18	19	20	21	22	23	24
$(1-B)X_t$	r_K	.47	.16	$-.06$	$-.20$	$-.23$	$-.29$	$-.31$	$-.28$	$-.14$	.14	.43	.52
	$\hat{a}_{KK}$	$-.17$	$-.07$	.01	.01	.07	$-.06$	.16	.00	$-.06$	$-.10$	.09	.11
	K	25	26	27	28	29	30	31	32	33	34	35	36
	r_K	.37	.16	$-.02$	$-.13$	$-.17$	$-.20$	$-.22$	$-.21$	$-.12$	.09	.28	.38
	$\hat{a}_{KK}$	$-.00$	.06	$-.04$	.04	.02	.03	.09	$-.00$	.01	$-.08$	$-.02$	.07

	K	1	2	3	4	5	6	7	8	9	10	11	12
	r_K	.94	.87	.80	.75	.71	.67	.63	.59	.55	.52	.49	.46
	$\hat{a}_{KK}$	.94	$-.18$	.09	.05	$-.01$	.07	$-.04$	$-.02$	.01	.03	$-.02$	$-.00$
	K	13	14	15	16	17	18	19	20	21	22	23	24
$(1-B^{12})X_t$	r_K	.44	.42	.40	.38	.37	.36	.35	.36	.36	.35	.34	.31
	$\hat{a}_{KK}$	.04	$-.01$	.02	.02	.02	$-.01$	.09	.02	.01	$-.02$	$-.04$	$-.13$
	K	25	26	27	28	29	30	31	32	33	34	35	36
	r_K	.27	.24	.20	.17	.14	.11	.09	.06	.03	.00	$-.03$	$-.07$
	$\hat{a}_{KK}$	$-.06$	$-.02$	$-.04$	$-.03$	$-.01$	$-.02$	$-.01$	$-.01$	$-.03$	$-.07$	$-.07$	$-.05$

	K	1	2	3	4	5	6	7	8	9	10	11	12
	r_K	.17	$-.10$	.02	.00	$-.21$	$-.01$	.07	$-.12$	$-.15$	.13	.08	$-.14$
	$\hat{a}_{KK}$	.17	$-.14$	.07	$-.03$	$-.20$	.08	.01	$-.13$	$-.09$	.12	.02	$-.13$
	K	13	14	15	16	17	18	19	20	21	22	23	24
$(1-B)(1-B^{12})X_t$	r_K	.01	$-.03$	$-.09$	$-.08$	$-.02$	$-.21$	$-.20$	$-.02$	.04	.07	.22	.14
	$\hat{a}_{KK}$	.03	$-.13$	.01	$-.07$	$-.11$	$-.22$	$-.13$	$-.07$	$-.08$	.06	.10	.03
	K	25	26	27	28	29	30	31	32	33	34	35	36
	r_K	.06	.06	.05	$-.03$	$-.05$	.05	$-.00$	.05	.03	.13	.01	$-.04$
	$\hat{a}_{KK}$	.10	.02	.01	.01	.00	.03	$-.02$	.08	$-.08$	.15	$-.01$	$-.08$

the first autocorrelation, which is moderate in size, may be of importance, and therefore the model

$$(1 - B)(1 - B^{12})X_t = (1 + b_1 B)\epsilon_t$$

will be carried forward for estimation.

EXAMPLE 3   The final example, a series of 64 quarterly values of construction begun in England and Wales, presents rather more difficulties, particularly with regard to the choice of an appropriate difference operator.

Examination of the sample autocorrelations of the raw data in Table 3.7 is not terribly suggestive. There is some evidence of seasonality, as can be seen from the relatively high values at lags 4 and 8. Nevertheless the overall picture is rather unclear, and it must be admitted that the evidence in favor of differencing is less than overwhelming. (Perhaps the seasonal effect—occurring as it does every four lags—is clouding the picture.) On the other hand, it can be seen that the series $(1 - B)X_t$ has very persistently high autocorrelations at lags that are multiples of 4, suggesting that the operator $1 - B^4$ should be employed. Now, for the series $(1 - B^4)X_t$, the autocorrelations are not terribly high for longer lags, but the *pattern* of these autocorrelations is rather smooth—typical of the behavior of an underdifferenced series. Rather surprisingly, in view of what has preceded, when one comes to look at the series $(1 - B)(1 - B^4)X_t$ the autocorrelations suggest very strongly the multiplicative first-order moving average model since these values are to be compared in magnitude with $2/\sqrt{59} \approx 0.26$. (It seems that, if nothing else, differencing has finally produced an autocorrelation pattern that is easy to interpret!) Accordingly the model

$$(1 - B)(1 - B^4)X_t = (1 + b_1 B)(1 + b_{1,s} B^4)\epsilon_t$$

will be tentatively entertained.

It must be admitted that the degree of differencing proposed for this series is not based solely on the evidence of the data. Generally speaking two other considerations are of importance. First, all other things being equal, we would favor differencing any time series when in doubt. Otherwise one is tying the series to a fixed mean, which will not be estimable with any great precision if the series in question is highly autocorrelated. Further, forecasts based on the model will of necessity be highly dependent on this imprecisely estimated mean, and hence potentially unreliable. Finally, our experience in analyzing a large number of economic time series has convinced us that, in those situations where the degree of differencing has been in doubt, it is generally the case that superior forecasts are obtained when the difference operator in question is included in the model.

Table 3.7 *Sample autocorrelations and partial autocorrelations for construction begun data*

	K	1	2	3	4	5	6	7	8	9	10
	r_K	.69	.52	.52	.60	.36	.17	.12	.26	.07	−.08
	$\hat{a}_{KK}$	.69	.09	.24	.32	−.42	−.17	−.07	.29	−.21	−.02
X_t											
	K	11	12	13	14	15	16	17	18	19	20
	r_K	−.07	.05	−.09	−.17	−.14	.00	−.09	−.13	−.09	.05
	$\hat{a}_{KK}$	.06	−.05	−.11	.06	.10	−.04	−.03	.11	−.09	.01
	K	1	2	3	4	5	6	7	8	9	10
	r_K	−.27	−.34	−.15	.65	−.14	−.28	−.19	.57	−.09	−.31
	$\hat{a}_{KK}$	−.27	−.44	−.55	.35	.14	.08	−.19	.10	.04	−.11
$(1-B)X_t$											
	K	11	12	13	14	15	16	17	18	19	20
	r_K	−.19	.55	−.12	−.26	−.15	.47	−.10	−.21	−.13	.44
	$\hat{a}_{KK}$	−.18	.06	−.10	−.06	−.04	.02	−.06	.00	.03	.07
	K	1	2	3	4	5	6	7	8	9	10
	r_K	.52	.38	.30	−.02	.12	.08	−.05	−.08	−.12	−.17
	$\hat{a}_{KK}$	.52	.15	.08	−.33	.28	−.02	−.10	−.22	.14	−.10
$(1-B^4)X_t$											
	K	11	12	13	14	15	16	17	18	19	20
	r_K	−.18	−.14	−.16	−.15	−.10	−.09	−.02	.05	−.02	−.03
	$\hat{a}_{KK}$	−.10	−.04	.07	−.08	−.03	.03	.10	.01	−.18	−.01
	K	1	2	3	4	5	6	7	8	9	10
	r_K	−.34	−.07	.21	−.46	.21	.08	−.06	−.01	.02	−.03
	$\hat{a}_{KK}$	−.34	−.16	.00	−.08	−.10	−.17	−.04	.10	−.04	.02
$(1-B)(1-B^4)X_t$											
	K	11	12	13	14	15	16	17	18	19	20
	r_K	−.06	−.06	−.05	−.04	.03	−.06	.01	.14	.07	.06
	$\hat{a}_{KK}$	−.04	−.16	.00	−.08	−.10	.17	.04	.10	.04	.02

Estimation

Estimation of seasonal time series involves no new principles, and proceeds along the lines of Section 3.5. The residuals ϵ_t are computed from

$$\left(1 - a_1 B - a_2 B^2 - \cdots - a_p B^p\right)\left(1 - a_{1,s} B^s - a_{2,s} B^{2s} - \cdots - a_{P,s} B^{Ps}\right) Y_t$$

$$= \left(1 + b_1 B + b_2 B^2 + \cdots + b_q B^q\right)$$

$$\times \left(1 + b_{1,s} B^s + b_{2,s} B^{2s} + \cdots + b_{Q,s} B^{Qs}\right)\epsilon_t$$

where Y_t is the appropriately differenced series $Y_t = (1 - B)^d (1 - B^s)^D X_t$. The sum of squared errors (or, if the unconditional approach is employed, the sum of squares of the conditional expectations of the errors) can then be minimized using the techniques described in Section 3.5. Again, approximate standard deviations are obtained for the coefficient estimates from the nonlinear regression program.

Returning now to the series discussed earlier in this section, estimates of the coefficients of the identified models are given below (together with estimated standard errors in brackets):

(i)   For the series of employment in manufacturing in Canada, the estimated model was

$$(1 - B)(1 - B^{12})X_t = (1 - 0.50B^{12})\epsilon_t$$
$$[0.10]$$

(ii)   For the data on unemployment in Belgium, the fitted model was

$$(1 - B)(1 - B^{12})X_t = (1 + 0.25B)\epsilon_t$$
$$[0.10]$$

(iii)   For the quarterly series of construction begun in England and Wales the fitted model was

$$(1 - B)(1 - B^4)X_t = (1 - 0.37B)(1 - 0.68B^4)\epsilon_t \qquad (3.7.13)$$
$$[0.13] \qquad\qquad [0.10]$$

Diagnostic Checking

For seasonal time series, the range of reasonable alternatives to the originally chosen model is likely to be even greater than in the nonseasonal case. Accordingly, the implementation of diagnostic checks takes on even greater importance here. As before, the two most useful techniques are likely to be the fitting of extra coefficients and study of the autocorrelations of the residuals from the fitted model.

The difficulties in identifying an appropriate model for a seasonal time series are potentially quite severe. Hence, it is generally a good idea to test the validity of the chosen model by adding extra coefficients to one of the operators $a(B)$ and $b(B)$ of (3.7.1) and to one of the operators $a_s(B^s)$ and $b_s(B^s)$. The identification procedure itself may contain useful suggestions in this regard.

For the series on employment in manufacturing in Canada the augmented model

$$(1 - B)(1 - B^{12})X_t = (1 + b_1B)(1 + b_{1,s}B^{12})\epsilon_t$$

was fitted, adding the moving average coefficient b_1 to the previously estimated model. Estimating the parameters in the usual way yielded

$$(1 - B)(1 - B^{12})X_t = (1 - 0.03B)(1 - 0.50B^{12})\epsilon_t$$
$$[0.10] \qquad\qquad [0.10]$$

Clearly there is no justification in retaining the extra coefficient in this case, and on the basis of this check one could not be unhappy with the adequacy of representation of the originally chosen model.

For the unemployment in Belgium series the model was again augmented on the moving average side to the form of a multiplicative first-order process. This produced the estimated model

$$(1 - B)(1 - B^{12})X_t = (1 + 0.28B)\ (1 - 0.15B^{12})\epsilon_t$$
$$\qquad\qquad\qquad\quad [0.10]\qquad\qquad [0.11]$$

The decision with regard to the extra coefficient is much less clearcut in this case. The estimate of that coefficient is about 1.36 times its estimated standard deviation, indicating statistical significance at approximately the 17% level. Since the extra coefficient chosen was an obviously sensible one to contemplate, we are inclined to regard this as sufficiently strong evidence for preferring the more elaborate model to the simpler form previously fitted.

In addition to these possibilities, it is often worthwhile to check the assumption of multiplicativity in the chosen model by the addition of extra coefficients. One should be prepared to drop the multiplicativity assumption if such a course is suggested by this check.

For the series on construction begun in England and Wales, the fitted model (3.7.13) can be written

$$(1 - B)(1 - B^4)X_t = (1 - 0.37B - 0.68B^4 + 0.2516B^5)\epsilon_t$$

To check the assumption of multiplicativity the model

$$(1 - B)(1 - B^4)X_t = (1 + b_1B + b_4B^4 + b_5B^5)\epsilon_t$$

was fitted to this series. The estimated equation which resulted was

$$(1 - B)(1 - B^4)X_t = (1 - 0.33B - 0.65B^4 + 0.29B^5)\epsilon_t$$
$$\qquad\qquad\qquad\quad [0.13]\qquad\quad [0.10]\qquad\quad [0.13]$$

The estimated coefficients differ very little from those predicted by the multiplicative model, providing no grounds on which to question the adequacy of representation of that model.

Finally, the residual autocorrelations can provide, as in the nonseasonal case, a useful guide to model adequacy. To illustrate, consider again the series on construction begun in England and Wales. The first 20 autocorrelations $r_\tau(\hat\epsilon)$ of the errors from the fitted model (3.7.13) are given in Table 3.8.

Since five observations are "lost" in the application of the difference filter $(1 - B)(1 - B^4)$, the effective sample size is $n = 59$. None of the residual sample autocorrelations lies outside the range $\pm 2/\sqrt{n}$ ($= 0.26$), and so these autocorrelations provide no basis on which to question the fitted model. However, for further verification the Box–Pierce test statistic (3.6.3) can be

Table 3.8 *Autocorrelations of errors from model (3.7.13) fitted to series on construction begun in England and Wales*

τ:	1	2	3	4	5	6	7	8	9	10
$r_\tau(\hat{\epsilon})$:	0.01	−0.03	0.03	−0.03	0.14	0.10	−0.14	0.04	−0.01	−0.10
τ:	11	12	13	14	15	16	17	18	19	20
$r_\tau(\hat{\epsilon})$:	−0.11	0.02	−0.07	−0.08	−0.07	−0.04	−0.02	0.08	−0.03	0.02

applied to the data in Table 3.8. The calculated statistic is

$$Q = 59 \sum_{\tau=1}^{20} r_\tau^2(\hat{\epsilon}) = 6.03$$

This should be compared with tabulated values of χ^2 for 18 degrees of freedom (since the model (3.7.13) contains two estimated parameters), from which it emerges that the Box–Pierce test provides no indication of departure from randomness in the residuals, leading to acceptance of the fitted model as providing an adequate representation of the behavior of the construction begun series.

The autocorrelations of the residuals may suggest that an appropriate model for these residuals is

$$a^*(B)a_s^*(B^s)(1-B)^{d^*}(1-B^s)^{D^*}\epsilon_t = b^*(B)b_s^*(B^s)\eta_t \qquad (3.7.14)$$

where η_t is white noise. Combining (3.7.1) and (3.7.14) suggests that an appropriate model might be

$$a(B)a^*(B)a_s(B^s)a_s^*(B^s)(1-B)^{d+d^*}(1-B^s)^{D+D^*}X_t$$
$$= b(B)b^*(B)b_s(B^s)b_s^*(B^s)\eta_t$$

although the difficulties mentioned in Section 3.6 should be borne in mind in this context. The iterative cycle of identification, estimation, and diagnostic checking is then continued until a suitable model is found.

3.8 Time Series Model Building—An Overview

The time series model building procedure just described constitutes an attempt to construct, from a given set of data, the underlying stochastic process that might have generated the given observations. Such a task is, of course, extremely formidable and, indeed, would remain so even if the samples generally available were very large indeed. A good deal of progress can be made, however, if only linear processes are considered—the hope being that in most practical situations there can be found a linear model that approximates well in all relevant aspects the properties of the true underlying process. Accordingly, a class of linear processes is examined and the objective is to select from this class a single process to describe a particular given time series. (Nonlinear models are considered in Chapter 9.)

A further restriction is imposed by the assumption that the series under consideration can be well represented, after appropriate differencing, by a stationary stochastic process, and the validity of this assumption will undoubtedly influence to some degree the validity of any inference made from the fitted model. However, the facility to remove certain kinds of nonstationarity by suitable differencing is of considerable importance in the study of economic time series. In our experience, differencing of such series is almost always required to produce stationarity and models based on the differenced series generally provide superior representation to the low-order autoregressive processes traditionally favored by econometricians. Further evidence for this contention is provided by the typical spectral shape of economic time series (see, for example, Granger [1966]). Of course other types of nonstationarity may also be important, and this topic is discussed in Chapter 9.

The model building procedure introduced in this chapter consists of an iterative cycle of identification, estimation, and diagnostic checking. Not surprisingly, the stage that causes the most difficulties in practical attempts at time series model building is identification. Here one is required to choose from a wide class of models a single process that might adequately describe a given time series. While some objective criteria are available on which a rational choice can be based, it remains the case that there does not exist a clearly defined procedure leading in any given situation to a unique identification. Rather, it is necessary to exercise a good deal of judgment at this stage. To some extent, experience with the procedures involved will increase the chances of successful identification, but nevertheless it must be expected that occasional difficulties will crop up. Before discussing particular problems, it is perhaps worth reiterating that in selecting a particular model for subsequent estimation one is not irrevocably committed to retaining it. The model chosen is subjected to checks on its validity, and the iterative nature of the model building process allows one the possibility of making appropriate modifications.

Two particular problems in model identification are worth mentioning. First, it is at times extremely difficult from a given set of data to identify any particular model from the general class in which one has a great deal of confidence. Up to a point, one finds this happening less frequently the more experience one has in using the techniques. Nevertheless, difficulties of this nature do on occasion still arise. It would be extremely rash, for example, to claim that on the basis of sample autocorrelations and partial autocorrelations from a series of a length that is likely to occur in practice one could immediately identify complex models such as

$$(1 - a_1B - a_2B^2 - a_3B^3 - a_4B^4)X_t = (1 + b_1B + b_2B^2 + b_3B^3)\epsilon_t \quad (3.8.1)$$

In situations where one suspects a complex structure, the best strategy might be to begin by constructing a fairly simple model which one may then be able to modify on the basis of the autocorrelation structure of the residuals from the fitted equation or by suitable overfitting. Thus the construction of models like (3.8.1) might require as many as three or four iterations of the cycle of identification, estimation, and diagnostic checking. It should also be added that, in our experience, such complex models occur very rarely in practice (or, at least, if they do we do not succeed in detecting them very often). It may also be profitable to look for extensions of the identification procedure outlined in Section 3.2 as a further aid in the detection of more complex models. In this context, a strategy suggested to us by Professor G. M. Jenkins [1974], developing the ideas behind the use of the ordinary partial autocorrelation function, deserves further study. It has been shown earlier that for an ARMA(p, q) process

$$X_t - a_1X_{t-1} - \cdots - a_pX_{t-p} = \epsilon_t + b_1\epsilon_{t-1} + \cdots + b_q\epsilon_{t-q}$$

the autocorrelations obey

$$\rho_\tau = \sum_{j=1}^{p} a_j\rho_{\tau-j}, \qquad \tau = q + 1, q + 2, \ldots \quad (3.8.2)$$

Consider, now, the quantities $a_{KK,\,M}$ obtained by solving the set of simultaneous linear equations in $a_{Kj,\,M}$

$$\rho_\tau = \sum_{j=1}^{K} a_{Kj,\,M}\rho_{\tau-j}, \qquad \tau = M + 1, M + 2, \ldots, M + K \quad (3.8.3)$$

Clearly the values $a_{KK,\,0}$ are simply the partial autocorrelations a_{KK} defined in (3.2.2). Now, it follows from (3.8.2) that if X_t follows an ARMA(p, q) process

$$a_{KK,\,q} = 0 \qquad \text{for all} \quad K > p \quad (3.8.4)$$

Hence, in principle, the values of p and q for a mixed process can be determined by calculating $a_{KK,\,M}$, $K = 1, 2, 3, \ldots$, $M = 0, 1, 2, \ldots, q^*$ (where q^* is chosen to be the highest value of q one might reasonably expect —typically $q^* = 4$ or 5 ought to be sufficient) and testing the conditions (3.8.4) in turn for successively higher values of M. In practice, of course, the population autocorrelations are unknown and hence $a_{KK,\,M}$ must be estimated by substituting sample autocorrelations for population values in (3.8.3). The usefulness of this additional identification tool then hinges crucially on the sampling properties of the estimates $\hat{a}_{KK,\,M}$. We have no practical experience in using this technique, but feel that it certainly warrants further investigation. Another possibly useful tool in identification is the inverse autocorrelation function proposed by Cleveland [1972]. Its possible practical value again deserves further study.

A related point has been made by Chatfield and Prothero [1973]. It is sometimes the case that multiple identifications are thought possible; that is, the identification procedure might suggest two or more models from the general class that could well represent a particular set of data. It is further contended that, on the basis of available data, it might prove impossible to distinguish at acceptable levels of statistical significance between the alternatives. One solution to this dilemma is to construct a general model that incorporates as subsets the possibilities being entertained. As a simple illustration, suppose for a time series of length 80 the first two sample autocorrelations were $r_1 = 0.32$, $r_2 = 0.07$ and that the remaining sample autocorrelations were small. Does one regard r_2 as being close to zero (in which case the appropriate identification is $X_t = \epsilon_t + b_1\epsilon_{t-1}$) or should r_2 be taken as being approximately equal to r_1^2, and $r_j \approx r_1^j$ for $j = 3, 4, 5, \ldots$ (in which case the appropriate identification is $X_t - a_1 X_{t-1} = \epsilon_t$)? Perhaps the best solution to this difficulty would be to fit the more general model $X_t - a_1 X_{t-1} = \epsilon_t + b_1\epsilon_{t-1}$. As well as providing a diagnostic check by fitting an extra coefficient, the statistical significance of the estimated coefficients of the more general model can help determine which (if any) of the simple models is appropriate. Of course, as noted by Box and Jenkins [1973], it may well be the case that the "different" models are in fact very similar to one another. If this is so, they will yield very similar forecasts and it is not terribly important to distinguish between them. This point is nicely illustrated by the series on help wanted advertisements examined in Sections 3.2, 3.5, and 3.6.

As a final point on model identification, it should be repeated that in order to have any reasonable hopes of success a moderately long series of observations is necessary. This should be obvious by now, for it is required of the procedure that it select a particular stochastic model from a wide class of models. In order to do this with a fair amount of objectivity, one must have a considerable amount of evidence on which a rational choice of model can be based. It would be foolish to expect a series of, say, 30 observations to supply sufficient evidence for such a task. That is not to say that the applied time series analyst ought to refuse to handle such data, but rather that his analysis ought to be less ambitious than that outlined earlier in this chapter. In such situations it is still possible to fit two or three fairly simple models, test the estimated coefficients for statistical significance, and compare the resulting error variance estimates. In this way one might well arrive at a model in whose forecasting ability it is possible to have at least a fair amount of confidence.

Since identification contains so many difficulties, it is tempting to seek a procedure that circumvents the need for this stage entirely. Why not, for example, simply fit a model that contains a large number of both autoregressive and moving average coefficients and thus includes as subsets any models that are likely to arise in practice? As has been seen, such a procedure is unworkable since it will in general lead to coefficient estimates with extremely high standard deviations due to multiple solutions. A more viable alternative might be to fit a very high order autoregressive process since this could capture the essential characteristics of simple mixed processes. However, the model building philosophy outlined in this chapter is predicated on the belief that in most practical situations the underlying process is likely to be a very simple one; that is, few coefficients will be required to provide an adequate description of the behavior of the particular time series. It follows that if such is the case, the fitting of high-order autoregressive processes will lead to the estimation of models in which most of the coefficients are redundant, with the consequence that the few nonredundant coefficients will be estimated with unnecessary imprecision. In Chapter 5 it is shown that a useful compromise can be achieved by fitting high-order autoregressive equations not directly but by stepwise regression techniques.

CHAPTER FOUR

THE THEORY OF FORECASTING

Time present and time past are both perhaps present in time future and time future contained in time past.

T. S. ELIOT

4.1 Some Basic Concepts

Information Set

Let X_t be some discrete-time stochastic process, which for the time being will be assumed to be stationary. Suppose that one is at time n ($\equiv$ now) and one wishes to forecast[1] h time units ahead to time $n + h$ ($h \equiv$ hence), so that attention is directed to the random variable X_{n+h}. Thus for $h = 1$, a one-step ahead forecasting situation arises. Any forecast procedure will have to be based on some *information set*, consisting of data together with knowledge, theories, or assumptions about the properties of the process *as available at time n*. For example, the information set could consist of a sample of previous values of the series X_{n-j}, $j = 0, 1, \ldots, N$, together with the knowledge that the series has zero mean and an assumption that the series is stationary. A different information set may contain past and present values of several series. It is often important to be precise about the information set being used, particularly when comparing different forecasting procedures. For the time being it will be assumed merely that a specific information set is to be used, and this will be denoted by I_n.

---

[1] The words *forecast* and *prediction* will be used interchangeably. In many situations a better description is *extrapolation*.

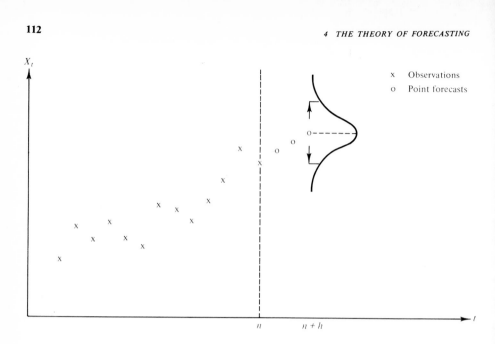

FIG. 4.1 *Conditional density of X_{n+h}, given information up to time n, point and interval forecasts.*

Conditional Variables

Since the variable to be forecast X_{n+h} is a random variable, it can be fully characterized only in terms of a probability density function or some equivalent function. However, since the information set I_n is to be utilized one needs to use a *conditional* density function, i.e.,

$$\text{Prob}(x < X_{n+h} \leqslant x + dx|I_n) = g_c(x)\,dx \qquad (4.1.1)$$

where the subscript c denotes "conditional." If $g_c(x)$ were available, then all other properties of X_{n+h}, such as the conditional mean $E_c(X_{n+h})$, could be immediately determined. However, in practice, it is generally rather too ambitious to hope to be able to characterize fully X_{n+h}, and so one attempts the less ambitious task of finding some confidence band for X_{n+h} or some single value, called a *point forecast*, that in some way "best" represents the random variable X_{n+h}. The forecast picture is depicted in Fig. 4.1, which illustrates a possible conditional density function for X_{n+h} given information available at time n, together with interval and point estimates based on that density function.

Cost Function

To obtain any kind of best value for a point forecast, one requires a criterion against which various alternatives can be judged. An intellectually satisfying way to proceed is to introduce the idea of a *cost function*. Suppose that one is not just forecasting in some vacuous situation, but rather that someone, such as a production manager of a firm whose product demand is being forecast, will actually wish to base decisions upon the forecasts one makes. Since forecast errors are virtually certain to occur in connection with a random process, suppose that one can cost the effect of an error of size e to be $C(e)$, with $C(0) = 0$. If in some fashion a point forecast $f_{n,h}$ is made of X_{n+h}, based on I_n, with resulting error

$$e_{n,h} = X_{n+h} - f_{n,h} \qquad (4.1.2)$$

then because the user has not made the optimal decision a cost $C(e_{n,h})$ will arise. It is now a fairly natural criterion, given such a cost function, to choose the point forecast $f_{n,h}$ so that the expected cost $E_c\{C(e_{n,h})\}$ is minimized, the expectation being conditional on the information set I_n being used. One particular cost function gives a rather tidy and easily used solution, and that is the function

$$C(e) = ae^2 \qquad (4.1.3)$$

where a is some positive constant. Classical statistical theory tells us that the point forecast that minimizes this cost function, which corresponds to a least-squares criterion, is just the conditional mean of X_{n+h}, i.e.,

$$f_{n,h} = E_c\{X_{n+h}\} \qquad (4.1.4)$$

As will be seen, this is a particularly useful result, but it does depend on the cost function being of the form (4.1.3). More general cost functions are considered in Section 4.2.

Linear Forecasts

The procedure just outlined, with whatever cost function is used, may well lead to a point forecast $f_{n,h}$ that is a nonlinear function of the data within the information set I_n. However, in practice one will rarely know the conditional density function sufficiently well for a complete solution to be possible. In these circumstances, the problem of choosing the best point forecast is greatly eased by putting restrictions on the form of forecast to be considered; in particular by assuming that $f_{n,h}$ is a linear function of the data available in I_n. If this restriction is made, only *linear forecasts* are considered. Since the benefits in ease of procedure are so considerable from this restriction, it will generally be taken to apply, although some consideration will be given to nonlinear forecasts in Chapter 9. Alternatively, theory may suggest certain specific nonlinear functional forms by which the available data might enter the forecasting mechanism. This is often the case in econometric forecasting, which will be discussed in Chapter 6.

If one assumes that all subsets of variables have normal distributions, called a Gaussian assumption, then it may be shown that only linear forecasts need be considered. However, linear forecasts can also be optimal in non-normal situations. An example is when a series X_t is generated by a linear model, such as an AR(1) model, but with an error series of independent nonnormally distributed deviates.

Deterministic Processes

Suppose for the moment that a satisfactory measure of the quality of a forecast procedure is the mean square of the one-step error, i.e., $V(1)$ $= E_c\{e_{n,1}^2\}$. The size of this quantity will depend both upon the properties of the series X_t being forecast and also on the contents of the information set I_n being used. If, for example, the information set was $I_n = \{x_{n-j}, j = 0, 1, \ldots, N, \text{model}\}$, i.e., a finite sample of past and present values of the series one is interested in, together with a knowledge of the generating process for the series, then $V(1)$ can vary between variance of x_t if X_t is a white noise series,[1] down to zero if $X_t = a \cos \lambda t$ for example, with a and λ known. In the first case X_t might be said to be *unpredictable* and in the second case X_t is *self-deterministic*. A more conventional definition of deterministic considers $\lim_{N \to \infty} V_N(1)$, where $V_N(1)$ is the mean squared error of one-step forecasts based on the previous N terms of the series. If this limit is zero, X_t is usually said to be deterministic. Thus, given an infinite amount of past values of the series, so that presumably a generating process or model can be determined without error, the next value of the series X_{n+1} can be forecast without error if it is deterministic. However, we prefer the expression "self-deterministic." It follows by iterative reasoning that if one can forecast one step ahead perfectly, then it is possible to forecast perfectly any number of steps ahead. The expression self-deterministic is used here because a series need not be deterministic for one information set yet can be deterministic with respect to a larger information set. Consider the example $X_t = Y_{t-1}$, where Y_t is a zero mean, white noise series. Then X_t is not self-deterministic since given past values of X_t one cannot predict future values at all, but if the information set is expanded to include past and present values of the Y_t series, then X_{n+1} is known without error. If, for some set I_n, X_t is perfectly predictable, it will be said to be deterministic with respect to I_n.

The concept of a series being deterministic need not be phrased in terms of the mean squared errors $V(1)$ but could be defined as the series being predictable with zero cost of error for any given cost function or, almost equivalently in practice, by saying that the conditional distribution function $G_c(x)$, corresponding to the density function $g_c(x)$ of (4.1.1), is a single-step function. This broadening of definition has little or no practical significance.

As a *statement of belief* we would suggest that virtually no series of any importance within the field of economics is deterministic for *any* information set I_n, however widely defined. Thus one would *always* expect to produce forecasts with error, however sophisticated one's forecasting procedure. This belief has implications when the problem of the evaluation of forecasts is discussed in Chapter 8. This statement does not exclude the possibility that a *component* of the series X_t is deterministic, so that Wold's decomposition (see Section 1.11) can apply.

[1] This variance is assumed to be finite.

Memory

It is convenient to attempt to discuss the *memory* of a series X_t when forecasts are based on an information set I_n. Such a definition is necessarily rather arbitrary, but the following two definitions seem to be sensible although not necessarily equivalent. Let X_t be purely nondeterministic, so that it contains no deterministic components, and let $f_{n,h}^{(N)}$ be the optimal linear forecast based on the information set $I_{n,N} = \{x_{n-j}, j = 0, \ldots, N, \text{model}\}$, i.e., a finite sample of past values of the series and knowledge of the true, single-series model. Define

$$V_N(h) = E\left\{\left(X_{n+h} - f_{n,h}^{(N)}\right)^2\right\} \quad \text{and} \quad V(h) = \lim_{N \to \infty} V_N(h)$$

The two suggested definitions of ϵ-memory, where ϵ is some small positive quantity, are:

(i) *The backward ϵ-memory* is the integer M_ϵ such that

$$V_N(1) - V(1) \leqslant \epsilon, \quad \text{all } N \geqslant M_\epsilon$$

so that adding more past terms to the information set does not add significantly to the ability to forecast one step ahead.

(ii) *The forward ϵ-memory* is the integer M_ϵ' such that

$$V(\infty) - V(h) < \epsilon, \quad \text{all } h \geqslant M_\epsilon'$$

If X_t is purely nondeterministic, then $V(\infty) = \text{var}(X_t)$, so that M_ϵ' effectively measures the length ahead that any worthwhile forecasting is possible with the information set $I_{n,\infty}$.

If X_t is an autoregressive process $AR(p)$, then $M_\epsilon \leqslant p$ for any ϵ. If X_t is a moving average process, and if one takes $\epsilon = 0$, then the backward memory is infinite, but with $\epsilon > 0$ one would hope that M_ϵ is finite. In fact, it will be necessary to assume in what follows that the backward memory of the process is much less than the length of data that is available. If X_t is an integrated process, as described in Section 1.13, then only the first definition is relevant, as $V(\infty)$ is not finite.

4.2 Generalized Cost Functions

Suppose that one wishes to form an optimal point forecast $f_{n,h}$ of X_{n+h} using the information set I_n and a cost function $C(e)$. The cost function will be assumed to have the properties that $C(0) = 0$, $C(e)$ is monotonic nondecreasing for $e > 0$ (i.e., $C(e_1) \geqslant C(e_2)$ for every $e_1 > e_2 > 0$) and monotonic nonincreasing for $e < 0$. Thus, if one forecasts without error, then no cost arises; but if there is an error, then the larger it is the greater will be the cost. However, $C(e)$ need not be symmetrical, so that $C(e)$ and $C(-e)$ need not be equal. Let $g_{c,h}(x)$ be the conditional probability density function (pdf) of

X_{n+h} given I_n, then the required optimal forecast $f_{n, h}$, which will be a function only of I_n, will be found by choosing $f_{n, h}$ so that the expected cost is a minimum, i.e., minimizing

$$J = \int_{-\infty}^{\infty} C(x - f_{n, h}) g_{c, h}(x) \, dx \qquad (4.2.1)$$

In the case when the error function is a quadratic, so that $C(e) = ae^2$ with $a > 0$, J becomes

$$J = \int_{-\infty}^{\infty} a(x - f_{n, h})^2 g_{c, h}(x) \, dx \qquad (4.2.2)$$

Define M_h to be the conditional mean of X_{n+h} given I_n, so that

$$M_h = \int_{-\infty}^{\infty} x \, g_{c, h}(x) \, dx \qquad (4.2.3)$$

Then some simple algebraic manipulation, combined with the fact that $\int_{-\infty}^{\infty} g_{c, h}(x) \, dx = 1$ as $g_{c, h}(x)$ is a pdf, shows that (4.2.2) may be written

$$J = a(M_h - f_{n, h})^2 + a \int_{-\infty}^{\infty} (x - M_h)^2 g_{c, h}(x) \, dx \qquad (4.2.4)$$

The quantity $f_{n, h}$ appears only in the first term of (4.2.4) and so J is minimized by taking

$$f_{n, h} = M_h = E_c\{X_{n+h}|I_n\} \qquad (4.2.5)$$

In the Gaussian situation, where every finite subset of variables in the set $\{I_n, X_{n+h}\}$ is normally distributed, this leads to the important result that the optimal least-squares prediction of X_{n+h} is a linear sum of the terms in I_n. It should also be noted that in this case $g_{c, h}(x)$ is a normal distribution with mean M_h and variance that does not depend on I_n, so that the expected or average cost is a constant if the optimal predictor (4.2.5) is used. In non-Gaussian situations, M_h need not be a linear function of the terms in I_n, and the expected cost may vary through time depending on the values actually found in I_n. Thus, without an assumption of normality, the usual restriction of prediction theory to linear predictors may be suboptimal. However, the restriction to linear predictors does ensure considerable computational simplicity and is generally thought to be a reasonable one unless specific information is available as to the kind of nonlinear form that should be used. This problem is discussed further in Chapter 9.

It is interesting to ask whether the predictor given by (4.2.5) is optimal for a wider class of cost functions than just the quadratic function. Two theorems proved by Granger [1969a] show that in fact it is. These theorems may be summarized as follows:

THEOREM   The optimal predictor of X_{n+h} given I_n is $f_{n, h} = M_h$ if

(i) $C(e)$ is symmetric about $e = 0$, the derivative $C'(e)$ exists almost everywhere and is strictly monotonically increasing on the whole range

$-\infty < e < \infty$, and also $g_{c,h}(x)$ is symmetric about $x = M_h$; or

(ii) $C(e)$ is symmetric about $e = 0$ and $g_{c,h}(x)$, symmetric about $x = M_h$, is continuous and unimodal.

For proofs, readers are referred to the original article. An example may also be found there showing that both $C(e)$ and $g_{c,h}(x)$ can be symmetric but the optimal $f_{n,h}$ need not be equal to M_h.

In practice, one rarely has any direct evidence about the properties of the conditional pdf $g_{c,h}(x)$, so an assumption of symmetry about the conditional mean M_h is likely to be an easy one to accept. It is not difficult to then go further and assume $g_{c,h}(x)$ to be a normal density function, although there is surprisingly little evidence for or against such an assumption in data originating either in industry or in many branches of economics. To a statistician, a Gaussian assumption is an extremely useful one since it leads to considerable simplification in techniques, tests, and interpretation. Such an assumption will underlie much of what follows, although this will not always be stated explicitly.

An assumption of symmetry for the cost function is much less acceptable. As examples of situations where nonsymmetric cost functions arise, consider two cases:

(i) A bank intends to purchase a computer to handle its current accounts. To determine the size of computer to buy, a prediction of future business is made. If this prediction is too high, the result will be that the computer will be underutilized and a cheaper machine could have been bought. If the prediction is too low, the result will be that part of the accounts will have to be handled by other means. There is no reason to suppose that cost of errors will be symmetric in these circumstances.

(ii) A firm is about to issue shares on a stock exchange for the first time. It approaches a bank to act as issuing agent. The bank guarantees to buy all shares not taken by the public but is allowed to choose the price at which the shares are offered. The problem is to predict public demand for each price and then pick the best price. If too low a price is chosen, the original firm makes less money than would otherwise have been possible. If the price is too high, the issuing bank will be left with many unsold stocks and could lose more than its commission. The cost to the bank, which makes the prediction, is clearly not symmetric. The result, in the U.K. at least, is usually that the price chosen is subsequently found to be too low.

In just a few cases, the optimal predictor can be found for specific nonsymmetric cost functions. Consider the linear cost function

$$C(e) = ae, \qquad e > 0, \quad a > 0$$
$$= 0, \qquad e = 0$$
$$= be, \qquad e < 0, \quad b < 0 \qquad (4.2.6)$$

Then, from (4.2.1), the expected cost with predictor $f \equiv f_{n,h}$ is

$$J = E_c\{C(X_{n+h} - f)|I_n\} = a\int_f^\infty (x - f)g_{c,h}(x)\, dx$$

$$+ b\int_{-\infty}^f (x - f)g_{c,h}(x)\, dx \tag{4.2.7}$$

Differentiating this with respect to f and equating to zero to find the minimum expected cost gives

$$G_{c,h}(f) = a/(a - b) \tag{4.2.8}$$

where $G_{c,h}(x)$ is the conditional cumulative distribution function of X_{n+h} given I_n, i.e.,

$$G_{c,h}(x') = \int_{-\infty}^{x'} g_{c,h}(x)\, dx \tag{4.2.9}$$

The second derivative of J is $(a - b)g_{c,h}(f)$, which will always be positive, so the predictor f found from solving (4.2.8) will correspond to a minimization of the expected cost. In the symmetric case $a = -b$, and so the optimal predictor $f \equiv f_{n,h}$ will be given by $G_{c,h}(f) = \frac{1}{2}$, so that f is the median of $g_{c,h}(x)$.

The linear cost function may well provide a good approximation to nonsymmetric cost functions that arise in practice. Consider the production of some highly perishable commodity, such as creamcakes. A producer decides to predict tomorrow's demand for his cakes, which cannot be stored from one day to another, and then to produce as many cakes as is predicted will be demanded. A low prediction of demand $(e > 0)$ will result in an underproduction and the cost will equal the lost profits, which might be approximately equal to the marginal profit (a) times the number of cakes that could have been sold if available. A high prediction $(e < 0)$ will result in a loss due to unsold cakes, this loss may be approximated by the marginal production cost $(-b)$ times the number of cakes unsold. There is no reason to suppose that marginal costs and marginal profits are equal, so a nonsymmetric cost function will arise.

Results such as (4.2.8) are of little practical importance since only very rarely will one have sufficient knowledge of the conditional distribution for solutions to be available. If the conditional distribution is assumed to be normal, so that the distribution of $X_{n+h} - M_h$ is independent of I_n, then it is easily shown that the optimal predictor is given by $f_{n,h} = M_h + \alpha$ where α depends only on the cost function being used and not on I_n. This may be seen as follows. Consider the expected cost J given by (4.2.1), but assume a Gaussian situation, so that $\bar{g}_{c,h}(x)$, the conditional pdf of $X_{n+h} - M_h$, is in fact independent of I_n. Then J may be written $J = \int_{-\infty}^\infty C(x - \alpha)\bar{g}_{c,h}(x)\, dx$ and α has to be chosen so that J is minimized. If the linear cost function (4.2.6) is used, α will be given by $\bar{G}_{c,h}(\alpha) = a/(a - b)$ where $\bar{G}_{c,h}(x)$ is the cumulative distribution function of $X_{n+h} - M_h$, that is of a normally distributed variable, but with zero mean. The optimal forecast is then $M_h + \alpha$.

The practical relevance of these results seems to be that a least-squares approach, corresponding to a quadratic cost function, is both more general and more defensible than might originally have been thought. With a normality assumption about one's data, the optimal least-squares predictor will also be optimal for all symmetric cost functions that are likely to arise in practice. If the cost function is not symmetric, then a generally acceptable procedure is to form the best least-squares predictor and then, at the decision stages, to allow for the cost function being used by adding the appropriate bias term. This is somewhat reassuring, since actual cost functions are unlikely to be quadratic, but nevertheless the classical least-squares prediction theory to be presented in later sections will still be usable.

It might also be noted that with the cost function $C(e) = ae^2$, $a > 0$, the actual value of the parameter a is unimportant. This is also a useful property since otherwise one would need to know the cost function at time $n + h$, which is the time at which the error will occur. For this, one would need to predict the cost function, and a circular situation would result, as well as throwing a difficult and undesirable prediction problem on the reluctant accountants who would need to produce the future cost function.

4.3 Properties of Optimal, Single-Series Forecasts

Let X_t be a zero-mean, stationary, and purely nondeterministic series with finite variance, therefore necessarily having an MA(∞) representation

$$X_t = \sum_{j=0}^{\infty} c_j \epsilon_{t-j}, \qquad c_0 = 1 \qquad (4.3.1)$$

where ϵ_t is a zero-mean, white noise process with variance $\sigma_\epsilon^2 < \infty$. Thus,

$$E\{X_t\} = 0 \qquad \text{and} \qquad \text{var}(X_t) = \sigma_\epsilon^2 \sum_{j=0}^{\infty} c_j^2 < \infty$$

It was noted in Section 1.11 that any stationary series could be represented in this fashion, possibly after removal of its deterministic components.

A further assumption about X_t is also required for the following theory, that the MA(∞) representation is invertible, so that a model of the form

$$X_t = \sum_{j=1}^{\infty} a_j X_{t-j} + \epsilon_t$$

also exists, i.e., if $a_N(z) = 1 - \sum_{j=1}^{N} a_j z^j$, then $\lim_{N \to \infty} a_N(z)$ exists. A necessary and sufficient condition for this to hold is that the roots of $c(z) = 0$ all lie outside the unit circle $|z| = 1$, where $c(z) = \sum_{j=0}^{\infty} c_j z^j$. The reason for requiring this invertibility condition should become clear within this section and is further discussed in Section 4.9.

The information set to be considered in this section is $I_n = \{x_{n-j}, j \geqslant 0,$ model} so that all past and present values of the series are known together with the generating model, which is equivalent to knowing the values $c_j, j \geqslant 0$, in the above formulation. In practice this second assumption is an unreasonable one but it does provide an obvious and important starting point for a theory of forecasting. Given that the model is known, knowledge of the complete past of the series could be replaced by knowledge of a finite sample prior to time n, provided the length of this sample is greater than the backward ϵ-memory for small enough ϵ. However, since an inherently unreal situation is being considered, there is little point in examining the slightly more difficult finite sample case.

With I_n as given above, the problem is to find a linear forecast for X_{n+h} of the form

$$f_{n,h} = \sum_{j=0}^{\infty} w_{j,h} X_{n-j} \tag{4.3.2}$$

using a least-squares criterion. Thus, the $w_{j,h}$ need to be chosen so that

$$J = E\left\{ \left(X_{n+h} - \sum w_{j,h} X_{n-j} \right)^2 \right\} \tag{4.3.3}$$

is minimized. In the Gaussian case, it is clear that the resulting $f_{n,h}$ will be the conditional mean M_h defined in the previous section.

In operator terms, (4.3.1) becomes

$$X_t = c(B)\epsilon_t \tag{4.3.4}$$

where

$$c(B) = \sum_{j=0}^{\infty} c_j B^j \tag{4.3.5}$$

and (4.3.2) gives

$$f_{n,h} = w_h(B)X_n \tag{4.3.6}$$

where

$$w_h(B) = \sum_{j=0}^{\infty} w_{j,h} B^j \tag{4.3.7}$$

Letting t take the value n in (4.3.4) and substituting into (4.3.6) gives

$$f_{n,h} = \phi_h(B)\epsilon_n \tag{4.3.8}$$

where

$$\phi_h(B) = \sum_{j=0}^{\infty} \phi_{j,h} B^j \tag{4.3.9}$$

and

$$\phi_h(B) = w_h(B)c(B) \tag{4.3.10}$$

Thus the forecast $f_{n,h}$ is expressed in terms of the white noise, or innovation, process ϵ_t.

By letting t take the value $n + h$ in (4.3.1) and using (4.3.8), it is seen that J of (4.3.3) may be expressed as

$$J = E \left\{ \left(\sum_{j=0}^{h-1} c_j \epsilon_{n+h-j} + \sum_{j=0}^{\infty} (c_{j+h} - \phi_{j,h}) \epsilon_{n-j} \right)^2 \right\} \tag{4.3.11}$$

which gives

$$J = \sigma_\epsilon^2 \sum_{j=0}^{h-1} c_j^2 + \sigma_\epsilon^2 \sum_{j=0}^{\infty} (c_{j+h} - \phi_{j,h})^2 \tag{4.3.12}$$

by using the fact that $E\{\epsilon_s \epsilon_t\} = 0$, $s \neq t$, since ϵ_t is a white noise series. Since the $w_{j,h}$ in (4.3.2) are at our choice, it follows that the $\phi_{j,h}$ are similarly at our choice. It is immediately clear that J in (4.3.12) is minimized by taking

$$\phi_{j,h} = c_{j+h} \tag{4.3.13}$$

so that

$$\phi_h(B) = \sum_{j=0}^{\infty} c_{j+h} B^j \tag{4.3.14}$$

and the optimal h-step forecast may then be written as

$$f_{n,h} = \sum_{j=0}^{\infty} c_{j+h} \epsilon_{n-j} \tag{4.3.15}$$

An occasionally useful notation may be introduced here as follows: consider a function $A(z) = \sum_j A_j z^j$, where j ranges over positive and negative values; then the notation $[A(z)]_+$ will be used to denote the sum over nonnegative values of j, i.e., $[A(z)]_+ = \sum_{j \geq 0} A_j z^j$. Using this notation, $\phi_h(z)$ may be expressed as

$$\phi_h(z) = [c(z)/z^h]_+ \tag{4.3.16}$$

However, the problem as originally posed was to find the parameters $w_{j,h}$ in (4.3.2). In theory at least, they can be found by equating coefficients of powers of z in the expression $\phi_h(z) = w_h(z)c(z)$ with $\phi_h(z)$ given by (4.3.16). Since $w_h(z)$ is now expressed in terms of the known function $c(z)$, a full solution has been reached. Before considering the form of the optimal predictor for some particular models, a number of general properties of considerable importance will be derived.

The h-step error when using an optimal predictor will be

$$e_{n,h} = X_{n+h} - f_{n,h} \tag{4.3.17}$$

but from (4.3.1), with t taking the value $n + h$, and from (4.3.13) it is immediately seen that

$$e_{n,h} = \sum_{j=0}^{h-1} c_j \epsilon_{n+h-j} \tag{4.3.18}$$

so that $e_{n,h}$ is an MA($h - 1$) process. In particular, the one-step error resulting from an optimal predictor is

$$e_{n,1} = \epsilon_{n+1} \tag{4.3.19}$$

and so is a white noise series. Alternatively, this result can be reversed, giving

$$\epsilon_n = X_n - f_{n-1,1} \tag{4.3.20}$$

so that the innovation process ϵ_n can be expressed in terms of the observed series X_n and the previous one-step optimal predictor, which is itself a linear function of $X_{n-j}, j \geq 1$.

From (4.3.18) it follows that the errors have mean zero and variances given by

$$V(h) = E\{e_{n,h}^2\} = \sigma_\epsilon^2 \sum_{j=0}^{h-1} c_j^2 \tag{4.3.21}$$

Further $\lim_{h\to\infty} V(h) = \text{var}(X_t)$. The sequence of error variances is seen to be monotonically nondecreasing since

$$V(h+1) - V(h) = \sigma_\epsilon^2 c_h^2 \geq 0 \tag{4.3.22}$$

It thus follows that, when using optimal forecasts, the further ahead one forecasts the worse one does, on average. An apparent counterexample to this is to consider the sequence of daily figures for car production in the U.K. There are certain days in the year for which car production can be predicted exactly, for example Christmas Day will certainly have zero production. However, this feature occurs because of a (multiplicative) deterministic component in the series and such components were assumed to have been removed before the above analysis was applied.

From (4.3.18) it is seen that

$$e_{n,h} = e_{n-1,h+1} - c_h \epsilon_n \tag{4.3.23}$$

Subtracting both sides of this expression from X_{n+h} and substituting for ϵ_n from (4.3.20), one gets a formula known as the *updating formula*:

$$f_{n,h} = f_{n-1,h+1} + c_h(X_n - f_{n-1,1}) \tag{4.3.24}$$

Clearly, given the sequence of forecasts made at time $n-1$ and the most recent value of the observed series, X_n, one can generate all of the new h-step forecasts. It will be seen in Chapter 5 that use of (4.3.24) can result in considerable computational saving in the calculation of forecasts.

Consider now the pair of forecasts $f_{n,h}$ and $f_{n,h+k}$ $(k > 0)$ of X_{n+h} and X_{n+h+k} made at time n. It follows immediately from (4.3.18) that

$$E(e_{n,h} \, e_{n,h+k}) = \sigma_\epsilon^2 \sum_{j=0}^{h-1} c_j c_{j+k} \tag{4.3.25}$$

Hence forecast errors from the same base n are typically correlated with one another.

Further properties of optimal forecasts and their errors that follow easily from (4.3.15), (4.3.17), and (4.3.18) are that

$$\text{var}(X_{n+h}) = \text{var}(f_{n,h}) + \text{var}(e_{n,h})$$

so that $\text{var}(X_{n+h}) > \text{var}(f_{n,\,h})$ unless the series X_t is deterministic, and

$$\text{cov}(f_{n,\,h},\,e_{n,\,h}) = 0$$

so that $\text{cov}(X_{n+h},\,f_{n,\,h}) = \text{var}(f_{n,\,h})$.

4.4 Optimal Forecasts for Particular Models

In this section, the optimal single-series forecasts derived above will be applied to the models considered in the first chapter.

A series X_t is a moving average of order q, MA(q), if it is generated by

$$X_t = \sum_{j=0}^{q} b_j \epsilon_{t-j}, \qquad b_0 = 1 \tag{4.4.1}$$

where ϵ_t is a zero-mean white noise series. It immediately follows from the theory of the previous section that the optimal h-step forecast is given by

$$f_{n,\,h} = \sum_{j=0}^{q-h} b_{j+h} \epsilon_{n-j} \tag{4.4.2}$$

which may be expressed as

$$f_{n,\,h} = \sum_{j=0}^{q-h} b_{j+h}(X_{n-j} - f_{n-j-1,\,1}) \tag{4.4.3}$$

Given the infinite past of the series, (4.4.3) can be used to generate the forecasts. In practice, there will be a problem in "starting-up" the iterative procedure. This problem will be discussed in Chapter 5. It should be noted that $f_{n,\,h} = 0$, $h > q$, and thus for a white noise series with zero mean, which is MA(0), the optimal forecast is always zero when the forecast is based just on the past of the series. This reflects the basic property of a white noise series that earlier terms contain no information about later terms.

If X_t is a first-order autoregressive process AR(1) generated by

$$X_t = aX_{t-1} + \epsilon_t, \qquad |a| < 1 \tag{4.4.4}$$

then it may be written

$$X_t = \sum_{j=0}^{\infty} a^j \epsilon_{t-j} \tag{4.4.5}$$

and so applying (4.3.13) one gets as the optimal h-step forecast

$$f_{n,\,h} = \sum_{j=0}^{\infty} a^{j+h} \epsilon_{n-j} \tag{4.4.6}$$

Thus

$$f_{n,\,h} = a^h \sum_{j=0}^{\infty} a^j \epsilon_{n-j} = a^h X_n \tag{4.4.7}$$

The same formula can, of course, also be derived directly from (4.3.16) by noting that $c(z) = 1/(1 - az)$. The variance of the h-step forecast error is immediately seen from (4.3.21) to be

$$V(h) = \frac{(1 - a^{2h})}{1 - a^2} \sigma_\epsilon^2 \tag{4.4.8}$$

Note that, as h tends to infinity, (4.4.8) tends to $\sigma_\epsilon^2/(1 - a^2)$, the variance of the process X_t.

Consider now the mixed ARMA(p, q) process X_t generated by

$$a(B)X_t = b(B)\epsilon_t \tag{4.4.9}$$

where

$$a(B) = 1 - \sum_{j=1}^{p} a_j B^j \quad \text{and} \quad b(B) = \sum_{j=0}^{q} b_j B^j, \quad b_0 = 1$$

The corresponding MA(∞) representation will be

$$X_t = c(B)\epsilon_t \tag{4.4.10}$$

where

$$b(z)/a(z) = c(z) \tag{4.4.11}$$

Thus

$$a(z)c(z) = b(z) \tag{4.4.12}$$

Equating coefficients of z^k gives

$$c_k - \sum_{j=1}^{p} a_j c_{k-j} = b_k \quad \text{with} \quad c_j \equiv 0, \quad j < 0, \quad k = 0, 1, 2, \ldots \tag{4.4.13}$$

Recall that

$$f_{n,h} = \sum_{i=0}^{\infty} c_{i+h}\epsilon_{n-i} \tag{4.4.14}$$

and, taking $f_{n,k}$ to be X_{n+k} for $k \leqslant 0$, consider

$$f_{n,h} - \sum_{j=1}^{p} a_j f_{n,h-j} = \sum_{i=0}^{\infty} \left\{ c_{i+h}\epsilon_{n-i} - \sum_{j=1}^{p} a_j c_{i+h-j}\epsilon_{n-i} \right\}$$

$$= \sum_{i=0}^{\infty} b_{i+h}\epsilon_{n-i}, \quad b_i \equiv 0, \quad i > q \tag{4.4.15}$$

as follows from (4.4.13). This is an important formula since it allows one to form the sequence of optimal forecasts for given n and increasing h. To illustrate this, suppose that X_t is an AR(p) process given by

$$X_t = \sum_{j=1}^{p} a_j X_{t-j} + \epsilon_t \tag{4.4.16}$$

Then (4.4.15) gives

$$f_{n,h} = \sum_{j=1}^{p} a_j f_{n,h-j} \tag{4.4.17}$$

with $f_{n, k} = X_{n+k}$ for $k \leqslant 0$. In particular, for $h = 1$,

$$f_{n, 1} = \sum_{j=1}^{p} a_j X_{n-j+1} \qquad (4.4.18)$$

Equation (4.4.18) gives the optimal one-step forecast in terms of known values and (4.4.17) gives an iterative formula for $f_{n, h}$, $h > 1$.

The more general formula (4.4.15) is similar in nature, but with the addition of terms involving ϵ_{n-i}, which may be replaced by $X_{n-i} - f_{n-i-1, 1}$ by (4.3.20). A useful interpretation of (4.4.15) may be derived as follows: consider the generating formula (4.4.9) for the observed series, with t taking the value $n + h$, i.e.,

$$X_{n+h} = \sum_{j=1}^{p} a_j X_{n+h-j} + \sum_{j=0}^{q} b_j \epsilon_{n+h-j} \qquad (4.4.19)$$

If one is at time n, replace every term in this expression either by its known value at time n or its optimal forecast made at time n. Thus X_{n+k} is replaced by $f_{n, k}$ for $k > 0$ or by X_{n+k} if $k \leqslant 0$, and ϵ_{n+k} is replaced by zero, its optimal forecast for $k > 0$, and by itself, or equivalently, by $X_{n+k} - f_{n+k-1, 1}$, for $k \leqslant 0$. This procedure gives (4.4.15), and so this formula is seen to be completely reasonable in commonsense terms.

So far attention has been restricted to stationary processes, but prediction of integrated processes with a known model is equally straightforward. Suppose X_t is ARIMA(p, 1, q), so that $Y_t = X_t - X_{t-1}$ is ARMA(p, q). The preceding methods provide forecasts of Y_{n+h}, denoted by $f_{n, h}^Y$. The optimal forecast of X_{n+h}, denoted by $f_{n, h}^X$, is given simply by

$$f_{n, h}^X = f_{n, h-1}^X + f_{n, h}^Y$$

with, in the particular case $h = 1$,

$$f_{n, 1}^X = X_n + f_{n, 1}^Y$$

Extension to processes that require a higher degree of differencing follows immediately.

All of the above theory has assumed that the series to be forecast are purely nondeterministic. Suppose that X_t does have deterministic components, and may be written $X_t = T_t + Y_t$ where T_t is deterministic and Y_t is purely nondeterministic. If T_t is known exactly, the optimal forecast of X_{n+h} is obviously given by

$$f_{n, h}^X = T_{n+h} + f_{n, h}^Y$$

Since T_t is deterministic and known it follows that T_{n+h} will also be known exactly. In practice things are not necessarily this simple since, although a series may have a deterministic component, the form of this component is rarely precisely known.

One property of the optimal forecast derived in these past two sections is worth noting particularly, and that is that the generating equations of $f_{n,h}$, the optimal forecast, and X_{n+h}, the series being forecast, are usually quite different and so the time series properties, and the plots through time, will be far from identical. Put another way, the spectrum of $f_{n,h}$ and X_{n+h} will not be the same. This property is particularly relevant when considering the evaluation of forecasts. This will be discussed in Chapter 8.

4.5 A Frequency-Domain Approach

The theory outlined in the previous two sections has involved analysis only in the time domain, and is largely due to Wold [1954a] and Kolmogorov [1939, 1941a]. The finding of the MA(∞) representation

$$X_t = \sum_{j=0}^{\infty} c_j \epsilon_{t-j}, \qquad c_0 = 1 \tag{4.5.1}$$

is clearly of major importance in this approach. In the frequency domain this representation may be written

$$\lambda_x(z) = c(z^{-1})c(z)\lambda_\epsilon(z) \tag{4.5.2}$$

where $\lambda_x(z)/2\pi \equiv s_x(\omega)$, $z = e^{-i\omega}$, is the spectrum of X_t, $\lambda_\epsilon(z)/2\pi \equiv s_\epsilon(\omega)$ is the spectrum of ϵ_t, and $c(z) = \sum_{j=0}^{\infty} c_j z^j$. Equation (4.5.2) follows from the fact that X_t may be regarded as a filtered version of the series ϵ_t. Since ϵ_t is white noise, with variance σ_ϵ^2, (4.5.2) becomes

$$\lambda_x(z) = \sigma_\epsilon^2 c(z)c(z^{-1}) \tag{4.5.3}$$

Considerable theoretical work has been concerned with the prediction problem given merely the spectrum of X_t rather than the generating model. However, since in practice it is most unusual for the spectrum to be known exactly, this work does not as it stands have a great deal of practical relevance. Nevertheless, a usable procedure based on the estimated spectrum has recently been proposed and studied in some detail by Bhansali [1973, 1974]. If one is given a spectral function, it is usually very difficult in practice to find a function $c(z)$ so that the decomposition (4.5.3) holds. As an example the reader might consider a series X_t that is the sum of two independent MA(∞) components with known parameters. In theory a decomposition can be achieved as follows. Suppose that $\log \lambda_x(z)$ has a power series expansion in z, with both negative and positive powers, of the form

$$\log \lambda_x(z) = \sum_{j=-\infty}^{\infty} d_j z^j \tag{4.5.4}$$

This is known as a Laurent expansion and is assumed to be valid in some annulus $\rho < |z| < \rho^{-1}$, $\rho < 1$. Thus

$$\lambda_x(z) = \exp\left\{ \sum_{j=-\infty}^{\infty} d_j z^j \right\} \tag{4.5.5}$$

and

$$c(z) = \exp\left\{ \sum_{j=1}^{\infty} d_j z^j \right\}$$ (4.5.6)

will then be an obvious choice for $c(z)$ since it will have a Taylor series expansion in nonnegative powers of z. By equating coefficients of z^0 in (4.5.5) one sees that

$$\sigma_\epsilon^2 = e^{d_0} = \exp\left\{ \frac{1}{2\pi} \int_{-\pi}^{\pi} \log 2\pi s_x(\omega) \, d\omega \right\}$$ (4.5.7)

It was shown in Section 4.3 that σ_ϵ^2 is the variance of the one-step prediction error when an optimal predictor is used. Thus (4.5.7) gives a general expression in terms of the spectrum of X_t for the minimum achievable one-step error variance when X_t is predicted from its own past and has a known spectrum. It follows directly that a necessary condition that X_t be nondeterministic is that

$$\int_{-\pi}^{\pi} \log 2\pi s_x(\omega) \, d\omega > -\infty$$ (4.5.8)

otherwise X_t is deterministic. This condition may also be shown to be a sufficient one. It further follows that if $s_x(\omega) = 0$ on a set of nonzero measure, for example if $s_x(\omega) = 0$ for all ω in (ω_0, ω_1), then X_t will be deterministic. Rosenblatt [1957] has discussed how one would form the optimal predictor in such cases. A practical use for Eq. (4.5.7) is discussed in Chapter 9.

A class of processes for which the decomposition (4.5.3) is particularly simple is the mixed ARMA class. If X_t is generated by

$$X_t = \sum_{j=1}^{p} a_j X_{t-j} + \sum_{j=0}^{q} b_j \epsilon_{t-j}$$

i.e., $a(B)X_t = b(B)\epsilon_t$, then it was shown in Chapter 2 that the spectrum of X_t is given by

$$\frac{\lambda_x(z)}{2\pi} \equiv s_x(\omega) = \frac{b(z) b(z^{-1})}{a(z) a(z^{-1})} \frac{\sigma_\epsilon^2}{2\pi}$$

and so, immediately $c(z) = b(z)/a(z)$. Further, writing $c(z)$ as

$$c(z) = \prod_{j=1}^{p} (1 - \gamma_j z) / \prod_{j=1}^{q} (1 - \theta_j z), \qquad |\gamma_j| < 1, \quad |\theta_j| < 1$$

which will always be possible, then since

$$\int_{-\pi}^{\pi} \log(1 - \gamma z) \, d\omega = 0, \qquad z = e^{-i\omega}, \quad |\gamma| < 1$$

as may be seen by expanding $\log(1 - \gamma z)$ as a power series in z, the truth of (4.5.7) is easily verified in this case by substitution for $s_x(\omega)$ in that equation.

These results indicate why the mixed ARMA model is of particular importance when considering the formation of optimal single series forecasts.

The formula for the optimal predictor (4.3.16), previously obtained by a time-domain analysis, can also be achieved by an argument in the frequency domain. Since this derivation, known as the Wiener–Hopf technique, has some considerable power, in the sense that a variety of situations can be handled by it, a brief outline of the approach will be given. (We follow Whittle [1963, p. 67].) One may start with the more general problem of how to best "explain" a series Y_t by a linear sum of past and present terms of a series X_t, both series being stationary and having zero means. Consider a representation

$$\hat{Y}_t = \sum_{j=0}^{\infty} w_j X_{t-j} \qquad (4.5.9)$$

Strictly, one should consider the sum over the range $j = 0, \ldots, N$ and then consider the limit as $N \to \infty$, but such subtleties will be ignored and all infinite sums will be assumed to exist. Using a least-squares criterion, the w_j have to be chosen so that

$$J = E\left\{ \left(Y_t - \hat{Y}_t \right)^2 \right\} \qquad (4.5.10)$$

is minimized. By considering $\partial J / \partial w_j = 0$, one gets

$$\sum_{k=0}^{\infty} \lambda_{j-k}^{(xx)} w_k = \lambda_j^{(yx)}, \qquad j = 0, 1, 2, \ldots \qquad (4.5.11)$$

where

$$\lambda_k^{(xx)} = E\{ X_t X_{t-k} \}, \qquad \lambda_k^{(yx)} = E\{ Y_t X_{t-k} \} \qquad (4.5.12)$$

Multiplying (4.5.11) by z^j and summing over all integers j, both positive and negative, gives

$$\lambda_x(z) w(z) = \lambda_{yx}(z) + h(z) \qquad (4.5.13)$$

where $\lambda_x(z)/2\pi \equiv s_x(\omega)$, $z = e^{-i\omega}$, is the spectrum of X_t, $\lambda_{yx}(z)/2\pi \equiv s_{yx}(\omega)$ is the cross spectrum between Y_t and X_t, $w(z) = \sum_{j=0}^{\infty} w_j z^j$, and $h(z)$ is some unknown series in negative powers of z. The left-hand side term of (4.5.13) arises because the left-hand side of (4.5.11) is a convolution, as discussed in Section 1.3. The function $h(z)$ is necessary because (4.5.11) holds only for positive j. Equation (4.5.11) can be derived from (4.5.13) by noting that

$$\lambda_x(z) = \sum_{j=-\infty}^{\infty} \lambda_j^{(xx)} z^j \qquad (4.5.14)$$

and

$$\lambda_{yx}(z) = \sum_{j=-\infty}^{\infty} \lambda_j^{(yx)} z^j \qquad (4.5.15)$$

and then equating coefficients of positive powers of z.

Assume now that X_t is purely nondeterministic and so its spectrum has the decomposition (4.5.3), then dividing both sides of (4.5.13) by $c(z^{-1})$ gives

$$\sigma_\epsilon^2 c(z)w(z) = \frac{\lambda_{yx}(z)}{c(z^{-1})} + \frac{h(z)}{c(z^{-1})} \qquad (4.5.16)$$

The term on the left-hand side of (4.5.16) contains no negative powers of z and the second term on the right-hand side contains only negative powers. It follows therefore that

$$\sigma_\epsilon^2 c(z)w(z) = \left[\frac{\lambda_{yx}(z)}{c(z^{-1})} \right]_+ \qquad (4.5.17)$$

using the notation introduced in Section 4.3.

Two particular cases will now be considered. First, suppose that Y_t, the series to be "explained," is just X_{t+h}, so that the problem considered is the prediction of X_{n+h} by $X_{n-j}, j \geqslant 0$, which is the single-series forecast problem considered in Section 4.3. Then

$$\lambda_k^{(yx)} = E\{X_{t+h}X_{t-k}\} = \lambda_{k+h}^{(xx)} \qquad (4.5.18)$$

and so

$$\lambda_{yx}(z) = z^{-h}\lambda_x(z) \qquad (4.5.19)$$

Further, since $\lambda_x(z) = \sigma_\epsilon^2 c(z)c(z^{-1})$, Eq. (4.5.17) becomes

$$c(z)w_h(z) = [c(z)/z^h]_+ \qquad (4.5.20)$$

which is identical to formula (4.3.16). Hence the optimal forecast has been derived through a frequency-domain approach.

As a second example, consider the case where one wishes to forecast Y_{n+h} by the information set $I_n = \{X_{n-j}, j \geqslant 0, \text{model}\}$, so that one series Y_t is to be forecast in terms of past and present values of another series X_t. If the cross spectrum between Y_t and X_t is $\lambda_{yx}^{(1)}(z)/2\pi$, then the optimal terms w_j in (4.5.9) are given by

$$w(z) = \frac{1}{\sigma_\epsilon^2 c(z)} \left[\frac{\lambda_{yx}^{(1)}(z)}{z^h c(z^{-1})} \right]_+ \qquad (4.5.21)$$

as follows from (4.5.17) by noting that $\lambda_{yx}^{(1)}(z) = z^h \lambda_{yx}(z)$ since, by definition, $\lambda_{yx}(z)/2\pi$ is the cross spectrum between Y_{t+h} and X_t.

Suppose now that the two series are actually related by an equation of the form

$$Y_t = \sum_{j=0}^{\infty} \delta_j X_{t-j} + U_t \qquad (4.5.22)$$

where U_t is some stationary series independent of X_t, while X_t is generated as before by the process $X_t = c(B)\epsilon_t$. Then

$$\lambda_{yx}^{(1)}(z) = \delta(z)\lambda_x(z)$$

where

$$\delta(z) = \sum_{j=0}^{\infty} \delta_j z^j \tag{4.5.23}$$

The optimal prediction of Y_{n+h} in terms of $X_{n-j}, j \geq 0$ is then given by

$$f_{n,h}^{(y)} = \sum_{j=0}^{\infty} w_{j,h} X_{n-j} \tag{4.5.24}$$

where

$$w_h(z) = \frac{1}{\sigma_\epsilon^2 c(z)} \left[\frac{\delta(z)\lambda_x(z)}{z^h c(z^{-1})} \right]_+ \tag{4.5.25}$$

from (4.5.21). Noting that $\lambda_x(z) = \sigma_\epsilon^2 c(z) c(z^{-1})$ this becomes

$$w_h(z) = \frac{1}{c(z)} \left[\frac{c(z)\delta(z)}{z^h} \right]_+ \tag{4.5.26}$$

and expanding $\delta(z)$ in the form (4.5.23), the expression for $w_h(z)$ may be written

$$w_h(z) = \sum_{j=0}^{h-1} \frac{\delta_j}{c(z)} \left[\frac{c(z)}{z^{h-j}} \right]_+ + \sum_{j=h}^{\infty} \delta_j z^{j-h} \tag{4.5.27}$$

In operator form (4.5.24) is $f_{n,h}^{(y)} = w_h(B)X_n$. Thus by noting from (4.5.20) that

$$\frac{1}{c(B)} \left[\frac{c(B)}{B^{h-j}} \right]_+ X_n = f_{n,h-j}^{(x)}, \quad h > j$$

where $f_{n,h}^{(x)}$ is the optimal forecast of X_{n+h} made at time n and B now operates on the index n of X_n, one finds that the optimal forecast of Y_{n+h} based on $X_{n-j}, j \geq 0$, may be written

$$f_{n,h}^{(y)} = \sum_{j=0}^{h-1} \delta_j f_{n,h-j}^{(x)} + \sum_{j=h}^{\infty} \delta_j X_{n+h-j} \tag{4.5.28}$$

Comparing this to (4.5.22) with t given the value $n + h$, the optimal forecast is seen to be the commonsense one, with all future X_t values replaced by their optimal forecasts made at time n, using past and present values of X_t and with U_{n+h} given the value zero, which is the optimal prediction if one only has information on the X_t series, which is independent of U_t. The model (4.5.22) is one frequently used in economics, known as the distributed lag model (see Dhrymes [1971]).[1] The optimal forecast of Y_{n+h} using just past and present X values may be rather a poor one if the U_t component of Y_t is of

---

[1] Strictly, the distributed lag formulation writes $\delta(z) = \beta(z)/\alpha(z)$, where $\alpha(z)$ and $\beta(z)$ are finite order polynomials in z.

greater importance than the component involving values of X_t. It is clear that a much improved forecast can be achieved in this case when the information set contains both $X_{n-j}, j \geqslant 0$, and $Y_{n-j}, j \geqslant 0$, or, equivalently, $U_{n-j}, j \geqslant 0$. It is particularly worth noting that the form of U_t does not at all affect the forecast (4.5.28). This could be taken as a criticism of the distributed lag approach. Put another way if, as is frequently done, it is assumed that the error term in a distributed lag model is white noise, this is equivalent to postulating a situation in which forecasts of future Y are based only on current and past X. It would seem more reasonable to expect that current and past values of Y would also be relevant, and these can be accounted for by incorporating into the model a more general time series structure for the error series. Procedures of this kind will be discussed further in Chapter 7.

4.6 Expectations and Forecasts

Economists frequently encounter expectations in their theories, that is, values of some economic variable that are thought likely to occur in the future, the expectations being made either by individuals or, implicitly by their actions, by some group of individuals, such as those making a market. An expectation may vary from a very naïve guess about the future based on little data and no analysis to a full-scale forecast. If the individuals involved are sufficiently sophisticated, it seems reasonable to consider the possibility that their expectations are, in fact, optimal forecasts based on a wide enough information set. This set need not consist merely of past and present numerical data but could also include subjective information, so that the forecasts will reflect concepts such as "experience" and "prejudices" which are difficult to quantify. Such information sets cannot be fitted into the foregoing theory without adopting a Bayesian viewpoint of some kind, and so attention will be limited here to numerical information sets. To ease exposition, the only case to be considered will be the single-series one, so that the information set will consist only of the past and present of the series to be forecast.[1]

The first example to be presented, showing how forecast theory can be used in economics, involves the term structure of interest rates. At any moment of time there exists not just one interest rate but many rates, depending on how long one wishes to borrow the money. Denote by $R_{k,t}$ the rate quoted at time t for a loan made at that time and to be repaid at the end of the kth year of the loan, so that $R_{1,t}$ represents the rate on a one-year loan and $R_{2,t}$ the rate on a two-year loan, both loans to be made at time t. At each time t one may also get quotes for loans to be made in the future, at least in theory. Let $r_{t,h}$ be the rate quoted at time t for a one year loan to start at time $t + h$. This will be a forward rate and one would naturally expect there to be a relationship between the two-year rate $R_{2,t}$ and the pair of one-year rates $R_{1,t}$ and $r_{t,1}$

[1] Results to be presented in Chapter 7 will show that no considerable changes result from using multiseries information sets.

since a two-year loan is equivalent to borrowing for one year now and for a further year in a year's time. The required formula is in fact

$$r_{t,1} = \frac{(1 + R_{2,t})^2}{1 + R_{1,t}} - 1 \tag{4.6.1}$$

These forward rates may not actually be quoted in the market but can be taken as implied or expected rates and can be estimated from (4.6.1), using the values of $R_{1,t}$ and $R_{2,t}$ actually observed in the money market.

A particular form of expectations hypothesis proposed by Meiselman [1962] is an error-learning model of the form

$$r_{t,h} - r_{t-1,h+1} = g_h(E_t) \tag{4.6.2}$$

where E_t is the error made in forecasting the current one-year rate, so that

$$E_t = R_{1,t} - r_{t-1,1} \tag{4.6.3}$$

Both $r_{t,h}$ and $r_{t-1,h+1}$ are expectations of the one-year rate to prevail in year $t + h$, but made in years t and $t - 1$, respectively. The function $g(x)$ is assumed to be linear, of the form

$$g_h(E_t) = A_h + D_h E_t \tag{4.6.4}$$

The resulting formula, obtained from (4.6.4) and (4.6.2), should be compared to the updating formula (4.3.24), which states that if $r_{n,h}$ is the optimal forecast of $R_{1,n+h}$ based on the information set $I_n = \{R_{1,n-j}, j \geqslant 0\}$ then

$$r_{n,h} - r_{n-1,h+1} = c_h(R_{1,n} - r_{n-1,1}) \tag{4.6.5}$$

where

$$R_{1,t} = \sum_{j=0}^{\infty} c_j \epsilon_{t-j}, \qquad c_0 = 1 \tag{4.6.6}$$

and ϵ_t is a white noise process. Thus the model suggested by Meiselman as a working hypothesis is seen to follow from the theory of optimal forecasting, assuming the market to be using optimal forecasts for its expectations, but with specific values for A_h and D_h in (4.6.4), namely $A_h = 0$, $D_h = c_h$ where c_h is given by (4.6.6).

Meiselman fitted his model to some U.S. data for the period 1901–1954 and estimated A_h and D_h. The values for A_h, $h = 1, 2, \ldots, 8$ were all very small, and none were significantly different from zero, in agreement with the optimal forecasting theory. The estimated D_h values were well fitted by the equation

$$D_h = \beta \alpha^h \tag{4.6.7}$$

with $\beta = 0.72$, $\alpha = 0.84$. Setting c_h in (4.6.6) equal to D_h of (4.6.7) this suggests the model for the one-year rates

$$R_{1,t} = \epsilon_t + \beta \sum_{h=1}^{\infty} \alpha^h B^h \epsilon_t \tag{4.6.8}$$

where B is the backward operator. This may be written as

$$R_{1,t} = \frac{1 - \alpha(1 - \beta)B}{1 - \alpha B} \epsilon_t \qquad (4.6.9)$$

Hence, if the forward rates constitute optimal forecasts of future one-year rates based on the information set consisting of just current and past one-year rates, and if the relation (4.6.7) is taken to hold exactly, the series $R_{1,t}$ must be generated by the ARMA(1, 1) process

$$(1 - 0.84B)R_{1,t} = (1 - 0.24B)\epsilon_t \qquad (4.6.10)$$

as follows by substituting for α and β in (4.6.9). In fact, Nelson [1972a] has fitted several models to the series $R_{1,t}$. For the ARMA process he obtained (with standard errors in brackets)

$$R_{1,t} - \underset{[0.05]}{0.94} R_{1,t-1} = 0.20 + \epsilon_t - \underset{[0.13]}{0.13} \epsilon_{t-1}$$

which is in reasonable agreement with the predicted form (4.6.10). A slightly better fit in terms of error variance was given by the IMA(1, 1) process

$$R_{1,t} - R_{1,t-1} = \epsilon_t - \underset{[0.12]}{0.16} \epsilon_{t-1}$$

which differs insignificantly from a random walk. Nelson concludes that the simple error-learning theory of forward rates is in itself an insufficient explanation of their behavior and that the level of interest rates and an index of business confidence should be incorporated in the model (4.6.5).

A second example of the use of expectations in economics is provided by what is known as the current valuation formula. Suppose a company has, or is contemplating, an investment in some asset from which it is reasonable to expect a flow of returns or dividends. An example would be an investment in a new machine or in a particular portfolio of stocks. A frequently used method of valuing the investment at time n is by the formula

$$V_n = \sum_{h=1}^{\infty} \frac{D^*_{n,h}}{(1+r)^h} \qquad (4.6.11)$$

where V_n is the current value, at time n, $D^*_{n,h}$ is the expected dividend to be paid in year $n + h$, and r is a positive discount factor, used to reflect a preference for dividends in the near future rather than those in the more distant future. The valuation formula merely says that current value is a discounted sum of all expected dividends.

This formula is much used by accountants and by financial economists. An account of its uses may be found in Van Horne [1971]. A problem of considerable importance and difficulty in practice is the choice of r, the discount factor. One suggested method is to put r equal to a "normally expected" rate plus a term reflecting a belief about the degree of risk involved with the investment. In the following analysis, r will be taken to be a known constant.

Suppose that one has available a flow of past dividends from the investment, $I_n = \{D_{n-j}, j \geqslant 0\}$, and that the company's expectations are optimal forecasts based on this information set. Suppose that the series D_t is stationary and has an MA(∞) representation

$$D_t = \sum_{j=0}^{\infty} c_j \epsilon_{t-j} \equiv c(B)\epsilon_t \qquad (4.6.12)$$

where ϵ_t is a white noise series. From (4.6.11) one can write

$$V_{n+1} + D_{n+1} - (1+r)V_n = D_{n+1} - D_{n,1}^* + \sum_{h=1}^{\infty} \frac{D_{n+1,h}^* - D_{n,h+1}^*}{(1+r)^h} \qquad (4.6.13)$$

From (4.3.19) and the updating formula (4.3.24) it follows that

$$V_{n+1} + D_{n+1} - (1+r)V_n = \sum_{h=0}^{\infty} \frac{c_h}{(1+r)^h} \epsilon_{n+1} \qquad (4.6.14)$$

$$= c(\mu)\epsilon_{n+1}, \qquad \mu = 1/(1+r) \quad (4.6.15)$$

The rate of return in the year $n + 1$ may be taken to be the change in the capital value of the investment $V_{n+1} - V_n$ plus the direct monetary return D_{n+1} divided by the value of the capital involved in the investment at the start of the period. This value may be taken to be V_n since, if other people agreed with the company's valuation, in theory the investment or asset could be sold at price V_n, so that V_n is the value of the investment in terms of alternative costs. The accounting rate of return is thus

$$R_{n+1} = \frac{V_{n+1} + D_{n+1} - V_n}{V_n}$$

and from (4.6.15), it is seen that $R_{n+1} = r + c(\mu)(\epsilon_{n+1}/V_n)$. Thus, the sequence of accounting returns R_{n+1} will be the sum of a constant r plus a white noise term since ϵ_{n+1} is white noise and is uncorrelated with V_n. The variance of the white noise term also depends on the value of r through the term $c(\mu)$ since $\mu = 1/(1 + r)$. The use of forecasting theory in conjunction with the valuation formula casts some doubt on the usefulness of this formula since the average return that results is equal to the discount factor inserted and so the argument becomes circular. Further discussion of this formula with a more general forecasting formulation can be found in Granger [1975].

4.7 Forecasting with Misspecified Models

The theory discussed in Sections 4.3 and 4.4 began with the unreasonable assumption that the generating model of the series being forecast was known exactly. In practice, one is usually given just a sample from the series and a plausible model then has to be identified and estimated, using the techniques of Chapter 3. The model so derived will generally not be identical to the true model, but nevertheless the derived or fitted model will be used to generate forecasts as though it were the correct one. The theory introduced earlier can be adapted to this situation and the consequences of using a misspecified model can be found. The only case considered in any detail here will be that of an AR(p) process, although mixed models can be handled in the same manner but with more complicated algebra.

Suppose that X_t is a zero-mean, stationary AR(p) series generated by

$$a(B)X_t = \epsilon_t \qquad (4.7.1)$$

where

$$a(B) = 1 - \sum_{j=1}^{p} a_j B^j \qquad (4.7.2)$$

and ϵ_t is zero mean white noise. Let the corresponding MA(∞) representation be

$$X_t = c(B)\epsilon_t = \sum_{j=0}^{\infty} c_j \epsilon_{t-j}, \qquad c_0 = 1 \qquad (4.7.3)$$

where

$$c(B)a(B) = 1 \qquad (4.7.4)$$

Now suppose that an AR(p') model is fitted to a sample from the process X_t, and that the model achieved is of the form

$$\alpha(B)X_t = \eta_t \qquad (4.7.5)$$

where

$$\alpha(B) = 1 - \sum_{j=1}^{p'} \alpha_j B^j \qquad (4.7.6)$$

with corresponding MA(∞) type representation

$$X_t = \gamma(B)\eta_t = \sum_{j=0}^{\infty} \gamma_j \eta_{t-j}, \qquad \gamma_0 = 1 \qquad (4.7.7)$$

where

$$\alpha(B)\gamma(B) = 1 \qquad (4.7.8)$$

For (4.7.1) and (4.7.5) to both hold, it is necessary that

$$\eta_t = \frac{\alpha(B)}{a(B)} \epsilon_t \qquad (4.7.9)$$

so that η_t will generally not be white noise although, presumably, the estimated values of η_t over the sample period will be indistinguishable from white noise over this sample period if some real effort has been made to fit the "best" model.

The model (4.7.5) will be the only one available and forecasts will be made from it as though it were the correct model, with η_t taken to be white noise.

Let $g_{n,h}$ represent the forecast of X_{n+h} made at time n from the model (4.7.5). The sequence $g_{n,h}$, $h = 1, 2, \ldots$, will be generated by the equation corresponding to (4.4.17), i.e.,

$$g_{n,h} = \sum_{j=1}^{p'} \alpha_j g_{n,h-j} \tag{4.7.10}$$

with $g_{n,j} = X_{n+j}$, $j \leq 0$. Substituting $n + h$ for t in (4.7.5) and subtracting (4.7.10) from this equation yields

$$e'_{n,h} = \sum_{j=1}^{p'} \alpha_j e'_{n,h-j} + \eta_{n+h} \tag{4.7.11}$$

where $e'_{n,h} = X_{n+h} - g_{n,h}$ are the errors made when forecasts are computed from (4.7.5), so that $e'_{n,h} = 0$, $h \leq 0$. Solving the difference equation (4.7.11) yields, by (4.7.8),

$$e'_{n,h} = \sum_{j=0}^{h-1} \gamma_j \eta_{n+h-j} \tag{4.7.12}$$

and so it follows, by setting $t = n + h$ in (4.7.7) and subtracting (4.7.12), that the forecasts are given by

$$g_{n,h} = \sum_{j=0}^{\infty} \gamma_{j+h} \eta_{n-j} \tag{4.7.13}$$

It follows directly from (4.7.12) that

$$e'_{n,h} = e'_{n-1,h+1} - \gamma_h \eta_n \tag{4.7.14}$$

Subtracting both sides of this expression from X_{n+h} then gives the updating formula corresponding to (4.3.24)

$$g_{n,h} = g_{n-1,h+1} + \gamma_h \eta_n \tag{4.7.15}$$

where

$$\eta_n = X_n - g_{n-1,1} \tag{4.7.16}$$

The updating formula is thus seen to be merely an identity resulting from the method of forming h-step forecasts and is in no way connected with the use of optimal forecasts, although it is generally formulated in such a context.

The h-step error may also be written in the form

$$e'_{n,h} = (X_{n+h} - f_{n,h}) + (f_{n,h} - g_{n,h}) \tag{4.7.17}$$

where $f_{n,h}$ is the optimal h-step forecast. Since the first term of this expression, given by (4.3.18), is a linear weighted sum of ϵ_{n+k}, $k > 0$, and the second term is an infinite linear weighted sum of ϵ_{n-k}, $k \geq 0$, it follows that the two terms are uncorrelated. Thus, the h-step error variance when using the incorrect model is

$$V'(h) = \text{var}(e'_{n,h}) = V(h) + G(h) \tag{4.7.18}$$

where $V(h)$ is given by (4.3.21) and

$$G(h) = \text{var}(f_{n,h} - g_{n,h}) \tag{4.7.19}$$

Clearly $V'(h) \geqslant V(h)$, as must occur, and $V'(h) \to \sigma_x^2$ as $h \to \infty$. As shown in Section 4.3, $V(h)$ is monotonic nondecreasing, but $G(h)$ is not necessarily monotonic nonincreasing or nondecreasing. This is seen by considering the particular case where the true model is $X_t = aX_{t-1} + \epsilon_t$ but the model used is $X_t = \alpha X_{t-1} + \eta_t$. Then

$$G(h) = \sigma_x^2 (\alpha^h - a^h)^2 \tag{4.7.20}$$

If one takes $\alpha = 0.7$, $a = 0.5$, then

$$
\begin{aligned}
\alpha^h - a^h &= 0.2, & h &= 1 \\
&= 0.24, & h &= 2 \\
&= 0.228, & h &= 3 \\
&= 0.1776, & h &= 4, \quad \text{etc.}
\end{aligned}
$$

The error variances $V'(h)$ are also found to be not necessarily monotonic nondecreasing. This is seen by considering the following particular case. Let the true model be

$$X_t = aX_{t-1} + \epsilon_t + b\epsilon_{t-1}, \qquad \sigma_\epsilon^2 = 1 \tag{4.7.21}$$

whereas the model used is

$$X_t = aX_{t-1} + \eta_t \tag{4.7.22}$$

η_t being treated as though it were white noise, whereas in fact

$$\eta_t = \epsilon_t + b\epsilon_{t-1} \tag{4.7.23}$$

From (4.7.22) it follows that the fitted model can be written

$$X_t = \sum_{j=0}^{\infty} \gamma_j \eta_{t-j}$$

where

$$\gamma_j = a^j, \qquad j = 1, 2, \ldots \tag{4.7.24}$$

Thus, from (4.7.12), the forecast error variance is

$$V'(h) = E\left[\left(\sum_{j=0}^{h-1} \gamma_j \eta_{n+h-j} \right)^2 \right] \tag{4.7.25}$$

Hence $V'(1) = \text{var}(\eta_t)$, so that, by (4.7.23), $V'(1) = 1 + b^2$. Further $V'(2) = \text{var}(\eta_t + a\eta_{t-1})$ by (4.7.24). Thus

$$V'(2) = (1 + a^2)V'(1) + 2aE[\eta_t \eta_{t-1}] \tag{4.7.26}$$

But by (4.7.23) $E[\eta_t \eta_{t-1}] = b$. Hence, from (4.7.26)

$$V'(2) - V'(1) = a^2(1 + b^2) + 2ab$$

Thus, for example, if $a = 0.5$ and $b = -0.3$, then $V'(2) - V'(1) = -0.0275$, and hence one can have, with an incorrectly specified model, $V'(h + 1)$ less that $V'(h)$. It further follows that if an incorrect model has been used, the h-step error variance sequence predicted by this model may be quite different from the actual h-step error variances.

This difference can become particularly pronounced when the true model is an integrated process but the fitted model is a stationary one. To illustrate this, consider an integrated autoregressive process of order p, X_t, such that the first difference series, $Y_t = X_t - X_{t-1}$ is generated by the AR(p) model $a(B)Y_t = \epsilon_t$ where ϵ_t is zero-mean, white noise. The model for X_t is thus

$$a(B)(1 - B)X_t = \epsilon_t \tag{4.7.27}$$

and the moving average representation is

$$X_t = \sum_{j=0}^{\infty} S_j \epsilon_{t-j} \tag{4.7.28}$$

where

$$S_j = \sum_{k=0}^{j} b_k \tag{4.7.29}$$

and

$$\frac{1}{a(B)} = \sum_{j=0}^{\infty} b_j B^j, \qquad b_0 = 1 \tag{4.7.30}$$

The optimal forecast theory of Section 4.4 may be applied and the optimal h-step forecast $f_{n,h}$ of X_{n+h} will obey an equation

$$f_{n,h} - f_{n,h-1} = \sum_{j=1}^{p} a_j(f_{n,h-j} - f_{n,h-j-1}) \tag{4.7.31}$$

with $f_{n,k} = X_{n+k}$ if $k \leqslant 0$. Using these optimal forecasts, the h-step errors will be

$$e_{n,h} = \sum_{j=0}^{h-1} S_j \epsilon_{n+h-j} \tag{4.7.32}$$

with variances

$$V(h) = \sigma_\epsilon^2 \sum_{j=0}^{h-1} S_j^2 \tag{4.7.33}$$

For sufficiently large h_0, $S_h \approx$ constant, all $h \geqslant h_0$, where the constant is $1/a(1)$; so that

$$V(h) \approx A + \frac{\sigma_\epsilon^2 h}{\{a(1)\}^2} \tag{4.7.34}$$

for h large and where A is a constant.

An alternative, and for our purposes a more useful, representation of an integrated process is obtained by noting that (4.7.28) may be written

$$X_t = \sum_{j=0}^{t-1} S_j \epsilon_{t-j} + X_0 \tag{4.7.35}$$

where $X_0 = \sum_{j=0}^{\infty} S_{t+j} \epsilon_{-j}$ so that (4.7.27) becomes

$$(1 - B)a(B)(X_t - X_0) = \epsilon_t \tag{4.7.36}$$

where, now $\epsilon_j = 0$ for all $j \leqslant 0$.

Suppose now that the true model is an integrated AR(p) process but that a stationary AR(p') model of the form

$$\alpha(B)(X_t - M) = \eta_t \tag{4.7.37}$$

has in fact been, incorrectly, identified and fitted to the data. In (4.7.37) $\alpha(B)$ is a polynomial in B of order p', with the alternative representation

$$(X_t - M) = \sum_{j=0}^{\infty} \gamma_j \eta_{t-j}, \qquad \gamma_0 = 1$$

M being a constant. It follows that

$$\eta_t = \frac{\alpha(B)}{(1 - B)a(B)} \epsilon_t + \mu, \qquad \epsilon_j = 0, \quad j \leqslant 0 \tag{4.7.38}$$

where

$$\mu = \alpha(1)(X_0 - M) \tag{4.7.39}$$

and with X_0 now viewed as a given constant. Let $g'_{n,h}$ be the forecast of $X_{n+h} - M$ formed by using the model in (4.7.37) as though it were the correct one, so that the forecast of X_{n+h} is

$$g_{n,h} = g'_{n,h} + M \tag{4.7.40}$$

The forecast error is found to be, using (4.7.39),

$$e'_{n,h} = (X_0 - M)D_h + \sum_{j=0}^{h-1} \gamma_j \bar{\eta}_{n+h-j} \tag{4.7.41}$$

where $\bar{\eta}_t = \eta_t - \mu$ and

$$D_h = \alpha(1) \sum_{j=0}^{h-1} \gamma_j \tag{4.7.42}$$

Thus, from (4.7.38), $E\{e'_{n,h}\} = (X_0 - M)D_h$, and the expectation of the square of the h-step error is

$$S'(n, h) = \{E(e'_{n,h})\}^2 + \text{var}\left(\sum_{j=0}^{h-1} \gamma_j \bar{\eta}_{n+h-j}\right)$$

and will depend both on n and h. However, for h large enough, providing

$\gamma_k \to 0$ sufficiently quickly as k increases,

$$\sum_{j=0}^{h-1} \gamma_j \, \bar{\eta}_{n+h-j} \approx \gamma(B)\, \bar{\eta}_{n+h} = \frac{1}{(1-B)a(B)} \, \epsilon_{n+h}$$

$$= X_{n+h} - X_0 = \sum_{j=0}^{n+h-1} S_j \epsilon_{n+h-j}$$

Hence

$$\mathrm{var}\left(\sum_{j=0}^{h-1} \gamma_j \, \bar{\eta}_{n+h-j} \right) \approx \sigma_\epsilon^2 \sum_{j=0}^{n+h-1} S_j^2 \approx \sigma_\epsilon^2 (n+h) S_\infty^2 + F$$

where F is a constant. It is seen that the expected value of the average squared h-step error taken over some prediction period will be, for large h,

$$S'(h) = \text{average } S'(n, h) \approx h \frac{\sigma_\epsilon^2}{[a(1)]^2} + H$$

by the reasoning which led to (4.7.34), where H is a constant. Thus, the estimate of the expected squared h-step error will again be approximately linearly increasing with h, quite contrary to what would be expected if the fitted model were taken to be true since then $S'(h) \to$ constant as h increases sufficiently. It seems therefore that the observed pattern of the "estimated variances of h-step errors" should indicate that an integrated model should have been fitted rather than the stationary model actually used. The converse also holds. If an integrated model is used, but the true model is a stationary one, then the actual error variances should settle down whereas the expected values would continually increase.

4.8 Unbiased Forecasts

A forecast $f_{n, h}$ may be thought of as being an estimate of the random variable being forecast X_{n+h}; and, as elsewhere in statistics, it seems reasonable to require that the estimate be unbiased. If $f_{n, h}$ is based on the information set I_n, then $f_{n, h}$ will be said to be "completely unbiased" if

$$E_c\{e_{n, h} | I_n\} = 0 \tag{4.8.1}$$

where E_c denotes conditional expectation and

$$e_{n, h} = X_{n+h} - f_{n, h} \tag{4.8.2}$$

is the forecast error. When using a least squares criterion, the only forecast that is completely unbiased is the optimal forecast since

$$f_{n, h} = E_c\{X_{n+h} | I_n\} \tag{4.8.3}$$

The forecast $f_{n, h}$ will be said to be "unbiased on average" (henceforth just "unbiased") if

$$E\{e_{n, h}\} = 0 \tag{4.8.4}$$

To illustrate these definitions, consider the AR(1) model $X_t = aX_{t-1} + \epsilon_t$ where ϵ_t is a white noise series, but with mean $E\{\epsilon_t\} = m$. Then $E\{X_t\} = \mu$ where $\mu = m/(1-a)$. Using $I_n = \{X_{n-j}, j \geq 0\}$, consider the two forecasts $f_{n,1} = aX_n + m$ with error $e_{n,1}$, and $g_{n,1} = \mu$ with error $e'_{n,1}$. Then

$$e_{n,1} = (X_{n+1} - aX_n - m) \quad \text{and} \quad E_c(e_{n,1}) = E_c(\epsilon_{n+1} - m) = 0$$

since ϵ_{n+1} has mean m and is independent of I_n. When using $g_{n,1}$, one has

$$e'_{n,1} = X_{n+1} - \mu = aX_n + \epsilon_{n+1} - \mu$$

and so

$$E_c\{e'_{n,1}\} = aX_n + m - \mu = a(X_n - \mu)$$

Thus, $g_{n,1}$ is not a completely unbiased forecast, but since $E\{e'_{n,1}\} = 0$, it *is* unbiased on average.

The example illustrates the fact that the completely unbiased property is a very strong one, but that the other definition of unbiasedness is rather a weak one, although it is the only one generally applied or considered. With model misspecification, which will be the rule rather than the exception, the resulting forecasts will generally not be completely unbiased.

It is common practice, with stationary series, to estimate the mean of the series and to then subtract this estimated mean from the data before model building. (A better approach is to estimate the mean and model parameters simultaneously.) This procedure can introduce a different type of model misspecification to that considered in the previous section. To investigate the consequences of misestimating a mean, consider a stationary series generated by the AR(p) model

$$a(B)X_t = \epsilon_t \tag{4.8.5}$$

but with $E\{X_t\} = \mu$, so that $E\{\epsilon_t\} = m = a(1)\mu$. Note that, since X_t is stationary, it necessarily follows that $a(1) > 0$. Equation (4.8.5) may also be written

$$a(B)(X_t - \mu) = \epsilon_t - m \tag{4.8.6}$$

Suppose now that a value for μ is determined in some way, either being assumed to take some value such as zero or being estimated from data, and let the value so determined be M. Further suppose that a model is fitted to the available data of the form

$$\alpha(B)(X_t - M) = \eta_t \tag{4.8.7}$$

where η_t is taken to be zero-mean white noise, whereas in fact from (4.8.6) and (4.8.7) it follows that

$$\eta_t = \frac{\alpha(B)}{a(B)}(\epsilon_t - m) + M' \quad \text{where} \quad E\{\eta_t\} = M' = \alpha(1)(\mu - M)$$

If $g_{n,h}$ is the forecast of $X_{n+h} - M$ using the model (4.8.7) as though it were the correct model, then with $1/\alpha(B) = \sum_{j=0}^{\infty} \gamma_j B^j$, the forecast error, using the results of the previous section, will be

$$e'_{n,h} = \sum_{j=0}^{h-1} \gamma_j \eta_{n+h-j} = (\mu - M)D_h + \sum_{j=0}^{h-1} \gamma_j \overline{\eta}_{n+h-j}$$

where $\overline{\eta}_t = \eta_t - M'$ and $D_h = \alpha(1)\sum_{j=0}^{h-1}\gamma_j$. Thus $E[e'_{n,h}] = (\mu - M)D_h$. It should be noted that, although $\lim_{h\to\infty} D_h = 1$, D_h can be less than or greater than unity for any finite h value, so that the extent of the bias depends on the form of the fitted model.

The expected squared h-step forecast error is thus

$$s'(h) = (\mu - M)^2 D_h^2 + V'(h)$$

where $V'(h)$ is as in Section 4.7, with $\overline{\eta}_t$ replacing the previous η_t. From the considerations of the previous section it is clear that $s'(h)$ need not be monotonically nondecreasing. This is so even if the correct model were fitted apart from the estimate of the mean, so that $\alpha(z) \equiv a(z)$ but $\mu \neq M$. As a simple example, consider the process

$$X_t + aX_{t-1} = \epsilon_t, \qquad 0 < a < 1, \quad \sigma_\epsilon^2 = 1$$

where the mean μ of X_t is, incorrectly, taken to be M. Then

$$s'(1) = (\mu - M)^2(1 + a)^2 + 1$$

$$s'(2) = (\mu - M)^2(1 + a)^2(1 - a)^2 + 1 + a^2$$

so that

$$s'(1) > s'(2) \qquad \text{if} \quad (\mu - M)^2 > \frac{a^2}{(1 + a)^2\left[1 - (1 - a)^2\right]}$$

4.9 Invertibility

In this section, the discussion of Section 3.4, in which it was noted that moving average representation is not generally unique, is elaborated. Consider a zero-mean, stationary series X_t with theoretical autocovariances $\lambda_\tau = E\{X_t X_{t-\tau}\}$ such that

$$\lambda_\tau = 0, \qquad \tau \geqslant q + 1 \tag{4.9.1}$$

It follows from Section 1.6 that provided the λ_τ obey certain restrictions X_t will be a moving average process, which may be represented by

$$X_t = b(B)\epsilon_t = \left(\sum_{j=0}^{q} b_j B^j\right)\epsilon_t, \qquad b_0 = 1 \tag{4.9.2}$$

It was also shown in Section 1.6 that if $\lambda(z)$ is the autocovariance generating function

$$\lambda(z) = \sum_{j=-q}^{q} \lambda_j z^j \tag{4.9.3}$$

then

$$\lambda(z) = \sigma_\epsilon^2 b(z)b(z^{-1}) \tag{4.9.4}$$

The model-building problem can be thought of as being given the λ_τ and being required to find the b_j. In theory at least one could proceed as follows: find the roots of

$$\lambda(z) = 0 \tag{4.9.5}$$

noting that as $\lambda_j = \lambda_{-j}$, and hence $\lambda(z) = \lambda(z^{-1})$, then if z_j is a root it follows that z_j^{-1} must also be a root. Denote those roots of (4.9.5) that lie outside the unit circle by $z_j, j = 1, \ldots, q$, so that $|z_j| > 1$; and it is assumed that no roots have modulus one. Then the $2q$ roots of (4.9.5) may be denoted

$$z_j, \; j = 1, \ldots, 2q, \quad \text{where} \quad z_{j+q} = z_j^{-1}, \; j = 1, \ldots, q$$

Let $z_j', j = 1, \ldots, q$ be a subset of size q of these roots, such that if z_j is contained in it, then z_j^{-1} is not. Then a possible solution for $b(z)$ is

$$b'(z) = \sigma_\epsilon^{-1/2} \prod_{j=1}^{q} (z - z_j') \tag{4.9.6}$$

and by expanding this as a power series in z and equating coefficients in (4.9.2), some possible values for the b_j result. Since there are at most[1] $M_q = 2^q$ ways of choosing the subset z_j', it follows that there are this many possible moving average MA(q) processes all with the same autocovariance sequence. However, only one of these models has the invertibility property, so that ϵ_t may be written as a convergent sum of past and present values of X_t. To see this, transform (4.9.2) to

$$\epsilon_t = \frac{1}{b(B)} X_t = \frac{1}{\displaystyle\prod_{j=1}^{q} (B - z_j')} \sigma_\epsilon X_t \tag{4.9.7}$$

Expanding the reciprocal of the product into partial fractions gives terms such as

$$\frac{A_j}{B - z_j'} = -\frac{A_j}{z_j'} \sum_{k=0}^{\infty} \frac{B^k}{(z_j')^k} \tag{4.9.8}$$

and for convergence one requires $|z_j'| > 1$. Thus, the only moving average model of those considered that is invertible is the one corresponding to the set of roots $z_j' = z_j, j = 1, \ldots, q$, such that all have modulus greater than one. Clearly this is unique since, by definition, if z_j is not in the subset, then z_j^{-1} must be in it.

To show the considerable importance of this invertibility condition consider an MA(1) process with

$$\lambda_0 = 1, \quad \lambda_1 = \rho \leqslant \tfrac{1}{2}, \quad \lambda_\tau = 0, \quad \tau > 1 \tag{4.9.9}$$

---

[1] The qualification "at most" is required since the b_j of (4.9.2) are required to be real. Hence, any complex roots must be taken in conjugate pairs.

The foregoing theory suggests that there are two alternative MA(1) representations, of the form

$$X_t = \epsilon_t - \beta\epsilon_{t-1}, \qquad |\beta| < 1 \tag{4.9.10}$$

and

$$X_t = e_t - \beta^{-1}e_{t-1} \tag{4.9.11}$$

where ϵ_t and e_t are both white noise series, only the first being invertible. By considering $E\{X_t X_{t-1}\}$ for both models one gets

$$\beta\sigma_\epsilon^2 = \beta^{-1}\sigma_e^2$$

so that

$$\sigma_e^2 = \beta^2\sigma_\epsilon^2 < \sigma_\epsilon^2 \tag{4.9.12}$$

This suggests the rather surprising result that if one could find an optimal forecast using the noninvertible form (4.9.11), *then* the variance of the one-step forecast error σ_e^2 would be less than the corresponding variance σ_ϵ^2 using the invertible model (4.9.10). However, it is quite impossible to form an optimal forecast based on just past and present X_t using (4.9.11) since the e_t are not observable given this information set, just because the model is not invertible. That is to say, the optimal forecast theory of Section 4.3 suggests, based on (4.9.11), the forecast of X_{n+1}

$$g_{n,\,1} = -\beta^{-1}e_n$$

However, because of noninvertibility, e_n cannot be determined in terms of current and previous values of the series X. It is therefore seen that if one had the slightly larger information set $I_n' = \{X_{n-j}, j \geqslant 0, e_{n-k}, \text{ any } k \geqslant 0\}$ then one could forecast X_t one step ahead very much better than with the information set $I_n = \{X_{n-j}, j \geqslant 0\}$. But the set I_n' is impossible to realize in any practical situation. One is therefore forced always to use invertible models, and, given only I_n, one cannot be better off using a noninvertible form.

This can be shown more formally as follows, using $p(X)$ to represent the pdf of the random variable X and $\mathbf{X}_n$ to denote the vector $X_{n-j}, j \geqslant 0$. Then, using a least squares criterion, the optimal one-step forecast is known to be

$$f_{n,\,1} = E\{X_{n+1}|\mathbf{X}_n\}$$

and an expression for this can be derived from

$$p(X_{n+1}\mid \mathbf{X}_n) = \int p(X_{n+1}, \epsilon_n \mid \mathbf{X}_n)\,d\epsilon_n = \int p(X_{n+1}\mid \epsilon_n, \mathbf{X}_n)p(\epsilon_n \mid \mathbf{X}_n)\,d\epsilon_n$$

Hence

$$f_{n,\,1} = E\{X_{n+1}\mid \mathbf{X}_n\} = \int\int X_{n+1}p(X_{n+1}\mid \epsilon_n, \mathbf{X}_n)p(\epsilon_n\mid \mathbf{X}_n)\,d\epsilon_n\,dX_{n+1}$$

$$= \int E\{X_{n+1}\mid \epsilon_n, \mathbf{X}_n\}\,p(\epsilon_n\mid \mathbf{X}_n)\,d\epsilon_n \tag{4.9.13}$$

But from (4.9.10)

$$E\{X_{n+1} \mid \epsilon_n, X_n\} = -\beta\epsilon_n \tag{4.9.14}$$

and for the invertible form, ϵ_n is assumed to be known exactly given X_n, so that (4.9.13) becomes

$$f_{n,1} = E\{X_{n+1} \mid X_n\} = -\beta\epsilon_n$$

where ϵ_n is some known function of X_n. Similarly using the noninvertible form, one gets

$$E\{X_{n+1} \mid X_n\} = \int E\{X_{n+1} \mid e_n, X_n\} p(e_n \mid X_n)\, de_n$$

and from (4.9.11)

$$E\{X_{n+1} \mid e_n, X_n\} = -\beta^{-1}e_n$$

But, for the noninvertible form, e_n is not known exactly given X_n, so that $p(e_n \mid X_n)$ takes a form such that

$$-\beta^{-1}\int e_n p(e_n \mid X_n)\, de_n = -\beta\epsilon_n = f_{n,1}$$

To illustrate this in a special case, suppose that the distributions are multivariate normal, so that $p(\epsilon_n, X_n) \sim N(0, \Sigma)$ where the covariance matrix Σ takes the form

$$\Sigma = \begin{bmatrix} \sigma_\epsilon^2 & \sigma_\epsilon^2 & 0 & 0 & 0 & \cdots \\ \sigma_\epsilon^2 & & & & & \\ 0 & & \Sigma_x & & & \\ 0 & & & & & \end{bmatrix}$$

where Σ_x is the covariance matrix of X_n. Then from classical multivariate normal regression theory $p(\epsilon_n \mid X_n)$ is normal with mean $\sigma_\epsilon^2 (\Sigma_x)_1^{-1} X_n$, where $(\Sigma_x)_1^{-1}$ denotes the first row of Σ_x^{-1}. Thus, from (4.9.13) and (4.9.14)

$$f_{n,1} = -\beta\sigma_\epsilon^2 (\Sigma_x)_1^{-1} X_n$$

In the same way, using the noninvertible form and considering $p(e_n, X_n)$, one gets

$$f'_{n,1} = -\beta^{-1}\sigma_e^2 (\Sigma_x)_1^{-1} X_n$$

but, as $\sigma_\epsilon^2 = \beta^2\sigma_e^2$, these two forecasts are seen to be identical.

Thus, in terms of forecasting ability, nothing can be gained from employing a noninvertible form. The invertible form is inevitably used in practice since the computation of coefficient estimates and of forecasts from the fitted model is considerably easier for invertible than for noninvertible models.

4.10 Types of Forecasts

The only kind of forecast to have been considered in any detail in the earlier sections has been the point forecast, where a best estimate is made of the future value of some variable. It would often be very much better if a more sophisticated type of forecast were available, such as a confidence interval with the property that the probability that the true value would fall into this interval takes some specified value. However, unless one has a very long series of forecast errors or one is able to make a realistic assumption about the distribution of the errors, such confidence intervals are impossible to provide. An assumption of normality for the error distribution is an easy one to make and to use but it is less easy to justify. Frequently one does, in fact, resort to a normality assumption in order to produce confidence intervals, the hope being that the distribution of the forecast errors is sufficiently close to normal for the probability content of the intervals to be but little affected. Occasionally the decision maker or forecast consumer wants some other quantity to be forecast rather than the conditional mean. One may, for example, be considering returns on a portfolio of risky investments and be asked to forecast the probability that the return will be negative since this probability is associated with an interesting definition of risk, known as safety-first risk. It is not clear how one would do this without knowing the conditional distribution of future return given one's information set.

An apparently different type of forecasting situation is that for which forecasts are required, not for a single point in the future, but for a whole trace of future points. Thus, one may require a vector of point forecasts $f'_{n,j}$, $j = 1, \ldots, h$, for the variables $X_{n+j}, j = 1, \ldots, h$, based on some information set I_n. In fact it may be shown that if expected cost is of the form

$$C = E\left(\sum_{k,j} C_{kj}(X_{n+k} - f_{n,k})(X_{n+j} - f_{n,j}) \right)$$

where $\sum_{k,j} C_{kj} x_k x_j > 0$ for all x_k, x_j when not all x_j are zero, then the optimal vector or trace of forecasts is just the sequence of optimal forecasts $f_{n,k}$, $k = 1, \ldots, h$, derived in Section 4.3. It is important to have such a trace of forecasts when making a decision about when to take some action. An example would be a buyer of some commodity such as cocoa. He would need to have forecasts of future cocoa prices over some period so that he could decide when to make a purchase on the commodity market. Further, it is straightforward to show that if a forecast of a linear combination of future values $L = \sum_{h=1}^{H} l_h X_{n+h}$ is required, the optimal least-squares forecast is given by $\hat{L} = \sum_{h=1}^{H} l_h f_{n,h}$.

A much more difficult forecasting situation occurs when one attempts to answer a question of the form, When, if ever, will some event occur? Examples are, When will a competitor announce a new product or new advertising campaign? When will the Government change the official interest rate? or When will a specific household buy a new car? A question of this type, of particular importance to economists, is, When will the next turning point in the economy occur? Apart from the purely definitional problem of what exactly constitutes a real turning point, it is also rather difficult to invent a cost function that emphasizes a particular feature of the series rather than some other feature. These types of forecasts may be called timing forecasts and might be tackled using the theory of point processes (see Lewis [1972]).

A further type of forecast, which may be called an event forecast, occurs when one knows that an event will occur in the future but the outcome of the event needs forecasting. For example, a baby is to be born, but what will its sex be, or an election is to occur, but which candidate will win? The obvious problem with this situation is that the future event may be unique, so that the usually assumed flow of past information need not be available. Clearly some information has to be collected for a forecast to be made, such as chemical tests on the pregnant wife or an opinion poll on the voting public.

The existence of these types of forecasting situations, and many others that can be constructed, merely indicates the very particular case considered by the classical statistical forecasting theory, as outlined in this chapter. At least there is comfort in the fact that the classical theory deals with what is generally recognized as being the most important type of forecasting problem.

CHAPTER FIVE

PRACTICAL METHODS FOR UNIVARIATE TIME SERIES FORECASTING

Chief Witch *Yes, that's right.*
MacBeth *I understand you can foretell the future.*

BBC RADIO PROGRAM
June 1968

5.1 Introduction

In this chapter, the problem of analyzing the information yielded in a given realization $x_1, x_2, \ldots, x_n$ of a particular time series is further considered, with attention now focused on the problem of predicting future values of the series. In forming the forecasts, only current and past values of the series to be predicted will be employed and so, in the terminology of Chapter 4, the information set used is a rather restricted one. Forecasts based on this information set will be termed *univariate forecasts*. It may surprise the reader that we should devote so much space to univariate forecasting methods. After all, the quantity of available information in the universe is vast, and it would be reasonable to expect the forecaster to make use of all relevant information in assessing the future. Nevertheless, we feel that univariate time series forecasting methods deserve consideration for a number of reasons:

(i) They are quick and inexpensive to apply, and may well produce forecasts of sufficient quality for the purposes at hand. The cost of making particular forecasting errors should always be balanced against the cost of *producing* forecasts, for it is hardly worth expending large resources to obtain a relatively small increase in forecast accuracy if the payoff, in terms of improved decision making, is likely to be only marginally beneficial.

(ii) Relevant extraneous information may be unavailable or available only at a prohibitively high cost.

(iii) Univariate forecasting procedures can be useful as a yardstick against which the success or otherwise of more elaborate forecasting exercises can be judged. This point will be amplified in Chapter 8.

(iv) Forecasts obtained in this manner can often be usefully combined with other forecasts in the production of a superior overall forecast, as will be seen in Chapter 8.

(v) Having produced univariate forecasts, one is in a position to assess how much of the variation in a particular quantity can be explained in terms of its own past behavior, and so form a clearer picture of what particular behavior patterns require consideration of extraneous factors for their explanation. This point will be discussed further in Chapter 7, where multivariate time series forecasting methods are considered.

(vi) For the vast majority of economic time series, the information considered here, while restrictive, *is* of great importance—a point all too often neglected in more traditional approaches to economic forecasting.

There exist a number of procedures, of varying degrees of complexity, whereby a time series can be forecast from its own current and past values. The more complex procedures do not produce forecasts either as quickly or as cheaply as do the simple ones, but it is to be expected that some compensation in terms of increased accuracy would obtain through their use since they allow for a more detailed investigation of the properties of the particular series under investigation. In the remainder of this chapter methods for univariate time series forecasting are outlined under three main headings: Box–Jenkins methods, exponential smoothing procedures, and stepwise autoregression. These procedures will be compared and contrasted and situations in which a particular approach is likely to prove valuable will be indicated.

5.2 Box–Jenkins Forecasting Methods

In Chapter 3 procedures for the fitting of autoregressive integrated moving average (ARIMA) models, and their seasonal variants, to a particular time series were described. It will be recalled that the fitting procedures consisted of an iterative cycle of identification, estimation, and diagnostic checking. Essentially, a particular model is chosen from the general ARIMA class, its coefficients estimated, and its adequacy of representation checked, possibly leading to the choice of an alternative form and a repeat of the model building cycle. In this section, it will be shown how forecasts can be generated from the fitted models, using some of the theory developed in Chapter 4. The whole process of constructing an ARIMA model and the generation of forecasts from that model will be referred to as the Box–Jenkins forecasting method since, although a number of elements in the methodology were well known before these authors wrote, it is their contribution that has allowed an integrated and well-defined approach to time series forecasting via model building, stimulating a good deal of practical application over a wide range of actual time series.

As a first step, a number of results derived in Chapter 4 are summarized.

(i)   Let X_t follow the stationary, invertible ARMA(p, q) process

$$X_t = \sum_{j=1}^{p} a_j X_{t-j} + \sum_{j=0}^{q} b_j \epsilon_{t-j}, \qquad b_0 = 1$$

Standing at time n, let $f_{n,h}$ be the forecast of X_{n+h} which has smallest expected squared error among the set of all possible forecasts which are linear in $X_{n-j}, j \geqslant 0$. Now write

$$X_{n+h} = \sum_{j=1}^{p} a_j X_{n+h-j} + \sum_{j=0}^{q} b_j \epsilon_{n+h-j}, \qquad b_0 = 1 \qquad (5.2.1)$$

Then a recurrence relation for the forecasts $f_{n,h}$ is obtained by replacing each element in (5.2.1) by its "forecast" at time n, as follows:

(a)   replace the unknown values X_{n+k} by their forecasts $f_{n,k}$ for $k > 0$;
(b)   "forecasts" of X_{n+k}, $k \leqslant 0$, are simply the known values x_{n+k};
(c)   since ϵ_t is white noise, the optimal forecast of ϵ_{n+k}, $k > 0$, is simply zero;
(d)   "forecasts" of ϵ_{n+k}, $k \leqslant 0$, are just the known values ϵ_{n+k}.

(ii)   The ARMA(p, q) process $a(B)X_t = b(B)\epsilon_t$ can be written as an infinite moving average $X_t = c(B)\epsilon_t$ where the elements of $c(B) = c_0 + c_1 B + c_2 B^2 + \cdots$ can be obtained by equating coefficients of $B^j, j = 1, 2, \ldots$, in $a(B)c(B) = b(B)$. Then the forecast errors are given by

$$e_{n,h} = X_{n+h} - f_{n,h} = \sum_{j=0}^{h-1} c_j \epsilon_{n+h-j} \qquad (5.2.2)$$

and hence the variances of the forecast errors are given by

$$V(h) = E\{e_{n,h}^2\} = \sigma_\epsilon^2 \sum_{j=0}^{h-1} c_j^2 \qquad (5.2.3)$$

(iii)   An "updating" formula for the forecasts is given by

$$f_{n,h} = f_{n-1,h+1} + c_h(X_n - f_{n-1,1}) \qquad (5.2.4)$$

These results hold for stationary time series. However, as was seen in Chapter 3, it is very often the case in dealing with economic time series that the integrated model is of importance. That is, for a particular process X_t, one may need to difference d times to produce a stationary series $Y_t = (1 - B)^d X_t$. In the case of seasonal time series, it is often the case that a multiplicative difference filter is required to produce stationarity. However, for the present purposes, this involves no new principle, and so for simplicity of exposition, attention will be restricted to the nonseasonal case. A sensible procedure, then, is to derive forecasts of the series X from those of the stationary series Y. Writing $f^x_{n,h}$ for forecasts of X and $f^y_{n,h}$ for forecasts of Y, an obvious formula for generating forecasts of X is then

$$f^y_{n,h} = (1 - B)^d f^x_{n,h} \qquad (5.2.5)$$

where here B operates on the index h, so that, for example, in the case $d = 1$,

$$f^x_{n,h} = f^x_{n,h-1} + f^y_{n,h}$$

Thus, forecasts could be obtained by a two-step procedure, where the stationary series Y is first forecast and then forecasts of X are obtained from (5.2.5). However, a moment's reflection should indicate that this is unnecessary. Write

$$\left(1 - A_1 B - A_2 B^2 - \cdots - A_p B^p\right) = a(B)(1 - B)^d$$

Then, corresponding to (5.2.1), one can write

$$X_{n+h} = \sum_{j=1}^P A_j X_{n+h-j} + \sum_{j=0}^q b_j \epsilon_{n+h-j}, \qquad b_0 = 1 \qquad (5.2.6)$$

and forecasts may be derived from this equation in the same way as in the stationary case.

Now, define $C(B) = 1 + C_1 B + C_2 B^2 + \cdots$ where

$$A(B)C(B) = b(B) \qquad (5.2.7)$$

Then it is clear that, corresponding to (5.2.2) and (5.2.3), the forecast errors in the integrated case are given by $e_{n,h} = \sum_{j=0}^{h-1} C_j \epsilon_{n+h-j}$, and hence the error variance is

$$V(h) = \sigma^2_\epsilon \sum_{j=0}^{h-1} C_j^2, \qquad C_0 = 1 \qquad (5.2.8)$$

Further, the updating formula (5.2.4) is now given by

$$f_{n,h} = f_{n-1,h+1} + C_h(X_n - f_{n-1,1}) \qquad (5.2.9)$$

All the necessary equipment for the efficient computation of point forecasts from a fitted ARIMA model, or its seasonal variant, is now at hand. Further, since an expression for the variance of forecast error has been derived, it is possible, provided distributional assumptions are made, to derive interval forecasts. In the remainder of this section, practical procedures for the computation of these forecasts are outlined and the methods involved illustrated with a few specific examples.

Initial Calculation of Point Forecasts

Suppose now that one has a set of observations $x_1, x_2, \ldots, x_n$ on a process X, and that an ARIMA model has been fitted. Then forecasts of future values of the series can be obtained from (5.2.6), substituting in that equation forecasts of each individual term. The forecasting formula can then be written as

$$f_{n,h} = \sum_{j=1}^{P} A_j f_{n,h-j} + \sum_{j=0}^{q} b_j \hat{\epsilon}_{n+h-j}, \qquad b_0 = 1 \qquad (5.2.10)$$

where

$$f_{n,k} = x_{n+k}, \qquad k \leq 0$$

$$\hat{\epsilon}_{n+k} = 0, \qquad k > 0 \qquad (5.2.11)$$

$$= \text{estimate of } \epsilon_{n+k}, \qquad k \leq 0$$

Equations (5.2.11) require some further explanation. The theoretical development of Chapter 4 concerned the forecasting of X_{n+h} given $X_{n-j}, j \geq 0$. That is, it was assumed that an infinite past record of the series to be forecast was available, in which case $\epsilon_{n-j}, j \geq 0$ would also be known. However in the practical situation, where only a finite run of data is available, the ϵ_{n-j} will not be known, but must be estimated. This problem was encountered in Section 3.5, when the estimation of ARIMA models was discussed. Suppose that the fitted model is

$$Y_t - a_1 Y_{t-1} - \cdots - a_p Y_{t-p} = \epsilon_t + b_1 \epsilon_{t-1} + \cdots + b_q \epsilon_{t-q}$$

where $Y_t = (1 - B)^d X_t$. It was seen in Section 3.5 that the estimation of the ϵ_t essentially depended on the treatment of the unknown starting values $X_0, X_{-1}, \ldots, \epsilon_0, \epsilon_{-1}, \ldots$. There are two possible lines of attack: a conditional approach in which these values were all set to zero, and an unconditional approach which is more precise, employing the method of "back-forecasting" these unknown values. In fact, since only estimates of the most recent values $\epsilon_n, \epsilon_{n-1}, \ldots, \epsilon_{n-q+1}$ are required, it makes very little difference in most practical situations which approach is used. Exceptions are in cases where:

    (a)   the autoregressive operator $a(B)$ has a root near the unit circle;
    (b)   the moving average operator $b(B)$ has a root near the unit circle;
    (c)   the available time series is relatively short—a point of particular importance for seasonal time series which have rather longer memories due to the inclusion of the operator B^s in the models.

Thus the quantities $\hat{\epsilon}_{n+k}, k \leq 0$, are the residuals from the fitted model, and estimation programs are generally written so as to produce these values, which are also required for some of the diagnostic checks described in Section 3.6.

    Having constructed a forecasting model, the procedure for deriving point forecasts is then quite straightforward. Equation (5.2.10) is employed one step at a time for $h = 1, 2, 3, \ldots$, substituting in appropriate values from (5.2.11). Thus, for $h = 1$, the forecast $f_{n,1}$ can be obtained immediately from (5.2.10).

Then, substituting $h = 2$ into (5.2.10), the forecast $f_{n,2}$ is obtained using $f_{n,1}$, which has already been calculated. Forecasts can then be obtained as far ahead as is required.

To illustrate these calculations, some of the examples introduced in Chapter 3 are further considered.

EXAMPLE 1   In Sections 3.2 and 3.5, a series of 108 monthly values of official gold and foreign exchange holdings in Sweden was examined. The model fitted was the ARIMA(1, 1, 0) process

$$(1 - 0.38B)(1 - B)X_t = \epsilon_t \qquad (5.2.12)$$

Transposing this equation to the form of (5.2.10) yields the forecast generating formula

$$f_{n,h} = 1.38f_{n,h-1} - 0.38f_{n,h-2} + \hat{\epsilon}_{n+h} \qquad (5.2.13)$$

Now, for $h \geqslant 1$, no values of $\hat{\epsilon}_{n+k}$, $k \leqslant 0$, occur in (5.2.13). Thus, the calculation of forecasts is particularly easy since all that is required are the two most recent values of the series. These, together with forecasts made up to twelve steps ahead are shown in Table 5.1. Substituting $h = 1$ in (5.2.13) yields

$$f_{n,1} = 1.38f_{n,0} - 0.38f_{n,-1} + \hat{\epsilon}_{n+1}$$

Hence, from (5.2.11), using the known values $x_n = 702$ and $x_{n-1} = 725$ given in Table 5.1,

$$f_{n,1} = (1.38)(702) - (0.38)(725) = 693.26$$

Again, from (5.2.13), the two steps ahead forecast is given by

$$f_{n,2} = (1.38)(693.26) - (0.38)(702) = 689.94$$

The first 12 forecasts are given in Table 5.1; they have been rounded to the nearest integer since the original series was available in this form.

In a sense, these forecasts appear rather dull, but they do illustrate the point that changes in the series after the first are not very predictable given only past observations on the time series. This is by no means an uncommon phenomenon in nonseasonal economic time series. Indeed, it is straightforward to see that, for any ARIMA process with $d = 1$, the quantity

Table 5.1   *Some actual and predicted values for official gold and foreign exchange holdings series*

h:	-1	0	1	2	3	4	5	6
$f_{n,h}$:	725	702	693	690	689	688	688	688
h:	7	8	9	10	11	12		
$f_{n,h}$:	688	688	688	688	688	688		

Table 5.2 *Some actual and predicted values for index of help wanted advertisements*

h:	-2	-1	0	1	2	3	4	5
$f_{n,h}$:	130	134	134	135	136	137	137	138
h:	6	7	8	9	10	11	12	
$f_{n,h}$:	138	139	139	139	139	139	140	

$f_{n,h} - f_{n,h-1}$ tends to zero as h increases. This follows since, for the stationary process $Y_t = X_t - X_{t-1}$, $f_{n,h}^y$ tends to zero as h tends to infinity.

EXAMPLE 2   As a second example, consider the series of help wanted advertisements in the U.S.A. In Chapter 3 the ARIMA(2, 1, 1) model

$$(1 - 0.45B - 0.31B^2)(1 - B)X_t = (1 - 0.36B)\epsilon_t$$

was fitted to the given data. Thus, in the formulation (5.2.10),

$$f_{n,h} = 1.45f_{n,h-1} - 0.14f_{n,h-2} - 0.31f_{n,h-3} + \hat{\epsilon}_{n+h} - 0.36\hat{\epsilon}_{n+h-1} \quad (5.2.14)$$

All that is needed to calculate the forecasts are the last three observations $x_n = 134$, $x_{n-1} = 134$, and $x_{n-2} = 130$, together with the estimate $\hat{\epsilon}_n = -0.628$ obtained from the estimation stage of the model building procedure. Substituting $h = 1$ in (5.2.14) and using (5.2.11) yields

$$f_{n,1} = (1.45)(134) - (0.14)(134) - (0.31)(130) + (0.36)(0.628) = 135.466$$

Similarly, substituting $h = 2$ in (5.2.14), it follows that

$$f_{n,2} = (1.45)(135.466) - (0.14)(134) - (0.31)(134) = 136.126$$

Forecasts for this series, made up to 12 steps ahead, are given, rounded to the nearest integer, in Table 5.2.

EXAMPLE 3   There is no difficulty in applying the forecast generating procedure to seasonal time series. To illustrate, consider again the data on construction begun in England and Wales, for which the fitted model was

$$(1 - B)(1 - B^4)X_t = (1 - 0.37B)(1 - 0.68B^4)\epsilon_t$$

Table 5.3 *Quantities required for forecasting construction begun*

t:	$n-4$	$n-3$	$n-2$	$n-1$	n
x_t:	89966	89566	103912	95504	101191
$\hat{\epsilon}_t$:	7211.12	10089.82	-10785.66	-7323.88	10804.23

Table 5.4   *Forecasts of construction begun series*

h:	1	2	3	4	5	6
$f_{n,h}$:	91747	115966	109824	106322	99596	123814
h:	7	8	9	10	11	12
$f_{n,h}$:	117673	114170	107444	131663	125522	122019

Transposing this to the form (5.2.10), it is seen that the forecast generating formula is given by

$$f_{n,h} = f_{n,h-1} + f_{n,h-4} - f_{n,h-5}$$

$$+ \hat{\epsilon}_{n+h} - 0.37\hat{\epsilon}_{n+h-1} - 0.68\hat{\epsilon}_{n+h-4} + 0.2516\hat{\epsilon}_{n+h-5} \qquad (5.2.15)$$

Table 5.3 shows relevant quantities for the computation of the forecasts. Substituting $h = 1$ in (5.2.15) yields

$$f_{n,1} = 101191 + 89566 - 89966 - (0.37)(10804.23) - (0.68)(10089.82)$$

$$+ (0.2516)(7211.12) = 91746.68$$

The forecast two steps ahead is now obtained by substituting $h = 2$ in (5.2.15), so that

$$f_{n,2} = 91746.68 + 103912 - 89566 + (0.68)(10785.66) + (0.2516)(10089.82)$$

$$= 115965.52$$

Forecasts of this series made up to twelve quarters ahead are shown in Table 5.4.

Calculation of Interval Forecasts

While it is almost certainly the case that more attention is paid to point forecasts than to any others, it is generally worthwhile to calculate wherever possible confidence intervals associated with these forecasts, if only to provide an indication of their likely reliability. Now, the variance of the error of the point forecast is given by (5.2.8), where the C_j are defined in (5.2.7). In fact this is an *underestimate* of the true variance since it assumes that the coefficients of the forecasting model are known, whereas in fact they must be estimated leading to a corresponding decrease in accuracy in the resulting forecasts. However, for moderately long time series, this factor will be of relatively small importance.

In addition, if one is prepared to assume that the forecast errors come from a normal distribution, it is possible to derive, in an obvious way, confidence intervals for the forecasts. Thus an approximate 95% interval is given by

$$f_{n,\,h} \pm 1.96\hat{\sigma}_\epsilon \sqrt{\sum_{j=0}^{h-1} C_j^2} \tag{5.2.16}$$

where $\hat{\sigma}_\epsilon$ is the estimated standard deviation of ϵ_t, obtained in estimating the coefficients of the fitted model. Similarly, approximate 75% intervals are given by

$$f_{n,\,h} \pm 1.15\hat{\sigma}_\epsilon \sqrt{\sum_{j=0}^{h-1} C_j^2} \tag{5.2.17}$$

To illustrate, consider again the three examples just discussed.

EXAMPLE 1   For the series on official gold and foreign exchange holdings, the fitted model is (5.2.12). To determine the C_j of (5.2.8), note that (5.2.7) takes the form

$$(1 - 1.38B + 0.38B^2)(1 + C_1 B + C_2 B^2 + \cdots) = 1$$

Equating coefficients in B yields

$$C_1 - 1.38 = 0, \qquad C_1 = 1.38$$

Equating coefficients in B^2 yields

$$C_2 - 1.38C_1 + 0.38 = 0 \qquad \text{and hence} \qquad C_2 = 1.5244$$

Equating coefficients in B^3

$$C_3 - 1.38C_2 + 0.38C_1 = 0 \qquad \text{and so} \qquad C_3 = 1.5793$$

Further values of the C_j are shown in Table 5.5.

Now, in estimating the model for this series, the estimated standard deviation of the errors was found to be $\hat{\sigma}_\epsilon = 20.24$. Thus, substituting the C_j of Table 5.5 and the forecasts of Table 5.1 into (5.2.16), it follows that for the one-step ahead forecast a 95% confidence interval is given by

$$693 \pm (1.96)(20.24) = 693 \pm 39.67$$

and a 75% interval by

$$693 \pm (1.15)(20.24) = 693 \pm 23.28$$

Table 5.5   *The C_j weights for official gold and foreign exchange holdings model*

j:	0	1	2	3	4	5	6	7	8	9	10	11
C_j:	1	1.38	1.52	1.58	1.60	1.61	1.61	1.61	1.61	1.61	1.61	1.61

Table 5.6 *Confidence intervals for forecasts of official gold and foreign exchange holdings*

Quantity forecast	95% Interval	75% Interval
X_{n+1}	653–733	670–716
X_{n+2}	622–758	650–730
X_{n+3}	598–780	636–742
X_{n+4}	577–799	623–753
X_{n+5}	561–815	613–763
X_{n+6}	546–830	605–771
X_{n+7}	532–844	596–780
X_{n+8}	519–857	589–787
X_{n+9}	508–868	582–794
X_{n+10}	497–879	576–800
X_{n+11}	486–890	570–806
X_{n+12}	477–899	564–812

Similarly for the two-steps ahead forecast the intervals are

$$95\% \text{ interval}: \quad 690 \pm (1.96)(20.24)\sqrt{1 + (1.38)^2} = 690 \pm 67.61$$

$$75\% \text{ interval}: \quad 690 \pm (1.15)(20.24)\sqrt{1 + (1.38)^2} = 690 \pm 39.67$$

Confidence intervals (with end points rounded to the nearest integer) for the first twelve forecasts are given in Table 5.6, where, for purposes of illustration, the C_j weights used here and throughout this section are correct to two decimal places, as given in the relevant tables.

EXAMPLE 2   The model

$$(1 - 0.45B - 0.31B^2)(1 - B)X_t = (1 - 0.36B)\epsilon_t$$

fitted to the series of help wanted advertisements in the U.S.A. may be written as

$$(1 - 1.45B + 0.14B^2 + 0.31B^3)X_t = (1 - 0.36B)\epsilon_t$$

Thus from (5.2.7) it follows that the C_j weights may be found by equating coefficients in

$$(1 - 1.45B + 0.14B^2 + 0.31B^3)(1 + C_1B + C_2B^2 + \cdots) = 1 - 0.36B$$

Equating coefficients in B yields

$$C_1 - 1.45 = -0.36, \qquad C_1 = 1.09$$

Table 5.7   *The C_j weights for help wanted advertisements model*

j:	0	1	2	3	4	5	6	7	8	9	10	11
C_j:	1	1.09	1.44	1.63	1.82	1.96	2.09	2.19	2.27	2.34	2.40	2.44

Table 5.8   *Confidence intervals for forecasts of index of help wanted advertisements*

Quantity forecast	95% Interval	75% Interval
X_{n+1}	128–142	131–139
X_{n+2}	125–147	130–142
X_{n+3}	122–152	128–146
X_{n+4}	117–157	125–149
X_{n+5}	114–162	124–152
X_{n+6}	110–166	122–154
X_{n+7}	107–171	120–158
X_{n+8}	103–175	118–160
X_{n+9}	99–179	116–162
X_{n+10}	96–182	114–164
X_{n+11}	92–186	111–167
X_{n+12}	90–190	110–170

Equating coefficients in B^2 yields

$$C_2 - 1.45C_1 + 0.14 = 0 \quad \text{and hence} \quad C_2 = 1.4405$$

Further values of the C_j are shown in Table 5.7.

The estimated standard deviation of the errors from the fitted model was found to be $\hat{\sigma}_\epsilon = 3.806$. Substituting the C_j of Table 5.7 and the forecasts of Table 5.2 into (5.2.16) and (5.2.17), interval forecasts can be obtained. Thus, for example, for the forecast of X_{n+5}, a 75% interval is given by

$$138 \pm (1.15)(3.806)\sqrt{1 + (1.09)^2 + (1.44)^2 + (1.63)^2 + (1.82)^2} = 138 \pm 14.00$$

Confidence intervals (with end points rounded to the nearest integer) for $X_{n+j}, j = 1, 2, \ldots, 12$, are given in Table 5.8. As in the previous example, the 95% intervals are rather wide, indicating the relatively high degree of certainty required by such an interval. For many forecasting situations we feel that 75% intervals are likely to prove rather more useful and informative.

EXAMPLE 3   As a final example, consider the quarterly seasonal series on construction begun in England and Wales, to which in Chapter 3 the model

$$(1 - B)(1 - B^4)X_t = (1 - 0.37B)(1 - 0.68B^4)\epsilon_t$$

Table 5.9 *The C_j weights for construction begun model*

j:	0	1	2	3	4	5	6	7	8	9	10	11
C_j:	1	0.63	0.63	0.63	0.95	0.83	0.83	0.83	1.15	1.03	1.03	1.03

Table 5.10 *Confidence intervals for forecasts of construction begun in England and Wales*

Quantity forecast	95% Interval	75% Interval
X_{n+1}	75716–107778	82341–101153
X_{n+2}	97018–134914	104848–127084
X_{n+3}	88343–131305	97220–122428
X_{n+4}	82596–130048	92401–120243
X_{n+5}	71398–127794	83051–116141
X_{n+6}	92634–154994	105520–142108
X_{n+7}	83784–151562	97789–137557
X_{n+8}	77748–150592	92800–135540
X_{n+9}	66629–148259	83497–131391
X_{n+10}	87626–175700	105824–157502
X_{n+11}	78488–172556	97925–153119
X_{n+12}	72179–171859	92776–151262

was fitted. The C_j weights are then found, as usual, by equating coefficients of powers of B, from

$$(1 - B - B^4 + B^5)(1 + C_1 B + C_2 B^2 + \cdots)$$

$$= 1 - 0.37 B - 0.68 B^4 + 0.2516 B^5$$

The weights are given in Table 5.9. The estimated standard deviation of the errors from the fitted model was $\hat{\sigma}_\epsilon = 8179$. Interval forecasts are obtained by substituting the C_j of Table 5.9 and the forecasts of Table 5.4 into (5.2.16) and (5.2.17). For example, a 95% interval for the forecast of X_{n+5} is given by

$$99596 \pm (1.96)(8179)\sqrt{1 + (0.63)^2 + (0.63)^2 + (0.63)^2 + (0.95)^2}$$

$$= 99596 \pm 16545$$

Confidence intervals for forecasts made up to twelve quarters ahead are shown in Table 5.10.

Updating the Forecasts

Once a forecasting equation has been built, it is generally not necessary to refit a model when a new piece of data becomes available. Neither is it necessary to employ the rather lengthy procedures just described to recompute forecasts. A convenient algorithm, based on (5.2.9), is available for the updating of previously computed forecasts. Writing $n + 1$ for n in that expression, produces

$$f_{n+1, h} = f_{n, h+1} + C_h(X_{n+1} - f_{n, 1}) \qquad (5.2.18)$$

where $X_{n+1} - f_{n, 1} = \epsilon_{n+1}$ is the error made in forecasting X_{n+1} at time n. In words, the forecast of X_{n+1+h} made at time $n + 1$ can be obtained by adding to the forecast of the same quantity, made at time n, a multiple of the error made in forecasting X_{n+1} at time n. Further, the weights C_h required in this expression will already be known since they will have been calculated for the derivation of interval forecasts. The forecast generating mechanism can thus be viewed as an "error learning process." That is to say, forecasts of future values of the series are modified in the light of past forecasting errors. It is not, of course, necessary to recompute estimates of the widths of appropriate confidence intervals since the estimated variance for an h-step ahead forecast does not change when another observation becomes available.

To illustrate, consider once again the series on construction begun in England and Wales. The forecast of X_{n+1} made at time n was $f_{n, 1} = 91747$. In fact, the actual value turned out to be $X_{n+1} = 86332$. Thus the point forecast turned out to be an overestimate (although, as can be seen from Table 5.10, the true value lies well within the 75% confidence interval.). The one-step ahead forecast error is

$$X_{n+1} - f_{n, 1} = 86332 - 91747 = -5415$$

Thus, using (5.2.18), the forecasts of Table 5.4 can be updated by the formula

$$f_{n+1, h} = f_{n, h+1} - 5415 C_h$$

where the C weights are those given in Table 5.9. Thus, for example, the revised forecast of X_{n+6} is

$$f_{n+1, 5} = f_{n, 6} - 5415 C_5 = 123814 - (5415)(0.83) = 119320$$

The updated forecasts for this series are given in Table 5.11.

Table 5.11   *Updated forecasts of construction begun series*

h:	1	2	3	4	5	6
$f_{n+1, h}$:	112555	106413	102911	94452	119320	113179
h:	7	8	9	10	11	
$f_{n+1, h}$:	109676	101217	126086	119945	116502	

Forecast Errors as a Diagnostic Check on Model Adequacy

In Section 3.6 it was shown how errors from a fitted model could be used as a check on the adequacy of representation of that model to a given set of data. It is also possible to obtain further checks through use of the forecast errors produced by the fitted model. As a simple example, it was noted in Section 4.3 that the h-step ahead errors for optimal forecasts constitute an MA($h - 1$) process. Given a sufficiently long run of such errors, this could be checked through examination of their sample autocorrelation functions.

Recently an alternative check has been proposed by Bhattacharyya and Andersen [1974]. Let $f_{n,h}$, $h = 1, 2, \ldots, H$ be forecasts of X_{n+h}, $h = 1, 2, \ldots, H$, all made at time n, with errors

$$e_{n,h} = X_{n+h} - f_{n,h}, \qquad h = 1, 2, \ldots, H$$

Then

$$E[e_{n,h}] = 0, \qquad h = 1, 2, \ldots, H$$

and, corresponding to (4.3.25),

$$E[e_{n,h}e_{n,h+k}] = \sigma_\epsilon^2 \sum_{j=0}^{h-1} C_j C_{j+k}, \qquad k \geqslant 0 \qquad (5.2.19)$$

Now, let

$$\sigma_{ij} = E[e_{n,i}e_{n,j}], \qquad i = 1, 2, \ldots, H, \quad j = 1, 2, \ldots, H$$

and Σ be the $H \times H$ matrix whose (i, j)th element is σ_{ij}. Then, if the $e_{n,h}$ are distributed as multivariate normal, the quantity

$$Q = \mathbf{e}'_H \Sigma^{-1} \mathbf{e}_H \qquad (5.2.20)$$

is distributed as χ^2 with H degrees of freedom, where

$$\mathbf{e}'_H = (e_{n,1}, e_{n,2}, \ldots, e_{n,H})$$

Thus the adequacy of the fitted model can be checked by comparing the statistic (5.2.20) with tabulated values of χ^2, the hypothesis of model adequacy being rejected for high values of the test statistic. In practice, σ_ϵ^2 of (5.2.19) is replaced by its estimate from the model fit, and the C_j of that equation are not the actual values, but are derived as before from the estimated model. However, Bhattacharyya and Andersen show that, if the sample to which the model is fit is moderately large, the validity of the test will be but little affected.

An Overview of the Box–Jenkins Method

We are now in a position to examine the whole framework of what has come to be called "the Box–Jenkins forecasting method." What is involved is first the definition of a class of models that might be thought to be capable of describing adequately the behavior of many practically occurring time series. The class of models considered is denoted as ARIMA, and a seasonal variant of this class can be introduced as appropriate. Next, for any given set of time series data, one attempts to fit a single ARIMA model, using the techniques described in Chapter 3. The model fitting framework is not sufficiently well defined to allow this to be done automatically. Rather it is necessary to employ judgment at various stages of the procedure, one's judgment being based on particular characteristics of the time series under study. The final stage is the computation of forecasts from the fitted model in the manner just described. Of course, as was seen in Chapter 4, if the time series actually were generated by the chosen model, the resulting forecasts would be the optimal linear forecasts of X_{n+h} given the information set X_{n-j}, $j \geq 0$, and, given further the Gaussian assumption, these forecasts would be optimum in the class of all forecasts based on that information set. However, it is not true, for three reasons, that the Box–Jenkins method will inevitably produce optimal forecasts. First, the class of models considered may not include the true generating process. Second, in individual cases, the actual model chosen from the class may not be the true one. Third, since the coefficients of the chosen model must be estimated from a finite run of data, they will in practice differ with probability one from the unknown true values.

The Box–Jenkins method does, however, postulate a wide class of potential generating models and defines a strategy that, provided the true model lies within this class, should in general give one a very good chance of achieving a reasonable approximation to that model, provided a sufficiently long run of data is available. There are also a number of ways in which the range of possible models can be extended. For example, a deterministic trend component could be removed from the series prior to a Box–Jenkins analysis. Again, one might consider some instantaneous transformation $T(X_t)$ of the series as being generated by a linear ARIMA model, rather than simply the series itself. Further elaborations, to allow for more complex forms of nonlinearity or nonstationarity are also possible, but present rather more practical difficulties. These points will be discussed in more detail in Chapter 9.

In any forecasting exercise, consideration of the cost of generating forecasts is of great importance and should be balanced against the expected gains to be obtained from having fairly accurate forecasts. Of course, the question of costs is a relative one, and while it is true that for very many purposes the Box–Jenkins approach could not be regarded as prohibitively expensive, it is often the case (for example, in routine sales forecasting) that less expensive, albeit less versatile, methods are required. While, on the average, one would not expect these methods to achieve as high a degree of accuracy as does Box–Jenkins, they may well be sufficiently precise for the purpose at hand. In the next three sections fully automatic forecasting procedures of this kind will be discussed.

5.3 Exponential Smoothing Methods

In the industrial context it is often the case that forecasts of future sales of several thousand products are required for inventory control or production planning. Typically these forecasts are needed for the short term (perhaps up to a year or so ahead) on the basis of monthly data. A number of procedures, which can be grouped under the heading "exponential smoothing," have been developed to handle routine forecasting problems of this kind. The primary requirement here is for an approach that produces forecasts of sufficient accuracy, but which at the same time is quick and inexpensive to operate. Ideally the forecasting method should be fully automatic so as to reduce as far as possible the need for skilled manpower in its operation. Exponential smoothing methods fall into this category, and so retain a good deal of popularity in industrial forecasting in spite of their theoretical limitations when compared with the more sophisticated Box–Jenkins procedure. In this section a number of variants of exponential smoothing that are in current use will be described. Although these methods were developed from an essentially pragmatic viewpoint, it is useful in analyzing them to consider in what circumstances they might be optimal, or to ask what particular view is being taken of the world in embracing these procedures. A great danger in employing a fully automatic predictor is that rather poor forecasts might result unless one exercises some control over the quality of forecasts produced. To meet this need a number of automatic monitoring or tracking systems have been developed for use in conjunction with exponential smoothing procedures. These provide a check on the quality of forecasts produced and can be useful in the modification of inadequate forecasts.

In fact, a wide range of forecasting models can be derived as special cases of a technique, long familiar to engineers, known as Kalman filtering (see, e.g., Kalman [1960, 1963]). However, only a few commonly used exponential smoothing procedures will be considered here.

Forecasting Procedures Based on Exponential Smoothing

Suppose that the available data consist of a series of observations $x_1, x_2, \ldots, x_n$ and that it is required to forecast X_{n+h} for some positive integer h, and denote a forecast of this quantity by $f_{n,h}$. The earliest version of exponential smoothing, called "simple exponential smoothing," regards a time series as being made up locally of its level and a residual (unpredictable) element. It is thus required to estimate the current level of the series, that is, the level at time n. This will then be the forecast of all future values of the series since it represents the latest assessment available of the single (constant) predictable element of the time series. How might the level at time t be estimated using information available up to that time? If the available data consisted of a random sample, then the obvious thing to do would be to take a simple average of the observations. However, in the time series context it is reasonable to give most weight to the most recent observation, rather less weight to the preceding observation, and so on. One way to achieve this is to employ a weighted average, with geometrically (exponentially) declining weights, so that the level of the series at time t is estimated by

$$\bar{x}_t = \alpha x_t + \alpha(1 - \alpha)x_{t-1} + \alpha(1 - \alpha)^2 x_{t-2}$$

$$+ \alpha(1 - \alpha)^3 x_{t-3} + \cdots, \qquad 0 < \alpha < 1 \qquad (5.3.1)$$

Substituting $t - 1$ for t in this expression and multiplying through by $1 - \alpha$ yields

$$(1 - \alpha)\bar{x}_{t-1} = \alpha(1 - \alpha)x_{t-1} + \alpha(1 - \alpha)^2 x_{t-2} + \alpha(1 - \alpha)^3 x_{t-3} + \cdots$$

Subtracting this from (5.3.1) then yields

$$\bar{x}_t = \alpha x_t + (1 - \alpha)\bar{x}_{t-1}, \qquad 0 < \alpha < 1 \qquad (5.3.2)$$

Equation (5.3.2) represents the basic algorithm for simple exponential smoothing, replacing the original x_t series by a "smoothed" series $\bar{x}_t$. In order to employ the algorithm, it is necessary to make some "starting up" assumption, the simplest being to set $\bar{x}_1$ equal to x_1. Equation (5.3.2) can then be used recursively for $t = 2, 3, \ldots, n$, the transient introduced by the assumed starting value being of little importance in the determination of $\bar{x}_n$ unless n is small. The quantity α is termed the "smoothing constant," and appropriate choices of its value will be discussed later. The forecasts of all future values of the series are given simply by the latest available smooth value, so that

$$f_{n,h} = \bar{x}_n \qquad (5.3.3)$$

The estimation procedure implied by (5.3.2) can be regarded as an updating mechanism, so that at time t the previous estimate of level $\bar{x}_{t-1}$ is updated in light of the new observation x_t. The new estimate of level $\bar{x}_t$ is then a weighted average of x_t and $\bar{x}_{t-1}$.

It was noted earlier that exponential smoothing procedures have their most common application in routine sales forecasting, where forecasts of future sales of a large number of products might be required. In this context, formulations of the type (5.3.2) have a great advantage from a computational point of view, for they do not require the storage of all past values of a time series, for all that is needed is the most recent smoothed value $\bar{x}_{t-1}$ and the current observation x_t. This saving can be extremely important when many series must be handled at the same time.

In practice, this simple version of exponential smoothing is rarely employed. A number of variants, which regard a time series as consisting locally of level, trend, and (possibly) a seasonal factor in addition to the unpredictable residual element, have been developed. Perhaps the most logical extension of the simple exponential smoothing algorithm is the approach devised by Holt [1957] and Winters [1960]. Initially the case of a nonseasonal time series, which is made up locally of the sum of level, linear trend, and residual, will be considered. Denote the estimate of level at time t by $\bar{x}_t$ and of trend by T_t, where

$$\bar{x}_t = Ax_t + (1 - A)(\bar{x}_{t-1} + T_{t-1}), \qquad 0 < A < 1 \qquad (5.3.4)$$

and

$$T_t = C(\bar{x}_t - \bar{x}_{t-1}) + (1 - C)T_{t-1}, \qquad 0 < C < 1 \qquad (5.3.5)$$

The updating formulas (5.3.4) and (5.3.5), as in the case of simple exponential smoothing, modify previous estimates in the light of a new piece of data. Thus the estimate of level at time $t - 1$, $\bar{x}_{t-1}$, in conjunction with the trend estimate T_{t-1}, would suggest a level $\bar{x}_{t-1} + T_{t-1}$ for time t. This estimate is modified in light of the new observation x_t, according to (5.3.4). At time $t - 1$ trend is estimated by T_{t-1}, but given the new observation x_t an estimate of trend as the difference between the two most recent estimates of level is suggested. The trend estimate at time t is then a weighted average, given by (5.3.5), of the previous estimate and the evidence provided by the new piece of data. Again one needs to make some assumption about "starting up" values in order to employ Eqs. (5.3.4) and (5.3.5). Perhaps the simplest approach is to set T_2 equal to $x_2 - x_1$ and $\bar{x}_2$ equal to x_2. The equations can then be used recursively for $t = 3, 4, \ldots, n$ to calculate the estimates $\bar{x}_n$ and T_n. Forecasts of future values of the series are then given by

$$f_{n,h} = \bar{x}_n + hT_n \qquad (5.3.6)$$

Consider now a seasonal time series with period s (so that $s = 4$ for quarterly data and $s = 12$ for monthly data). The most commonly employed variant of the Holt–Winters method regards the seasonal factor F_t as being multiplicative (while trend remains additive) so that this quantity is estimated as

$$F_t = D(x_t/\bar{x}_t) + (1 - D)F_{t-s}, \qquad 0 < D < 1 \qquad (5.3.7)$$

(It is assumed that $x_t > 0$ for all t.) The level $\bar{x}_t$, which can be thought of as a "deseasonalized" level, is estimated now by

$$\bar{x}_t = A(x_t/F_{t-s}) + (1 - A)(\bar{x}_{t-1} + T_{t-1}), \qquad 0 < A < 1 \qquad (5.3.8)$$

The trend component is again estimated using (5.3.5). In order to employ Eqs. (5.3.5), (5.3.7), and (5.3.8) it is again necessary to specify "starting up" values. A very simple way to accomplish this is to take

$$F_j = x_j \bigg/ \left(\frac{1}{s} \sum_{k=1}^{s} x_k \right), \quad j = 1, 2, \ldots, s, \qquad \bar{x}_s = \frac{1}{s} \sum_{k=1}^{s} x_k, \qquad \text{and} \qquad T_s = 0$$

The three updating equations are then used recursively for $t = s + 1, s + 2, \ldots, n$. Since trend is taken to be additive and seasonality multiplicative, forecasts of future values are given by

$$f_{n, h} = (\bar{x}_n + hT_n)F_{n+h-s}, \qquad h = 1, 2, \ldots, s$$
$$= (\bar{x}_n + hT_n)F_{n+h-2s}, \qquad h = s + 1, s + 2, \ldots, 2s$$
$$\vdots \tag{5.3.9}$$

The Holt–Winters approach can easily be modified to deal with situations in which the seasonal factor is thought to be additive rather than multiplicative. In this case Eqs. (5.3.7) and (5.3.8) are replaced by

$$F_t = D(x_t - \bar{x}_t) + (1 - D)F_{t-s}, \qquad 0 < D < 1 \tag{5.3.10}$$

and

$$\bar{x}_t = A(x_t - F_{t-s}) + (1 - A)(\bar{x}_{t-1} + T_{t-1}), \qquad 0 < A < 1 \tag{5.3.11}$$

The forecasting equation (5.3.9) is now replaced by

$$f_{n, h} = \bar{x}_n + hT_n + F_{n+h-s}, \qquad h = 1, 2, \ldots, s$$
$$= \bar{x}_n + hT_n + F_{n+h-2s}, \qquad h = s + 1, s + 2, \ldots, 2s$$
$$\vdots$$

It remains only to discuss the choice of the smoothing constants A, C, and D employed in the Holt–Winters algorithms (and similarly the choice of α in the simple exponential smoothing formula (5.3.2)). Clearly, the lower the values of these constants the more steady will be the final forecasts since the use of low values implies that more weight is given to past observations and consequently any random fluctuations in the present will exert a less strong effect in the determination of the forecast. One possiblity is to choose the smoothing constants according to one's assessment of the characteristics of the particular series under consideration. In general, the more random the series the lower the values of the optimal smoothing constants. If the smoothing constants are chosen in this way, considerable savings in storage and computation time can be effected, proving particularly valuable if many series are to be predicted simultaneously. A more objective approach, proposed by Holt and Winters themselves, is to select those values that would have best "forecast" the given observations. However, even here there remains an element of arbitrariness, for one must decide on a criterion of accuracy (cost of error function) for the forecasts and on how many steps ahead one is evaluating forecast accuracy. The most common procedure is to assume a quadratic cost function and to seek the smoothing constants that provide the best one-step ahead forecasts. The procedure is to choose a grid of possible values of A, C, and D, and to calculate the one-step ahead forecasts $f_{t, 1}$, $t = m, m + 1, \ldots, n - 1$, for each set of values of the smoothing constants. That particular set for which the sum of squared errors

$$S = \sum_{t=m+1}^{n} (x_t - f_{t-1, 1})^2$$

is smallest is then used to calculate actual forecasts of all future values of the series. The starting point m for this procedure is taken to be an integer sufficiently large as to allow the effects of the choice of initial "starting up" values to have died down. While this approach to the choice of smoothing constants adds greatly to the number of calculations required, it is very easily incorporated into a computer program and the forecast generating mechanism remains, of course, fully automatic. It should be added, however, that this approach also greatly increases storage of data requirements, and hence may not always be feasible.

An alternative procedure, known as general exponential smoothing, is due to Brown [1962]. Assume that locally the time series under study is the linear combination of k known deterministic functions of time, plus a residual, which for convenience can be written

$$x_i = \sum_{j=1}^{k} a_j f_j (i - t) + e_i, \qquad i = 1, 2, \ldots, t \qquad (5.3.12)$$

The k functions f_j are generally taken to be polynomials, exponentials, and mixtures of sine and cosine terms. The model (5.3.12) is assumed to hold only locally, so that a natural estimation procedure is to minimize the discounted sum of squared errors

$$S = \sum_{i=1}^{t} \beta^{t-i} \left(x_i - \sum_{j=1}^{k} a_j f_j (i - t) \right)^2, \qquad 0 < \beta < 1 \qquad (5.3.13)$$

Let

$$\mathbf{x} = \begin{bmatrix} x_1 \\ x_2 \\ \vdots \\ x_t \end{bmatrix}, \qquad \mathbf{a} = \begin{bmatrix} a_1 \\ a_2 \\ \vdots \\ a_k \end{bmatrix}$$

$$\mathbf{H} = \begin{bmatrix} f_1(1-t) & f_2(1-t) & \cdots & f_k(1-t) \\ f_1(2-t) & f_2(2-t) & \cdots & f_k(2-t) \\ \vdots & \vdots & & \\ f_1(0) & f_2(0) & \cdots & f_k(0) \end{bmatrix} = \begin{bmatrix} \mathbf{f}'(1-t) \\ \mathbf{f}'(2-t) \\ \vdots \\ \mathbf{f}'(0) \end{bmatrix}$$

$$\mathbf{W} = \begin{bmatrix} \beta^{(t-1)/2} & & & \\ & \beta^{(t-2)/2} & & \mathbf{0} \\ & & \ddots & \\ \mathbf{0} & & & 1 \end{bmatrix}$$

The quantity to be minimized (5.3.13) is then

$$S = (\mathbf{Wx} - \mathbf{WHa})'(\mathbf{Wx} - \mathbf{WHa})$$

The minimum, obtained by differentiating with respect to $\mathbf{a}$, occurs at the value given by

$$\hat{\mathbf{a}}(t) = (\mathbf{H'W'WH})^{-1}\mathbf{H'W'Wx}$$

where $\hat{\mathbf{a}}(t)$ is now a function of t. This can be written as

$$\hat{\mathbf{a}}(t) = \mathbf{F}^{-1}(t)\mathbf{g}(t) \tag{5.3.14}$$

where

$$\mathbf{F}(t) = \mathbf{H'W'WH} = \sum_{i=1}^{t} \beta^{t-i}\mathbf{f}(i-t)\mathbf{f}'(i-t) \tag{5.3.15}$$

and

$$\mathbf{g}(t) = \mathbf{H'W'Wx} = \sum_{i=1}^{t} \beta^{t-i}x_i\mathbf{f}(i-t) \tag{5.3.16}$$

Now, it follows immediately from (5.3.15) that

$$\mathbf{F}(t) = \mathbf{F}(t-1) + \beta^{t-1}\mathbf{f}(1-t)\mathbf{f}'(1-t) \tag{5.3.17}$$

Suppose, now, that there exists a nonsingular matrix $\mathbf{L}$ such that

$$\mathbf{f}(t) = \mathbf{Lf}(t-1)$$

for all values of t. This restriction is not too prohibitive since it allows any polynomial, exponential, and sinusoidal functions. From (5.3.16), it now follows that

$$\mathbf{g}(t) = \beta\mathbf{L}^{-1}\mathbf{g}(t-1) + \mathbf{f}(0)x_t \tag{5.3.18}$$

Further, for functions that do not die out too quickly, it follows from (5.3.15) that, for moderately large t, $\mathbf{F}(t)$ will converge to some steady state matrix $\mathbf{F}$, so that $\mathbf{a}(t)$ can be written as

$$\mathbf{a}(t) = \mathbf{F}^{-1}\mathbf{g}(t)$$

Thus, the forecast of X_{n+h} is given by

$$f_{n,h} = \hat{\mathbf{a}}'(n)\mathbf{f}(h) \qquad \text{or} \qquad f_{n,h} = \mathbf{g}'(n)\mathbf{F}^{-1}\mathbf{f}(h)$$

where $\mathbf{g}(n)$ is calculated recursively from (5.3.18), the steady state matrix $\mathbf{F}$ is obtained by applying (5.3.17) recursively until convergence obtains and $\mathbf{f}(h)$ is a vector of known constants. It remains only to determine a value for the discount parameter β. This can be achieved either by visual inspection of the characteristics of the series under consideration or by optimizing over a grid of possible values, as for the Holt–Winters predictor. Brown suggests choosing β so that β^k lies in the range 0.75–0.95, the actual value used depending on the stability of the series, while Harrison [1965] proposes that a value in the neighborhood of $\beta^k = 0.8$ would frequently be appropriate. As before, savings in the storage of data result if the discount factor is chosen judgment-ally.

Two particular problems should be mentioned with regard to general exponential smoothing. First, Reid [1969] notes that the errors from the fitted model are very often serially correlated, suggesting strongly that suboptimal forecasts will be produced. He suggests fitting a first-order autoregressive model (possibly estimated by discounted least-squares) to these residuals. Forecasts produced by the model may then be modified accordingly, and it was noted that substantial improvement in forecast performance very often results from such a procedure. The second difficulty concerns seasonal time series. In the first place it is difficult to know how many harmonics to fit, and a certain amount of experimentation may be required. Perhaps the best strategy would be to add further terms and test for an improvement in fit. Thus, Reid notes that for one of his series he first tried the model

$$X_t = a_1 + a_2 t + a_3 \cos\left(\frac{2\pi t}{12}\right) + a_4 \sin\left(\frac{2\pi t}{12}\right)$$

$$+ a_5 \cos\left(\frac{2\pi t}{6}\right) + a_6 \sin\left(\frac{2\pi t}{6}\right) \tag{5.3.19}$$

but that a better fit was achieved by adding

$$a_7 \cos\left(\frac{2\pi t}{4}\right) + a_8 \sin\left(\frac{2\pi t}{4}\right)$$

However, when large numbers of coefficients are to be fitted, estimation becomes less efficient, and Reid concludes that "it is normally desirable to keep the number of fitting functions as low as possible provided they still adequately describe the time series." A more serious objection, however, arises from the use in the Brown method of only a single smoothing para-meter β. In order to estimate models like (5.3.19) with any precision, β should be fairly large so as to allow a fair number of data points to have appreciable weight, otherwise one is estimating with very few degrees of freedom. On the other hand one very often requires an exponential smoothing procedure to adapt very quickly so as to put most weight on the few most recent observations. These two mutually incompatible requirements should make one suspicious on a priori grounds of the performance of Brown's method for seasonal time series. Put rather bluntly, is it reasonable to expect the Brown predictor to do with a single parameter what requires three parameters for Holt–Winters?

Harrison [1965] has proposed modifications of both the Holt–Winters and the Brown predictors for dealing with seasonal time series. A difficulty with the Holt–Winters approach is that, according to (5.3.7), each seasonal factor is updated only once every year. Harrison's procedure, known as "seatrend," smooths the factors $F_t, F_{t-1}, \ldots, F_{t-s+1}$ by Fourier analysis. The Fourier coefficients are estimated by

$$a_k = \frac{1}{2s} \sum_{j=1}^{s} F_{t-s+j} \cos(ku_j)$$

$$b_k = \frac{1}{2s} \sum_{j=1}^{s} F_{t-s+j} \sin(ku_j), \qquad k = 1, 2, \ldots, \frac{s}{2}$$

where $u_j = [2(j-1)\pi/s] - \pi$. The amplitudes $(a_k^2 + b_k^2)^{1/2}$ are then tested for statistical significance, and smoothed seasonal estimates obtained as

$$\overline{F}_{t-s+j} = 1 + \sum_{\text{sig } k} \left(a_k \cos(ku_j) + b_k \sin(ku_j) \right), \qquad j = 1, 2, \ldots, s$$

where $\sum_{\text{sig } k}$ denotes summation over those harmonics for which statistically significant amplitudes were found. Forecasts can now be obtained by replacing F_t by $\overline{F}_t$ in (5.3.9).The approach can be made more "streamlined" by replacing F_{t-s} in (5.3.7) by $\overline{F}_{t-s}$ to obtain initial seasonal estimates for new observations. Harrison's second method, known as "doubts," constitutes a double application of Brown's procedure. At the first step the seasonal cycle is removed by applying a moving average, and the smoothed series is then predicted using a simple Brown trend model. Deviations of the observations about the trend are taken, and Brown's general exponential smoothing is applied to fit sine and cosine terms, representing the seasonal and the harmonics, to these deviations. Different smoothing parameters β can be applied at the two stages, thus avoiding the difficulty mentioned earlier.

Approximate confidence limits can be derived for forecasts based on exponential smoothing methods if one is prepared to make distributional assumptions. Typically it is assumed that the forecast errors are either normal or log normal, and hence confidence limits can be derived using the errors from the fitted model in the usual way.

Optimality of Exponential Smoothing

Up to this point, exponential smoothing predictors have been presented as rather ad hoc, though intuitively reasonable, forecast generating mechanisms. Their theoretical justifications will now be examined and, in particular, an attempt will be made to find the stochastic processes for which these predictors are optimal. Since it is the generating stochastic process that is of interest, X_t will be written for x_t in the following derivations.

Consider the Holt–Winters seasonal predictor in its additive form. This can be written as

$$\overline{X}_t = A(X_t - F_{t-s}) + (1 - A)(\overline{X}_{t-1} + T_{t-1})$$

$$T_t = C(\overline{X}_t - \overline{X}_{t-1}) + (1 - C)T_{t-1} \qquad (5.3.20)$$

$$F_t = D(X_t - \overline{X}_t) + (1 - D)F_{t-s}$$

Denote the one-step ahead forecast of X_t by $f_t \equiv f_{t-1,\,1}$. Then

$$f_t = \overline{X}_{t-1} + T_{t-1} + F_{t-s}$$

and the forecast error is

$$e_t \equiv e_{t-1,\,1} = X_t - f_t = X_t - \left(\overline{X}_{t-1} + T_{t-1} + F_{t-s}\right) \qquad (5.3.21)$$

Now, from (5.3.20),

$$\overline{X}_t - \overline{X}_{t-1} = T_{t-1} + A\left(X_t - \overline{X}_{t-1} - T_{t-1} - F_{t-s}\right)$$

and so by (5.3.21)

$$\overline{X}_t - \overline{X}_{t-1} = T_{t-1} + Ae_t \qquad (5.3.22)$$

Again, from (5.3.20), for the trend term

$$T_t - T_{t-1} = C\left(\overline{X}_t - \overline{X}_{t-1}\right) - CT_{t-1}$$

and hence, from (5.3.22),

$$T_t - T_{t-1} = ACe_t \qquad (5.3.23)$$

For the seasonal factor, it follows from (5.3.20) that

$$F_t - F_{t-s} = D\left(X_t - \overline{X}_t - F_{t-s}\right)$$

or, by (5.3.22),

$$F_t - F_{t-s} = D\left(X_t - \overline{X}_{t-1} - T_{t-1} - F_{t-s} - Ae_t\right)$$

and so, from (5.3.21),

$$F_t - F_{t-s} = (1 - A)De_t \qquad (5.3.24)$$

Now, introducing the back-shift operator B, it follows from (5.3.22) and (5.3.23) that

$$(1 - B)^2 \overline{X}_t = A[1 - (1 - C)B]e_t \qquad (5.3.25)$$

and from (5.3.23) and (5.3.24) that

$$(1 - B)T_t = ACe_t \qquad (5.3.26)$$

and

$$(1 - B^s)F_t = (1 - A)De_t \qquad (5.3.27)$$

Combining (5.3.25), (5.3.26), and (5.3.27) yields

$$(1 - B)^2(1 - B^s)\left(\overline{X}_{t-1} + T_{t-1} + F_{t-s}\right)$$

$$= \left[AB(1 - B^s)[1 - (1 - C)B] + ACB(1 - B)(1 - B^s)\right.$$

$$\left. + (1 - A)DB^s(1 - B)^2\right]e_t$$

Then, from (5.3.21),

$$(1 - B)^2(1 - B^s)X_t = \left[(1 - B)^2(1 - B^s) + AB(1 - B^s)[1 - (1 - C)B]\right.$$

$$\left. + ACB(1 - B)(1 - B^s) + (1 - A)DB^s(1 - B)^2\right]e_t$$

Now, if the forecasts are optimal, the errors e_t will constitute a white noise process ϵ_t. Thus, if the Holt–Winters additive seasonal predictor is to produce optimal forecasts, it follows that the series X_t must be generated by a process of the form

$$(1 - B)^2(1 - B^s)X_t = \left(1 + b_1B + b_2B^2 + b_sB^s + b_{s+1}B^{s+1} + b_{s+2}B^{s+2}\right)\epsilon_t$$

where the five coefficients b_1, b_2, b_s, b_{s+1}, and b_{s+2} are all functions of the three smoothing constants A, C, and D. Thus, use of this exponential smoothing predictor is optimal for a process generated by a particular member of the class of seasonal models considered by Box and Jenkins.

One can show in the same fashion that the simple exponential smoothing predictor derived from

$$\overline{X}_t = \alpha X_t + (1 - \alpha)\overline{X}_{t-1}, \qquad 0 < \alpha < 1 \tag{5.3.28}$$

is optimal if and only if X_t is generated by the ARIMA(0, 1, 1) process $(1 - B)X_t = [1 - (1 - \alpha)B]\epsilon_t$ as was originally shown by Muth [1960].

Similarly for the Holt–Winters nonseasonal predictor generated by

$$\overline{X}_t = AX_t + (1 - A)\left(\overline{X}_{t-1} + T_{t-1}\right), \qquad T_t = C\left(\overline{X}_t - \overline{X}_{t-1}\right) + (1 - C)T_{t-1}$$

$$\tag{5.3.29}$$

it is straightforward to show that optimality of forecasts implies that X_t is generated by an ARIMA(0, 2, 2) process. This was demonstrated by Harrison [1967], who further showed that (5.3.28) and (5.3.29) are special cases of an early predictor introduced by Box and Jenkins [1962], which itself yielded optimal forecasts only if X_t was generated by an ARIMA(0, k, k) process.

Recently, Cogger [1974] has considered the nonseasonal Brown predictor, obtained by estimating a polynomial in time of degree m, showing that optimality of this predictor implies that the underlying process is generated by an ARIMA(0, $m + 1$, $m + 1$) model. In fact, optimality requires a very specific subclass of these models since all the moving average parameters must be functions of the single discount coefficient β of (5.3.13).

An alternative justification for the use of exponential smoothing predictions—in terms of their optimality for certain generating models—is considered by Theil and Wage [1964], Nerlove and Wage [1964], and Harrison [1967]. A particularly simple model might regard the actual value at time t, X_t, as the sum of a "true value" X_t^* and a white noise error $\epsilon_{1,t}$. Further, the true value might be updated from one time period to the next according to a random walk. A plausible generating mechanism then might be

$$X_t = X_t^* + \epsilon_{1,t}, \qquad X_t^* = X_{t-1}^* + \epsilon_{2,t}$$

where $\epsilon_{1,t}$ and $\epsilon_{2,t}$ are independent white noise processes. It can be shown that the simple exponential smoothing procedure (5.3.28), with α determined by the ratio of variances of the white noise processes, produces optimal forecasts for this model. An elaboration of this simple model allows for linear growth in the "true values" of the series, so that a trend term T_t^* is added to the updating equation. If, further, the trend term is taken to be a random walk, a plausible model could be

$$X_t = X_t^* + \epsilon_{1,t}, \qquad X_t^* = X_{t-1}^* + T_t^* + \epsilon_{2,t}, \qquad T_t^* = T_{t-1}^* + \epsilon_{3,t} \quad (5.3.30)$$

where $\epsilon_{1,t}$, $\epsilon_{2,t}$, and $\epsilon_{3,t}$ are independent white noise processes. It can be shown that the Holt–Winters nonseasonal approach (5.3.29), with A and C determined by the relative variances of the white noise processes, produces optimal forecasts for this model.

Monitoring Forecast Performance

The great virtue of exponential smoothing for routine sales forecasting is, as previously noted, that it allows forecasts of a large number of series to be generated rapidly and inexpensively. However, as has been shown, particular exponential smoothing methods are optimal only for corresponding underlying stochastic processes, which in many cases are simply subsets of the general class of models considered in a Box–Jenkins analysis. Furthermore, if the true generating process is different from that implicitly assumed, the forecasts produced could be very far from optimal. Thus, in spite of its great convenience, exponential smoothing should be treated with a degree of caution. It would be particularly valuable to have a check on forecast performance, which was itself fully automatic, so that forecasts could be produced routinely until an indication is given that a built-in safeguard in the system has been violated. The forecaster may then devote his attention to those few time series that are causing difficulty, while forecasts of the remaining series continue to be generated in the usual routine manner.

An early proposal in this context is due to Harrison and Davies [1964] who suggest the use of cumulative sum (cusum) techniques for the control of routine forecasts. Let f_t be a (one-step ahead) forecast of X_t with error e_t. Then, if the system begins to produce forecasts for time $t = 1$, the cumulative sums of the forecast errors are

$$C_1 = e_1, \qquad C_j = C_{j-1} + e_j, \qquad j = 2, 3, \ldots$$

These cumulative sums when plotted on a chart—called a cusum chart—can be expected to indicate any tendency toward bias in the forecasts. Harrison and Davies suggest a backward sequential test procedure, based on the sum of the most recent forecast errors, so that one defines at time t

$$S_1 = e_t = C_t - C_{t-1}$$
$$S_2 = e_t + e_{t-1} = C_t - C_{t-2}$$

$$\vdots$$

$$S_k = e_t + e_{t-1} + \cdots + e_{t-k+1} = C_t - C_{t-k}$$

The values S_i are calculated as each new observation occurs and are tested against corresponding control limits $\pm L_i$. If a limit is broken, lack of control is signaled. However, in short term forecasting applications, Harrison and Davies note that "the cusum scheme does not, in general, provide more than a vague indication of the way in which the forecasting scheme needs to be adjusted . . . ." Harrison and Davies show that, if the magnitude of the limits L_i is a linear function of the number of observations comprising the sum to be tested, considerable simplification in terms of storage of information is achieved. They suggest that if the forecast errors are independent, appropriate limits can be found through a nomogram of Ewan and Kemp [1960]. Alternatively, appropriate values can be derived by simulation.

An alternative proposal, due to Trigg [1964], computes a "tracking signal" as the ratio of an exponentially weighted estimate of mean error and an exponentially weighted estimate of mean absolute error. Denote the "smoothed error" at time t by E_t, where

$$E_t = \gamma e_t + (1 - \gamma)E_{t-1}, \qquad 0 < \gamma < 1 \tag{5.3.31}$$

and the "mean absolute deviation" by D_t, where $D_t = \gamma|e_t| + (1 - \gamma)D_{t-1}$. Then at time t, tracking signal $= E_t/D_t$, and $-1 \leqslant$ tracking signal $\leqslant 1$. The distribution of this quantity is extremely difficult to find since the numerator and denominator are correlated with one another. As an approximation, note that it is well known that, for a wide range of distributions, the standard deviation is approximately equal to 1.2 times the mean absolute deviation. Further, if the available sample is moderately large and γ is taken to be quite small, D_t can be taken as a close approximation to the true population mean absolute deviation. Then

$$\sigma_e \approx 1.2 D_t \tag{5.3.32}$$

where σ_e is the standard deviation of the error series. Now, from (5.3.31), one can write

$$E_t = \sum_{j=0}^{\infty} \gamma(1 - \gamma)^j e_{t-j}$$

and so, on the assumption that the errors are uncorrelated,

$$\text{Var}(E_t) = \sigma_e^2 \gamma^2 \sum_{j=0}^{\infty} (1 - \gamma)^{2j} = \frac{\sigma_e^2 \gamma}{2 - \gamma}$$

Thus, from (5.3.32), the standard error of the tracking signal is approximately equal to $1.2\gamma/\sqrt{(2\gamma - \gamma^2)}$. Two standard error limits are thus given by

$$\pm 2.4\gamma \Big/ \sqrt{(2\gamma - \gamma^2)} \tag{5.3.33}$$

Trigg derives the cumulative distribution of the tracking signal by simulation (assuming a normal distribution for the errors) and it emerges that for γ equal to 0.1—the value recommended by Trigg as being suitable for most general purposes—(5.3.33) provides a good approximation for a 5% level test. However, the validity of the test does rest crucially on the assumption that the forecasting mechanism is producing errors that are not serially correlated—an assumption that is frequently violated by exponential smoothing predictors. This point is demonstrated in a particular case by Batty [1969]. In a later paper, Trigg and Leach [1967] propose that the tracking signal be employed to modify exponential smoothing systems. Consider the simple exponential smoothing system defined by

$$\overline{X}_t = \alpha X_t + (1 - \alpha)\overline{X}_{t-1}$$

They note that the simplest way to modify a system which has "gone out of control" is to increase the value of the smoothing constant α, so as to give more weight to recent observations, and propose to choose this constant so that α = modulus of tracking signal. They also suggest extending this technique to deal with the modification of forecasts generated by other members of Brown's generalized exponential smoothing family of functions.

A more thorough approach to the problem of changing structure in a given time series is due to Harrison and Stevens [1971]. The model (5.3.30) can be modified to take account of a seasonal factor F_t^* by writing

$$X_t = X_t^* F_t^* + \epsilon_{1,t}, \qquad X_t^* = X_{t-1}^* + T_t^* + \epsilon_{2,t}, \qquad T_t^* = T_{t-1}^* + \epsilon_{3,t}$$

where $\epsilon_{1,t}$, $\epsilon_{2,t}$, and $\epsilon_{3,t}$ have respective variances σ_1^2, σ_2^2, and σ_3^2. Harrison and Stevens envisage the system generating the time series as being in one of four possible states, which may be characterized as follows:

(i)   The steady state or no change state—that is, level and trend remain constant. In this situation one would have σ_1^2 taking a "normal" value with σ_2^2 and σ_3^2 being zero.

(ii)   A step change, where the series changes level at a particular point. For this case σ_1^2 takes a "normal" value, σ_2^2 is large, and σ_3^2 is zero.

(iii)   A slope change at some point in the system. This would imply a "normal" value for σ_1^2, a value zero for σ_2^2, and a large value for σ_3^2.

(iv)   A transient, or outlier, in the system. In this case, σ_1^2 assumes a "large" value, while σ_2^2 and σ_3^2 are zero.

Given initial global probabilities of the four states, these are modified in light of data via Bayes' theorem and estimates of the expected level and slope corresponding to each state are calculated. (Seasonal factors are generally treated outside the system.) Thus forecasts are calculated, not on the basis of an assumption that the system is in one particular state, but rather through the use of estimates derived relative to all four states. The weights given to each of these estimates depend on the estimated probabilities of the system being in each particular state. Harrison and Stevens discuss the practical implementation of a forecasting system along these lines and give an example in which, for generated data, the system performs very well in the presence of the four states described.

5.4 Stepwise Autoregression

The basic exponential smoothing procedures discussed in the previous section generally postulate a single model from which forecasts are to be generated, and thus do not possess the great virtue of the Box–Jenkins approach, whereby the eventual form of the forecast function is dictated, through the processes of identification and diagnostic checking, by the data itself. Of course, there is some room for experimentation within Brown's generalized exponential smoothing framework, but even here there does not exist any clear-cut identification procedure. (In addition, the restriction to a single parameter renders this approach overly parsimonious in many situations.) The identification and diagnostic checking phases of the Box–Jenkins cycle require manual intervention, however, and it would be desirable for some routine forecasting purposes to eliminate such a requirement. A compromise might be achieved through the design of a forecasting procedure which, while remaining fully automatic, contained a mechanism for discriminating among various possible forms of forecast function. That is, one would like a system to contain an identification procedure which was itself fully automatic. One method for achieving this, briefly introduced by Newbold and Granger [1974], is via stepwise autoregression. The objective is to construct autoregressive models to describe the behavior of given time series. However, for economic data, it is preferable, for reasons discussed in Chapters 2 and 3, to work with changes $Y_t = X_t - X_{t-1}$ rather than with levels of the series. Consider, now, the general kth order autoregressive model

$$Y_t = \sum_{j=1}^{k} a_j Y_{t-j} + \epsilon_t \tag{5.4.1}$$

Typically, models of the form (5.4.1) can, as has been seen, easily be fitted to a given set of data. However, unless k is taken to be quite small, it is likely that the resulting model will be overparametrized. One way out of this dilemma is to employ the technique of stepwise regression. This has been studied in great detail by Payne [1973], and the treatment given here depends heavily on Payne's work.

One way to proceed is to first select the value Y_{t-j} which, on the criterion

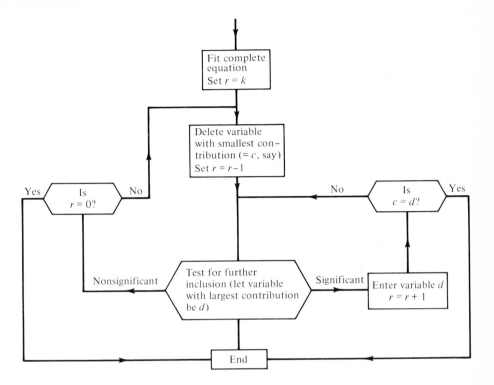

FIG. 5.1 *Payne's scheme for stepwise regression procedure.*

of residual sum of squares, contributes most toward "explaining" Y_t. At the second step, the lagged value that most improves the fit of the regression equation obtained at step one is added, and so on until addition of further variables produces no significant improvement in the fit of the regression. Variables, entered at an earlier stage, which cease to contribute significantly, can be dropped. An alternative, favored by Payne on the basis of his experience with a number of simulation experiments, is to proceed in the reverse direction, having initially fitted the complete model (5.4.1). At the first step, the lagged value contributing least to overall explanation of Y_t is dropped from the regression. At the second step the lagged value that contributes least in the model so achieved is dropped, and so on until deletion of further terms significantly worsens the fit of the regression equation. Variables, dropped at an earlier stage, can later be added if doing so would produce a significant improvement in the fit of the achieved regression. The procedure is set out schematically in Fig. 5.1. One might also include a constant term in the formulation (5.4.1). The constant could be treated as any other variable within the stepwise framework, or alternatively a decision as to its inclusion or exclusion could be made on purely subjective grounds.

Having decided to undertake the fitting of an autoregressive model by stepwise regression methods, three decisions must be taken:

(i)   A value k for the maximum permitted lag in (5.4.1) must be chosen.

(ii)   A significance level for testing for inclusion or exclusion of further variables must be decided upon.

(iii)   An appropriate hypothesis test to determine suitable stopping rules needs to be determined.

Choice of the maximum contemplated lag k is likely to be dictated in part by the nature of the time series under study and by the amount of data available. Our experience indicates that for all nonseasonal series, for quarterly seasonal series and shorter monthly seasonal series, a value $k = 13$ is generally adequate. For longer monthly seasonal time series, a value $k = 25$ is preferable. The question of choosing a suitable significance level for testing variable inclusion or exclusion is by no means a trivial one from a theoretical viewpoint. Looked at in this light, it might be desirable to reflect in one's choice preconceived notions as to how simple a model (in terms of number of parameters) is likely to provide reasonable forecasts. It is possible that one would like the significance level to vary according to how many lagged values have already been included in the model. Notwithstanding these considerations, however, we have found use of a constant 5% level to be adequate (in terms of forecasting accuracy of the resulting model) for most general purposes.

Testing of hypotheses presents one critical difficulty. Suppose that a stage has been reached where r terms are included in the regression. Clearly if each of the remaining $k - r$ terms was tested for futher inclusion at the 5% level, the probability of finding at least one term that apparently significantly improved the fit would be greater than 0.05 even if the true associated coefficient values were all zero. Payne suggests a number of procedures for overcoming this problem, and prefers use of the statistic

$$F' = \frac{m - r}{k - r} \frac{V_r}{V_k} - \frac{m - k}{k - r}$$

where

$$V_j = \frac{\text{residual sum of squares when } j \text{ terms are included in the regression}}{m - j}$$

and m is the effective number of observations for regression, so that if the original sample is $x_1, \ldots, x_n$, one observation is "lost" by differencing and a further k by formulation of the autoregressive model (5.4.1), so that $m = n - k - 1$. Under the null hypothesis that the coefficients on the excluded variables are all zero, the statistic F' is distributed approximately as Fisher's F with $k - r$ and $m - k$ degrees of freedom. (One is justified asymptotically in employing the usual normal theory regression tests in the context of autoregressive models as a result of Mann and Wald [1943].)

Stepwise autoregression, then, would appear to provide a reasonable alternative to exponential smoothing as a fully automatic forecasting technique. Its great advantage lies in the wide class of models contemplated, together with a built-in identification structure. Calculation of forecasts from the fitted model is straightforward along the lines described in Section 5.2.

5.5 A Fully Automatic Forecasting Procedure Based on the Combination of Forecasts

While either an exponential smoothing predictor or stepwise autoregression might produce optimal or near-optimal forecasts for particular underlying stochastic models, neither procedure can be expected to do so over the whole range of models likely to occur in practice. Given that both procedures may frequently produce suboptimal forecasts, it is worthwhile to attempt to improve the resulting predictions.

Let $F_{1,t}$ be a forecast of X_t based on an exponential smoothing predictor and let $F_{2,t}$ be a forecast of the same quantity derived through stepwise autoregression. Rather than restrict attention to just one of the forecasts, it might be profitable to consider the combined forecast $C_t = k_t F_{1,t} + (1 - k_t)F_{2,t}$ which is simply a weighted average of the two individual forecasts. The subscript on the weight k_t indicates that the appropriate weight for combining may vary through time. Although the computation of the predictor advocated here is quite lengthy, the forecasting procedure remains fully automatic since a single computer program can be written to generate forecasts without manual intervention. Thus calculation of the combined exponential smoothing—stepwise autoregressive predictor is quicker than that of the Box–Jenkins and requires less use of skilled manpower. The idea of combining individual forecasts in the production of an overall forecast was introduced by Bates and Granger [1969]. This concept will be discussed in more detail in Chapter 8, where appropriate choices of combining weights will be considered, and practical experience with the procedure surveyed.

5.6 Comparison of Univariate Forecasting Procedures

Given the availability of so many univariate forecasting techniques, it is natural to attempt to assess their relative worth and to try to come to some conclusion as to which methods are most likely to be successful in particular situations. In our view questions of this kind cannot be usefully answered through abstract consideration of the characteristics of the various procedures, but rather must be attacked through examination of the forecasting methods in action. The only reasonable test would seem to be an evaluation of forecast accuracy when the techniques are applied to "real" data in order to calculate "real" forecasts. This can be achieved by taking a particular time series and dividing it into two parts. A forecasting model is built, using only data in the first part of the series, and this is used to calculate "forecasts" of the known values in the second part, thus allowing forecast accuracy to be assessed. Comparison of forecast performance of the various techniques over a large number of time series should then give an indication of their relative worth, as well as of their various strengths and weaknesses. Of course, since it is impossible to draw a random sample of all time series, or of all economic time series, it will not be possible to make well-defined probabilistic conclusions in the usual hypothesis testing fashion. However, such an exercise is likely to yield valuable insights, although the conclusions drawn must retain an element of subjectivity. It remains only to decide on a standard for assessing forecast accuracy, and here we employ the criterion of mean squared error, although, as noted in Chapter 4, other cost functions may on occasion be appropriate.

Of course, assessment of forecast accuracy is only one side of the coin since strictly speaking this ought to be balanced against the cost of generating the forecasts. The question of cost is indeed very complex, but one on which we have very little to say. A priori, it would be reasonable to expect generation of Box–Jenkins forecasts to be more costly than that of exponential smoothing forecasts. However, the relative costs of the various procedures will depend crucially on the resources of the user and also on his requirements—for example, a computer program that is to be used many times will have a relatively low average cost per job, but might still be prohibitively expensive for a situation where only a few runs are required. Thus, while recognizing the importance of cost considerations, we can think of no general way to evaluate them, and so in the remainder of this section discussion is limited to evelution of forecast accuracy.

The earliest exercise of any reasonable size attempting to assess the relative merits of univariate predictors on real data was carried out by Wagle *et al.* [1968]. These authors examined the application of four procedures—Holt–Winters, Harrison's seatrend, Brown's multiple smoothing, and a simple linear regression on time—to sales of 20 products. They found Holt–Winters to be best 14 times, seatrend twice, and Brown's approach 4 times. Unfortunately very little detail is given in their report of this study, and in particular it is difficult to assess whether the series under consideration afforded a sufficient variety of characteristics as to make a fair comparison possible.

A much larger study, by Reid [1969], allowed for the first time a comparison of the performance of Box–Jenkins and various exponential smoothing predictors over a large set of real data. Reid assembled a collection of 113 macroeconomic time series and generated forecasts, employing methods he deemed reasonable, to each series. Both the Box–Jenkins and Brown's generalized exponential smoothing predictors were applied to every series in the collection, Holt–Winters was applied to 69 series, and Harrison's seatrend to 47. The Brown predictor modified to take account of the possibility of first-order autoregression in the one-step forecast errors was also evaluated.

Reid found for one-step ahead prediction, Box–Jenkins outperformed Brown's method on 88% of occasions, outperformed Holt–Winters 70%, and Harrison 77% of the time. Brown did better than modified Brown 24% of the time, better than Holt–Winters 20%, and better than Harrison 15% of occasions. Modified Brown forecast better than Holt–Winters and Harrison on 28 and 21% of occasions respectively, while Holt–Winters outperformed Harrison on 43% of occasions. Reid noted the large number of labor statistics in his collection tended to bias these results somewhat in favor of the Harrison approach. Reid also noted a tendency for Box–Jenkins to lose some of its advantage over exponential smoothing methods when forecasts over longer lead times are considered. This probably results from the fact that, while the past history of a typical nonseasonal economic time series may well contain valuable information about the next change, it is less likely to do so about subsequent changes. Thus, as one attempts to forecast further ahead, the substantive content in an efficient predictor becomes swamped by the element of uncertainty about distant future changes.

In a further large study, Newbold and Granger [1974] analyzed a collection of 106 time series, 80 of which were monthly and 26 quarterly. The collection included both seasonal and nonseasonal macroeconomic series and micro sales data. The Box–Jenkins, Holt–Winters, and stepwise autoregression methods were applied to every series in the collection. The stepwise autoregression program used in this study employed an arbitrary F ratio of 4.0 to determine whether or not a variable should be added to or dropped from a regression. Subsequently, Payne [1973] has reanalyzed many of the series in this collection, using the superior cutoff criterion described in Section 5.4. He achieved one-step ahead forecasts which, on average, were slightly superior to those obtained in this study. Table 5.12 shows the percentage of occasions one method outperforms another for forecasts made up to eight periods ahead. It is clear from this table that Box–Jenkins performs better than its competition on a sizable majority of series for all lead times. The Box–Jenkins versus Holt–Winters comparison for one-step ahead forecasts is very similar to that obtained by Reid, and here again it is noticed that over longer forecast horizons Box–Jenkins loses some of its relative advantage. The one-step ahead forecast errors are examined in more detail in Table 5.13, which gives an idea of the potential gains in accuracy which can be achieved through using the Box–Jenkins method. It can be seen from this table that Box–Jenkins very often considerably outperforms the two fully automatic procedures, particularly the Holt–Winters method. A feature of this table, which is of some practical relevance, is the wide spread of ranges containing an appreciable number of series in the Holt–Winters versus stepwise autoregression comparison. This suggests that each method contains useful features absent in the other, and consequently that the combined forecast introduced in Section 5.5 may well constitute a worthwhile improvement over the individual procedures. Summary statistics for the information in Table 5.13

Table 5.12 *Comparison of Box–Jenkins (B–J), Holt–Winters (H–W), and stepwise autoregressive (S–A) forecasts: percentage of series for which first named method outperforms second for various lead times*

	Forecast lead times							
Comparisons	1	2	3	4	5	6	7	8
B–J: H–W	73	64	60	58	58	57	58	58
B–J: S–A	68	70	67	62	62	61	63	63
H–W: S–A	48	50	58	57	55	56	58	59

Table 5.13 *Comparison of average square forecast errors (ASE) of Box–Jenkins, Holt–Winters, and stepwise autoregression in terms of ratios of average squared errors:* **number of forecasted series (in a total of 106 series) in various ranges**

	Method A Method B	Box–Jenkins Holt–Winters	Box–Jenkins Stepwise	Holt–Winters Stepwise
	0.1–0.2	1	0	0
	0.2–0.3	3	2	0
ASE Method A	0.3–0.4	4	4	2
ASE Method B	0.4–0.5	8	4	4
	0.5–0.6	5	2	4
when A is better	0.6–0.7	10	5	6
	0.7–0.8	13	14	13
	0.8–0.9	19	17	10
	0.9–1.0	14	23	12
Two methods identical		0	2	0
	0.9–1.0	11	14	5
	0.8–0.9	8	13	14
ASE Method B	0.7–0.8	2	3	10
ASE method A	0.6–0.7	3	1	12
	0.5–0.6	4	0	5
when B is better	0.4–0.5	0	2	2
	0.3–0.4	1	0	3
	0.2–0.3	0	0	4
	0.1–0.2	0	0	0

are provided by the geometric means of the various ratios of average squared forecast errors. The values obtained for these geometric means were:

$$\frac{\text{average squared error Box–Jenkins forecasts}}{\text{average squared error Holt–Winters forecasts}} = 0.80$$

$$\frac{\text{average squared error Box–Jenkins forecasts}}{\text{average squared error stepwise autoregressive forecasts}} = 0.86$$

$$\frac{\text{average squared error stepwise autoregressive forecasts}}{\text{average squared error Holt–Winters forecasts}} = 0.93$$

For one-step ahead prediction, the performance of various combined forecasts was also examined by Newbold and Granger. The combined Holt–Winters–stepwise autoregressive forecast was found to outperform Box–Jenkins on 46.25% of the monthly series in the sample. The geometric mean of the ratio of average squared error of the Box–Jenkins forecast to that of the combined forecast was found to be 0.99, using an appropriate procedure for choosing the combining weights. Thus a fully automatic forecast that performs very well indeed can be achieved. It was also found that combination of one or both of the fully automatic procedures with Box–Jenkins produced forecasts which, on the average, were a small improvement over Box–Jenkins taken alone. Detailed discussion of the results on combining is postponed until Chapter 8.

In addition to the forecasting methods that were applied to all the data in the collection, the Brown generalized exponential smoothing and Harrison seatrend procedures were applied to small subsets of the series, in order to obtain some insight into their merits. We noted, as did Reid, that seatrend was generally no better (and indeed, on occasion was much worse) than Holt–Winters, except on those series that had both a very strong seasonal factor and fairly large random (unpredictable) variation. Brown's method did not perform terribly well, particularly on seasonal time series, and we can imagine no circumstances in which we would prefer it to Holt–Winters.

It should be emphasized that the smoothing constants for exponential smoothing were chosen on the basis of goodness of fit over the sample period. It would be reasonable to expect forecast performance to deteriorate further if more arbitrary procedures were employed.

On the basis of the material described here and our general experience in applying univariate forecasting procedures, we now offer tentative guidelines. These should most certainly not be followed religiously but might provide a useful basis for deciding on a forecasting method in any given situation:

(a) For very short time series (with less than 30 observations), there is little alternative to use of an exponential smoothing predictor, and generally we would prefer Holt–Winters in such circumstances.

(b)   For moderately long series (at least 30 and no more than 40–50 observations), Box–Jenkins is still rather difficult to apply (more so for seasonal than nonseasonal series), but stepwise autoregression becomes feasible, and its use in combination with Holt–Winters seems the most promising approach. For longer series, if for some reason it is impracticable to use Box–Jenkins, then this combined forecast should again prove generally satisfactory.

(c)   For longer time series (at least 40–50 observations), both its versatility and its success in actual applications would argue for the use of Box–Jenkins in preference to any other single procedure. Very often this will yield forecasts of sufficient accuracy for the purpose at hand, but should it fail to do so, combination with either Holt–Winters or stepwise autoregression or both can be tried. The extra versatility of Box–Jenkins makes it particularly valuable in situations where either a time series has proved difficult to predict by routine methods or one is meeting a particular kind of series for the first time and is uncertain about its characteristics.

(d)   For those series that are strongly seasonal and exhibit large random variations, substitute Harrison's seatrend for Holt–Winters in the above.

(e)   Although we have little experience of the Harrison–Stevens Bayesian predictor, it appears that this approach might well prove useful in the situations for which it was designed—cases where it is suspected that the available data exhibit specific kinds of nonstationarities. Such series might crop up quite frequently in sales forecasting.

The above guidelines should certainly not be followed blindly. It is often the case that, in practice, one has valuable information about the particular series under study. This information should, if at all possible, be injected into the forecast generating mechanism, and may well influence any decision as to what procedure to employ.

CHAPTER SIX

FORECASTING FROM REGRESSION MODELS

Predict, v.   To relate an event that has not occurred, is not occurring and will not occur.
Prophecy, n.   The art and practice of selling one's credibility for future delivery.

Ambrose Bierce
"The Enlarged Devil's Dictionary", Doubleday, 1967

6.1 Introduction

Up to this point only methods of forecasting a time series given just its current and past values have been considered. As noted in Chapter 5, such procedures are of considerable practical value. Nevertheless, it will often be the case that forecasts of higher quality can be obtained through the use of a wider information set. One might include in such a set, for example, current and past values of related time series and perhaps also any relevant nonquantifiable information. This latter possibility makes rigorous treatment of the forecasting problem rather difficult. For example, the forecaster may know of the possiblility of an impending strike in the coal industry. However, he will generally not know for sure that the strike is to take place and almost certainly will be unsure of its duration. Thus, in addition to assessing the likely impact of the strike on the variable of interest, the forecaster must inject into his forecast an assessment of the probability of the strike taking place at all and also its likely duration. Presumably this kind of subjective judgment could be incorporated into a formal Bayesian treatment of the forecasting problem. However, the methods used in practice are of a more ad hoc nature.

Consideration of a wider information set for forecasting purposes introduces two additional concepts to the previous analysis. First it is necessary to examine and make use of any theory that postulates relationships determining the variable of interest, and second one might attempt to extend the univariate models of Chapter 3 to the multivariate case. In this chapter, traditional econometric approaches to forecasting, whereby economic theory is employed in the construction of equations to be used in prediction, will be discussed. Consideration of multivariate time series methods is postponed until Chapter 7.

6.2 Single Equation Models

Suppose that a variable Y is related to K variables $X_1, X_2, \ldots, X_K$ in such a way that the conditional expectation of Y given the X_j is linear, so that

$$E(Y|X_1, X_2, \ldots, X_K) = \beta_1 X_1 + \beta_2 X_2 + \cdots + \beta_K X_K$$

and let the residual be $\epsilon = Y - E(Y|X_1, X_2, \ldots, X_K)$. Suppose, now, that n equally spaced observations through time are observed on this process. One can write

$$Y_t = \mathbf{x}_t' \boldsymbol{\beta} + \epsilon_t, \qquad t = 1, 2, \ldots, n \tag{6.2.1}$$

where $\mathbf{x}_t'$ constitutes a vector of observations on $X_1, X_2, \ldots, X_K$, made at time t, so that $\mathbf{x}_t' = (x_{1t}, x_{2t}, \ldots, x_{Kt})$ and $\boldsymbol{\beta}$ is a vector of constant coefficients

$$\boldsymbol{\beta}' = (\beta_1, \beta_2, \ldots, \beta_K).$$

The set of equations (6.2.1) may be written more compactly as

$$\mathbf{y} = \mathbf{X}\boldsymbol{\beta} + \boldsymbol{\epsilon} \tag{6.2.2}$$

where $\mathbf{y}' = (Y_1, Y_2, \ldots, Y_n)$, $\boldsymbol{\epsilon}' = (\epsilon_1, \epsilon_2, \ldots, \epsilon_n)$, and

$$\mathbf{X} = \begin{bmatrix} x_{11} & x_{21} & \cdots \cdots & x_{K1} \\ x_{12} & x_{22} & \cdots \cdots & x_{K2} \\ \vdots & \vdots & & \vdots \\ x_{1n} & x_{2n} & \cdots \cdots & x_{Kn} \end{bmatrix} = \begin{bmatrix} \mathbf{x}_1' \\ \mathbf{x}_2' \\ \vdots \\ \mathbf{x}_n' \end{bmatrix}$$

Equation (6.2.2) formulates the (multiple) linear regression model. It should be emphasized that "linearity" refers here to the unknown coefficients $\boldsymbol{\beta}$. There is no reason to preclude consideration of transformed variables from the list of regressors $X_1, X_2, \ldots, X_K$. For example, in explaining change in wage rates it is common to include the reciprocal of unemployment as an explanatory variable. Again, a "dummy variable" taking a value 1 under particular special circumstances (for example, during government imposed wage control periods) and zero otherwise can be included among the X_j. Frequently it is desired to include a constant (intercept) term in the regression equation. This can be achieved by setting $x_{1t} = 1$, $t = 1, 2, \ldots, n$ whereupon the coefficient β_1 denotes the intercept.

In order to analyze the model (6.2.2), it is necessary to make various assumptions, although these may need to be relaxed in specific situations. The standard assumptions for linear regression are:

(i) $E(\epsilon) = 0$;
(ii) $E(\epsilon\epsilon') = \sigma^2 I$, where I is the identity matrix;
(iii) X is distributed independently of ϵ and, for any realization of the process, $X'X$ is nonsingular with probability 1, i.e., almost certainly.

Now consider the analysis of (6.2.2) for a given set of observations X, it being understood in what follows that expectations are taken conditional on X being fixed. The Gauss–Markov theorem then states that the estimator

$$\hat{\beta} = (X'X)^{-1}X'y \tag{6.2.3}$$

is best linear unbiased for β. That is, in the class of estimators that are linear in y and unbiased for β, $\hat{\beta}$ has smallest variance. It is straighforward to show that $E[(\beta - \hat{\beta})(\beta - \hat{\beta})'] = \sigma^2(X'X)^{-1}$, and that an unbiased estimate of σ^2 is given by

$$s^2 = (y - X\hat{\beta})'(y - X\hat{\beta})/(n - K) \tag{6.2.4}$$

Finally, if ϵ is assumed to be normally distributed, it can be shown that $\hat{\beta}$ is distributed independently of s^2, $\hat{\beta}$ being normal and $(n - K)s^2/\sigma^2$ distributed as χ^2 with $n - K$ degrees of freedom. Hence the quantity

$$(\hat{\beta}_j - \beta_j)/s\sqrt{a_{jj}} \tag{6.2.5}$$

where a_{jj} is the jth diagonal element of $(X'X)^{-1}$, is distributed as Student's t with $n - K$ degrees of freedom, thus allowing tests of significance of individual regression coefficients to be made.

Suppose now that the regression equation contains a constant term, so that

$$E(Y|X_2, X_3, \ldots, X_K) = \beta_1 + \beta_2 X_2 + \cdots + \beta_K X_K$$

It would be useful to have an indication of how much knowledge of the X_j contributes toward "explaining" Y. One could argue as follows: from (6.2.4) the sum of squared residuals about the regression is $(n - K)s^2$, whereas the variation of Y about its mean in the given sample is $\sum_{i=1}^{n}(Y_i - \bar{Y})^2$. Thus the regression "explains" the amount $\sum_{i=1}^{n}(Y_i - \bar{Y})^2 - (n - K)s^2$ and hence, in proportionate terms, the amount "explained" by the regression is

$$R^2 = 1 - (n - K)s^2 \left[\sum_{i=1}^{n} (Y_i - \bar{Y})^2 \right]^{-1} \tag{6.2.6}$$

This quantity is called the *coefficient of multiple correlation*. Now, it might be objected that R^2 can overstate the value of a regression fit since the quantity $(n - K)s^2$ can be reduced simply by adding further variables X_j, even if they are not relevant in "explaining" Y. An alternative is to employ a quantity that corrects for degrees of freedom in estimating the residual variance and the varance of Y. Thus one could calculate

$$\overline{R}^2 = 1 - s^2 \left[\sum_{i=1}^{n} \left(Y_i - \overline{Y} \right)^2 / (n - 1) \right]^{-1} \tag{6.2.7}$$

This is called the *corrected coefficient of multiple correlation*. It should be noted that, in cases of particularly bad fit, $\overline{R}^2$ can be negative. It is often required to test the hypothesis H_0: $\beta_2 = \beta_3 = \cdots = \beta_K = 0$, that is $X_2, X_3, \ldots, X_K$, taken as a set, contribute nothing (in the linear regression sense) toward explanation of Y. This can be accomplished by noting that, under H_0, the statistic

$$F = \frac{R^2}{1 - R^2} \frac{n - K}{K - 1}$$

is distributed as Fisher's F with $K - 1$ and $n - K$ degrees of freedom. The hypothesis H_0 is rejected for high values of the test statistic.

Suppose that the regression equation (6.2.1) continues to hold for some future time period $n + h$, so that

$$Y_{n+h} = x'_{n+h} \beta + \epsilon_{n+h} \tag{6.2.8}$$

If it is required to forecast Y_{n+h} for given x_{n+h}, a natural predictor is obtained by setting ϵ_{n+h} equal to its expected value of zero and substituting $\hat{\beta}$ of (6.2.3) for β in (6.2.8). It will now be proved that this predictor is best linear unbiased. Consider the set of all possible linear predictors $l'y$. Then

$$Y_{n+h} - l'y = (x'_{n+h} - l'X)\beta + \epsilon_{n+h} - l'\epsilon$$

by (6.2.2) and (6.2.8). Taking expectations (conditional on the fixed $x's$)

$$E(Y_{n+h} - l'y) = (x'_{n+h} - l'X)\beta$$

For the forecast to be unbiased, this quantity must be zero for all β, and hence

$$l'X = x'_{n+h} \tag{6.2.9}$$

The variance of the forecast is

$$E\left[(Y_{n+h} - l'y)^2 \right] = E\left[(\epsilon_{n+h} - l'\epsilon)^2 \right] = \sigma^2 (1 + l'l) \tag{6.2.10}$$

Thus, to obtain the best linear unbiased predictor, it is necessary to choose l such that $l'l$ is minimum subject to the restriction (6.2.9). This can be achieved using Lagrange multipliers. Define the multiplier $\lambda' = (\lambda_1, \lambda_2, \ldots, \lambda_K)$ and the function $F = l'l - \lambda'(X'l - x_{n+h})$. Then

$$\partial F / \partial l = 2l - X\lambda, \qquad \partial F / \partial \lambda = X'l - x_{n+h}$$

Setting these partial derivatives equal to zero yields

$$\left[\begin{array}{c|c} 2I & -X \\ \hline X' & 0 \end{array} \right] \left(\begin{array}{c} l \\ \hline \lambda \end{array} \right) = \left(\begin{array}{c} 0 \\ \hline x_{n+h} \end{array} \right)$$

Solving this set of equations by inverting the partitioned matrix then yields $l' = x'_{n+h}(X'X)^{-1}X'$. Hence the best linear unbiased predictor of Y_{n+h} for given x_{n+h} is given by

$$f_{n,h} = x'_{n+h}\hat{\beta} \qquad (6.2.11)$$

where $\hat{\beta}$ is given by (6.2.3). Further, it follows from (6.2.10) that the variance of this predictor is given by

$$E\left[(Y_{n+h} - f_{n,h})^2\right] = \sigma^2\left(1 + x'_{n+h}(X'X)^{-1}x_{n+h}\right)$$

An unbiased estimate for this variance is obtained by substituting s^2 of (6.2.4) for σ^2.

The forecasting problem solved here is one of *conditional* prediction, that is, the approach yields optimal forecasts of Y_{n+h} for given future values of $X_1, X_2, \ldots, X_K$. In itself this constitutes a worthwhile object since, for example, a policy maker may require estimates of the future values of the variable of interest under various policy options. The problem of *unconditional* prediction—forecasting what Y_{n+h} will actually be—can only be solved if forecasts of $X_{j, n+h}, j = 1, 2, \ldots, K$, are substituted for x_{n+h} in (6.2.11). Clearly the regression equation (6.2.2) is of no help here; these forecasts must be obtained from some other source—possibly through judgmental considerations or possibly through the application of some univariate or multivariate forecasting method.

The predictor (6.2.11) is, as just shown, optimal given the assumptions made. However, it will often be the case that the assumption concerning the variance–covariance matrix of the errors is untenable. In general, let these errors be written

$$u = y - X\beta \qquad (6.2.12)$$

reserving the symbol ϵ to denote white noise residuals. Suppose, now, that

$$E(u) = 0, \qquad E(uu') = \sigma^2\Omega \qquad (6.2.13)$$

where Ω is some positive definite symmetric matrix. Goldberger [1962] shows, through an approach similar to that employed in deriving (6.2.11), that the best linear unbiased predictor of Y_{n+h} is given by

$$f_{n,h} = x'_{n+h}b + W'\Omega^{-1}e \qquad (6.2.14)$$

where

$$b = (X'\Omega^{-1}X)^{-1}X'\Omega^{-1}y, \qquad e = y - Xb, \qquad W = E(u_{n+h}u) \quad (6.2.15)$$

Note that b is Aitken's generalized least squares estimator, that is, the best linear unbiased estimate of β in (6.2.12) under the specification (6.2.13). Unfortunately the result (6.2.14) is of little practical value as it stands since it requires knowledge of Ω and W. Typically, in practice, one must specify some plausible form of the covariance structure of the errors and estimate this along with the parameters β of the model. Goldberger also shows that if Ω is wrongly assumed to be the identity matrix I, the loss in prediction efficiency can in certain circumstances be extremely severe.

When dealing with time series data, perhaps the most frequently encountered problem is autocorrelation in the residuals from the regression equation. In the presence of autocorrelated errors the least-squares estimator (6.2.3) remains unbiased since

$$E(\hat{\beta}) = (\mathbf{X'X})^{-1}\mathbf{X'}E(\mathbf{y}) = (\mathbf{X'X})^{-1}\mathbf{X'}E(\mathbf{X}\beta + \mathbf{u})$$

from (6.2.12). Hence $E(\hat{\beta}) = \beta$ by (6.2.13). However, the best linear unbiased estimator is given by (6.2.15), which in general will differ from (6.2.3). More seriously, the usual formula for the variance–covariance matrix of the least-squares estimators is incorrect since now $\hat{\beta} - \beta = (\mathbf{X'X})^{-1}\mathbf{X'u}$ and hence by (6.2.13)

$$E\left[(\beta - \hat{\beta})(\beta - \hat{\beta})'\right] = \sigma^2(\mathbf{X'X})^{-1}\mathbf{X'\Omega X}(\mathbf{X'X})^{-1}$$

The usual tests of hypotheses, based for example on (6.2.5) and (6.2.6), are invalid in the presence of serially correlated errors—a point that will be taken up again in Section 6.4—and, as already noted, the forecast (6.2.11) is generally inefficient.

One approach to the problem of autocorrelated errors would be to assume some model from the general autoregressive integrated moving average class of processes $a(B)(1 - B)^d u_t = b(B)\epsilon_t$ introduced in Chapters 1 and 3. In principle, the data could be employed to suggest an appropriate model from this class, which could then be incorporated into the regression equation for estimation and forecasting. However, standard practice in applied economic work appears to be rather less general in terms of the alternative error structures that are contemplated. A white noise error structure is assumed as a null hypothesis, which is generally tested against some simple alternative—almost invariably a first-order autoregressive model—which is then adopted for purposes of estimation and forecasting should the test indicate significant autocorrelation. Occasionally higher order autoregressive processes for the errors are entertained (see, e.g., Wallis [1972]), but this seems to be the exception rather than the rule in current applied work.

The usual assumption, then, is a model of the form $\mathbf{y} = \mathbf{X}\beta + \mathbf{u}$ with

$$u_t = a_1 u_{t-1} + \epsilon_t \tag{6.2.16}$$

The first step in the analysis is to calculate the ordinary least-squares estimate $\hat{\beta}$ of (6.2.3) and to test the null hypothesis that a_1 of (6.2.16) is zero. Denote the residuals from the fitted regression as $\hat{u}_t = Y_t - \mathbf{x}_t'\hat{\beta}$, $t = 1, 2, \ldots, n$. The test generally employed is due to Durbin and Watson [1950, 1951], who propose the statistic

$$d = \sum_{t=2}^{n} (\hat{u}_t - \hat{u}_{t-1})^2 \Big/ \sum_{t=1}^{n} \hat{u}_t^2 \tag{6.2.17}$$

This statistic is available on virtually every standard linear regression program package, and is generally quoted along with R^2 or $\overline{R}^2$ in applied econometric work. Now, writing

$$\sum_{t=1}^{n} \hat{u}_t^2 \approx \sum_{t=2}^{n} \hat{u}_t^2 \approx \sum_{t=2}^{n} \hat{u}_{t-1}^2$$

it follows that

$$d \approx \left(2 \sum_{t=1}^{n} \hat{u}_t^2 - 2 \sum_{t=2}^{n} \hat{u}_t \hat{u}_{t-1} \right) / \sum_{t=1}^{n} \hat{u}_t^2$$

that is,

$$d \approx 2(1 - \hat{\rho}_1) \tag{6.2.18}$$

where $\hat{\rho}_1$ is the first-order sample autocorrelation of $\hat{u}_t$. Thus, in contrast to the methodology described in Chapter 3, no sample autocorrelations beyond the first are considered in assessing the time series structure of the residuals. Of course, this follows from the fact that only the formulation (6.2.16) is entertained as a possible model. It follows from (6.2.18) that if $\hat{\rho}_1$ is zero, then d is equal to 2. Positive values for $\hat{\rho}_1$ imply $0 < d < 2$ and negative values $2 < d < 4$, the larger is $|\hat{\rho}_1|$ the further from 2 is d. Unfortunately the distribution of the Durbin–Watson statistic d depends on the matrix $\mathbf{X}$. However, Durbin and Watson show that this distribution always lies between that of two other statistics d_L and d_U, which they tabulate for particular values of K and n. The null hypothesis of no autocorrelation is then rejected against the alternative of positive autocorrelation if $d < d_L^*$ and not rejected if $d > d_U^*$, where asterisks indicate tabulated values at appropriate significance levels. If $d_L^* < d < d_U^*$ the test is inconclusive. Tests against the alternative of negative autocorrelation proceed in the same way, except that the appropriate statistic is now $4 - d$. A number of procedures for resolving inconclusive test results are examined by Durbin and Watson [1971] and some alternative tests are described and compared by L'Esperance and Taylor [1975].

Suppose, now, that a model with first order autoregressive errors is assumed. This can be written

$$Y_t - a_1 Y_{t-1} = (\mathbf{x}'_t - a_1 \mathbf{x}'_{t-1})\boldsymbol{\beta} + \epsilon_t \tag{6.2.19}$$

The most direct (though not the most computationally simple) method of estimation is to estimate a_1 and $\boldsymbol{\beta}$ simultaneously through a nonlinear least squares routine, as described in Chapter 3. A number of alternative procedures are discussed in Chapter 13 of Malinvaud [1966]. Given the least-squares estimates $\hat{a}_1$ and $\hat{\boldsymbol{\beta}}$, one can write, approximately, from (6.2.19)

$$Z_t - \hat{a}_1 Z_{t-1} = \epsilon_t \qquad \text{where} \quad Z_t = Y_t - \mathbf{x}'_t \hat{\boldsymbol{\beta}}$$

Thus future values of Z, and hence of Y conditional on $\mathbf{X}$, can be forecast using the procedure described in Section 5.2.

As a final point on single equation models, it should be noted that lagged values of the variable Y can be included in the list of explanatory.variables X in the model (6.2.1). Thus, as a simple example, one might have

$$Y_t = \beta_1 + \beta_2 Y_{t-1} + \beta_3 x_t + \epsilon_t \qquad (6.2.20)$$

Application of ordinary least squares to models of this type yields an estimate $\hat{\beta}$ which is biased. However, provided the error term is nonautocorrelated, the usual least-squares estimators for β and the variance estimators are consistent —that is, they tend to the true values as the sample size tends to infinity. This is not generally the case if the error term is autocorrelated, however. A more thorough discussion of these results is given in Chapter 6 of Goldberger [1964]. A further difficulty with models of the form (6.2.20) is that the usual Durbin–Watson test is invalid, the test statistic being biased toward 2 (i.e., against rejection of the null hypothesis of no autocorrelation in the errors), as demonstrated by Nerlove and Wallis [1966]. A test for autocorrelated errors, which is valid in large samples, is given by Durbin [1970]. Forecasts of future values of Y are obtained in an obvious fashion from models of the form (6.2.20), along the lines described in Section 5.2.

To summarize, economists are frequently led to fit regression equations of the form (6.2.2), based on some particular theory that suggests an appropriate list of regressors $X_1, X_2, \ldots, X_K$. However, interpretation and analysis of these equations are dependent on the time series structure of the error terms —a point about which economic theory has little or nothing to say.

6.3 Simultaneous Equation Models

The analysis of the previous section was limited to consideration of the case where a single dependent variable Y is influenced by a set of independent variables $X_1, X_2, \ldots, X_K$, it being assumed that the values taken by these independent variables do not in turn depend on values taken by Y. Such an assumption is frequently untenable in economics. For example, the rate of change of money wages may well depend (among other things) on the rate of change of retail prices. However, it would also be reasonable to hypothesize a dependence of prices on wages, along with other factors. Thus it is often required to analyze in applied economic work a system of simultaneous equations, depicting the various interactions that might be taking place.

An econometric simultaneous equation system is a set of M equations that simultaneously determine the values of M "endogenous" variables Y_1, $Y_2, \ldots, Y_M$ in terms of K "exogenous," or "predetermined," variables. The endogenous variables can be thought of as the quantities of interest; one might want to explain their behavior or to forecast their future values. The exogenous variables $X_1, X_2, \ldots, X_K$ are quantities that are *not* themselves dependent on the endogenous variables, but that *do* influence the behavior of the endogenous variables. The exogenous variables may not necessarily be of any great interest in themselves, but they help explain the behavior of the endogenous variables, and may help in their prediction. Now, it should be obvious that any system of equations linking economic variables will not hold exactly at all times. Rather, it is necessary to add stochastic error terms, in which case one can think of the system as determining simultaneously the conditional expectations of the endogenous variables given the exogenous. The structures of the individual equations in a system are determined, as in the single equation case, by appeal to economic theory.

Suppose, for the time being, that a linear system is appropriate. Then, with the values $x_1, x_2, \ldots, x_K$ of the exogenous variables determined outside the system, one can write

$$Y_{1,t} = \gamma_{21} Y_{2,t} + \gamma_{31} Y_{3,t} + \cdots + \gamma_{M1} Y_{M,t} + \beta_{11} x_{1,t}$$

$$+ \cdots + \beta_{K1} x_{K,t} + u_{1,t}$$

$$Y_{2,t} = \gamma_{12} Y_{1,t} + \gamma_{32} Y_{3,t} + \cdots + \gamma_{M2} Y_{M,t} + \beta_{12} x_{1,t}$$

$$+ \cdots + \beta_{K2} x_{K,t} + u_{2,t}$$

$$\vdots$$

$$Y_{M,t} = \gamma_{1M} Y_{1,t} + \gamma_{2M} Y_{2,t} + \cdots + \gamma_{M-1M} Y_{M-1,t}$$

$$+ \beta_{1M} x_{1,t} + \cdots + \beta_{KM} x_{K,t} + u_{M,t}$$

for $t = 1, 2, \ldots, n$. In matrix notation this system can be expressed as

$$\mathbf{y}_t' \mathbf{\Gamma} = \mathbf{x}_t' \mathbf{B} + \mathbf{u}_t', \qquad t = 1, 2, \ldots, n \tag{6.3.1}$$

where

$$\mathbf{y}_t' = (Y_{1,t}, Y_{2,t}, \ldots, Y_{M,t}), \qquad \mathbf{u}_t' = (u_{1,t}, u_{2,t}, \ldots, u_{M,t})$$

$$\mathbf{x}_t' = (x_{1,t}, x_{2,t}, \ldots, x_{K,t})$$

$$\mathbf{\Gamma} = \begin{bmatrix} 1 & -\gamma_{12} & -\gamma_{13} & \cdots & -\gamma_{1M} \\ -\gamma_{21} & 1 & -\gamma_{23} & \cdots & -\gamma_{2M} \\ -\gamma_{31} & -\gamma_{32} & 1 & \cdots & -\gamma_{3M} \\ \vdots & \vdots & \vdots & \cdots & \vdots \\ -\gamma_{M1} & -\gamma_{M2} & -\gamma_{M3} & \cdots & 1 \end{bmatrix}$$

$$\mathbf{B} = \begin{bmatrix} \beta_{11} & \beta_{12} & \cdots & \beta_{1M} \\ \beta_{21} & \beta_{22} & \cdots & \beta_{2M} \\ \vdots & \vdots & & \vdots \\ \beta_{K1} & \beta_{K2} & \cdots & \beta_{KM} \end{bmatrix}$$

It is generally assumed that Γ is nonsingular and that

$$E(\mathbf{u}_t) = \mathbf{0}, \quad t = 1, 2, \ldots, n, \qquad E(\mathbf{u}_t\mathbf{u}_t') = \Sigma \text{ (nonsingular)}, \quad t = 1, 2, \ldots, n$$

Frequently it is further assumed that the errors u_t are not autocorrelated, so that

$$E(u_{i,t}u_{j,t+k}) = 0 \qquad \forall i, j \quad \text{and} \quad \forall k \neq 0$$

The set of equations (6.3.1) may be written more compactly as

$$\mathbf{Y}\Gamma = \mathbf{X}\mathbf{B} + \mathbf{U} \tag{6.3.2}$$

where

$$\mathbf{Y} = \begin{bmatrix} \mathbf{y}_1' \\ \mathbf{y}_2' \\ \vdots \\ \mathbf{y}_n' \end{bmatrix}, \qquad \mathbf{X} = \begin{bmatrix} \mathbf{x}_1' \\ \mathbf{x}_2' \\ \vdots \\ \mathbf{x}_n' \end{bmatrix}, \qquad \mathbf{U} = \begin{bmatrix} \mathbf{u}_1' \\ \mathbf{u}_2' \\ \vdots \\ \mathbf{u}_n' \end{bmatrix}$$

Equation (6.3.2) is called the structural form—the structure being imposed by economic theory. Typically, lagged values of the endogenous variables are included in the matrix of regressors $\mathbf{X}$ along with exogenous variables, so that the set $\mathbf{x}_t$ will be referred to as predetermined variables, that is, variables determined outside of the system of simultaneous equations (6.3.1) at time t.

Multiplying through (6.3.2) by the matrix Γ^{-1} yields

$$\mathbf{Y} = \mathbf{X}\mathbf{B}\Gamma^{-1} + \mathbf{U}\Gamma^{-1}$$

or

$$\mathbf{Y} = \mathbf{X}\Pi + \mathbf{V} \qquad \text{where} \quad \Pi = \mathbf{B}\Gamma^{-1}, \quad \mathbf{V} = \mathbf{U}\Gamma^{-1} \tag{6.3.3}$$

Equation (6.3.3), which constitutes the solution of the system for the endogenous variables in terms of the predetermined variables and residuals, is called the reduced form. The quantity $\mathbf{X}\Pi$ denotes the conditional expectation of the set of endogenous variables, given the predetermined variables. In general, however, knowledge of this conditional distribution is insufficient to determine $\mathbf{B}$, Γ, and Σ uniquely. To proceed further, prior restrictions must be imposed on the values of these matrices. The simplest approach to this problem of "econometric identification" is to assume, following the specifications of economic theory, that particular elements of $\mathbf{B}$ and Γ are zero. Thus, a condition for econometric identification of an equation in the system (6.3.2) is that the number of predetermined variables excluded from the equation must be at least as great as one less than the number of endogenous variables included in the equation. This condition is sufficient for identification provided that the matrix Π is not ill-conditioned. (For a proof, see Chapter 12 of Johnston [1972].) A discussion of other types of restrictions and resulting criteria for econometric identification is contained in Fisher [1966].

Application of ordinary least squares to an equation in the system (6.3.2) produces, in general, coefficient estimates that are inconsistent. A number of procedures for obtaining consistent estimators have been proposed, and many of these are discussed in standard econometrics textbooks (see, e.g., Johnston [1972], Chapter 13). These procedures are generally justified on asymptotic grounds, and their behavior in finite samples is extremely difficult to derive. A number of exact results have been obtained for simple specific models (see, e.g., Richardson [1968], Sawa [1969], and Anderson and Sawa [1973]), and the various estimation methods have often been evaluated and compared by Monte Carlo simulations, a survey of which is contained in Chapter 13 of Johnston [1972].

Suppose, now, that the system of equations has been estimated, yielding estimates $\hat{\mathbf{B}}$ and $\hat{\mathbf{\Gamma}}$. $\mathbf{\Pi}$ is then estimated as $\hat{\mathbf{\Pi}} = \hat{\mathbf{B}}\hat{\mathbf{\Gamma}}^{-1}$. An alternative, of course, would be to estimate $\mathbf{\Pi}$ directly from (6.3.3), but this would ignore the restrictions imposed by economic theory on the structural equations (6.3.2) and hence, as Klein [1960] has argued, would be less efficient, provided the economic specification was correct. Forecasts are most easily calculated through the reduced form (6.3.3), so that if the x's are truly exogenous variables $\mathbf{y}'_{n+h}$ is forecast conditional on the exogenous variables by

$$\mathbf{f}'_{n,h} = \mathbf{x}'_{n+h}\hat{\mathbf{\Pi}} \qquad (6.3.4)$$

If lagged values of the endogenous variables are included in the model, then forecasts can be obtained recursively, beginning with $h = 1$, in the usual way, substituting forecasts of endogenous variables for unknown future values in (6.3.4). Econometricians would generally consider such an approach naive, feeling, as will be seen, that judgment ought to be incorporated into the mechanism generating forecasts since no model can hope to take account of all the factors (many of which may be nonquantifiable) affecting the variables to be forecast.

The above discussion of the multivariate forecasting problem is extremely formal and inadequate to describe what actually goes on in an econometric forecasting exercise. To illustrate the practical considerations involved, attention is restricted to macroeconomic forecasting models, noting that essentially similar methodology is employed in microeconomic forecasting—though the models considered are often smaller. (As an example, see Houthakker and Taylor [1966], where an attempt is made to forecast all items of U.S. private consumption six years ahead.)

A good account of the principles involved in constructing a macroeconomic forecasting model (on an annual basis) is given by Suits [1962], who first presents a grossly oversimplified formulation for purposes of illustration. The first step is an appeal to economic theory to postulate various interrelationships in an economy. For example, consumption might depend on disposable income, that is, on gross income less taxes. Investment can be taken as a function of income, but generally with a time lag. In this oversimplified formulation, one might think of taxes as being just a proportion of income. Finally, income is the sum of consumption, investment, and government expenditure. These considerations must then be transposed to algebraic form, and the parameters of the model estimated. Suppose that this simple model, when estimated and ignoring error terms, is of the form

$$Y_{1,t} = 20 + 0.7(Y_{4,t} - Y_{3,t}), \qquad Y_{2,t} = 2 + 0.1Y_{4,t-1}$$

$$Y_{3,t} = 0.2Y_{4,t}, \qquad Y_{4,t} = Y_{1,t} + Y_{2,t} + x_t \qquad\qquad (6.3.5)$$

where Y_1 is consumption, Y_2 is investment, Y_3 is taxes, Y_4 is income, and x_t is government expenditure. Solving these equations yields the reduced form

$$
\begin{bmatrix} Y_{1,t} \\ Y_{2,t} \\ Y_{3,t} \\ Y_{4,t} \end{bmatrix} =
\begin{bmatrix} 1 & 0 & 0.7 & -0.7 \\ 0 & 1 & 0 & 0 \\ 0 & 0 & 1 & -0.2 \\ -1 & -1 & 0 & 1 \end{bmatrix}^{-1}
\begin{bmatrix} 20 & 0 & 0 \\ 2 & 0.1 & 0 \\ 0 & 0 & 0 \\ 0 & 0 & 1 \end{bmatrix}
\begin{bmatrix} 1 \\ Y_{4,t-1} \\ x_t \end{bmatrix}
$$

$$(6.3.6)$$

Now, for example, if income in year n is $Y_{4,n} = 100$ and government expenditure in year $n + 1$ is projected to be $x_{n+1} = 20$, then substitution in (6.3.6) yields the forecasts 86.2 for $Y_{1,n+1}$, 12 for $Y_{2,n+1}$, 23.7 for $Y_{3,n+1}$, and 118.2 for $Y_{4,n+1}$ when t is set equal to $n + 1$ in that equation. For year $n + 2$, forecasts are obtained by setting $t = n + 2$ in (6.3.6), inserting a projected value for x_{n+2}, and replacing $Y_{4,n+1}$ on the right-hand side of the equation by its forecasted value of 118.2. In this fashion, forecasts can be made as far into the future as required, given only projections of future government expenditure and a faith in the model continuing to represent adequately the behavior of the system under study. The forecasts so obtained can be viewed in two ways. The forecasts for year $n + 1$ may be thought of as conditional forecasts of the endogenous variables given that the exogenous variable—government expenditure—will actually be $x_{n+1} = 20$. Alternatively, they can be viewed as unconditional forecasts on the assumption that $x_{n+1} = 20$ is a forecast of future government expenditure, which has been obtained in some way from considerations outside of the model. It is a feature of econometric models, insofar as they contain truly exogenous variables, that they are not in themselves sufficient for the production of unconditional forecasts. Estimates of future values of exogenous variables must first be made outside of the model. Of course, it might be argued (see, e.g., Klein [1971a]) that, in many applications, conditional forecasts are of far more value than unconditional. This may well be the case, but what is generally required in such situations are conditional forecasts of endogenous variables given *some* of the exogenous variables (those over which the policy maker can exert some control, for example). In this situation one is still left with the problem of forecasting the remaining exogenous variables.

Moreover, for conditional forecasting of this type the exogenous variables must be truly exogenous. However, many "exogenous" variables in macroeconomic models are instruments of government policy. It would seem to us to be reasonable to assert that government policy is frequently determined in response to prevailing economic conditions, and hence, when building models, should be taken as endogenous, thus requiring the addition of further equations to the system. In that case, as just noted, it would now no longer be possible to use the model to obtain forecasts conditional on the future values of such variables.

Now the set of equations (6.3.5), and the method of generating forecasts just described, constitute a gross oversimplification of macroeconometric forecasting methodology. In particular, four points require further elaboration:

(i)   The vast majority of models in current use have far more equations than does this simple system.

(ii)   In order to depict economic theory faithfully, models frequently are nonlinear in the endogenous variables.

(iii)   A good deal of effort is put into the specification of individual equations to secure inclusion of appropriate variables, and to some extent as regards determination of lag and error structure.

(iv)   Forecasts obtained from models are generally modified judgmentally. In fact, there is often a good deal of adjustment to the model involved in the forecast generating process, which is typically not purely mechanistic.

The simple model (6.3.5) can be expanded in two directions. First, further equations would be required to depict behavior in other sections of the economy—for example, trade, employment, and price level might be explained by the addition of further equations. Moreover, it is generally felt that disaggregation of the quantities explained would allow a more realistic representation and hence lead to more accurate forecasts. Thus, for example, the forecasting model of the U.S. economy presented by Suits [1962] includes four consumption equations, explaining separately demand for automobiles and parts, other durable goods, nondurable goods, and services. The first attempt at large scale macroeconometric model building was due to Tinbergen [1939], further impetus being given by the work of Klein and Goldberger [1955]. Development of the computer allowed the treatment of far more sophisticated models, which began to proliferate in the 1960s. It is now held by an eminent authority in the field (see Klein [1971a]) that a minimum of 50 equations are required to represent adequately the behavior of a national economy. Indeed, many current models are far bigger than this. For several years now, there have been models containing 200 or more equations (see, e.g., Fromm *et al.* [1972]), and recently even larger models have been reported. For example, Eckstein *et al.* [1974] describe a model of 698 equations. However, Klein [1971a,b] would regard these as mere stepping stones along the path of disaggregation, leading eventually to models of a thousand equations for a national economy. In case the mind of a time series analyst, reading of these heroic efforts for the first time, remains insufficiently boggled, it should be added that yet further complexity is envisaged through the linking of various national models in the production of an international model. Indeed, work along these lines has been in progress for some time (see Ball [1973] and Hickman [1975]).

It has already been seen that faithfulness to economic theories can be a hard taskmaster for the aspiring model builder. Further difficulties follow from the fact that the relationships postulated by theory are often nonlinear in the endogenous variables. The chief implication for forecasting is that it is no longer possible to obtain closed-form solutions like (6.3.6) for the reduced form equations. Instead, the structural form equations are generally solved numerically, using an iterative technique, such as the Gauss–Seidel method (see, for example, Green *et al.* [1972a]).

It is often emphasized by econometric model builders that the construction of appropriate model equations is by no means a simple one–off exercise. For example, attention must be paid to the historical record concerning the sample data to be analyzed. One would hardly expect constant coefficient models to be appropriate if the period of observation includes such exogenous shocks as wars or strikes not specifically accounted for in the model specification. Use of dummy variables over the relevant periods can be of some assistance in such circumstances. It is very often the case that, in addition to postulating a structural form, economic theory specifies the signs of many coefficients in the model. Furthermore, the econometrician will generally have strong feelings as to what magnitudes are or are not "reasonable" for many coefficients. Now, in principle there exists a well-defined framework, using Bayes' theorem, whereby such prior beliefs are modified at the model estimation stage in light of the given data. The final outcome would represent a view of probable coefficient values taking into account both prior belief and the evidence in the data. It must be admitted that, for large models, the computational effort involved in such a procedure would be formidable. Further, its validity rests on the assumption that the model chosen is the correct one. In practice, the procedure followed is far more informal. Standard estimation procedures are employed, and results that do not accord with prior expectations noted. These may well lead to a respecification of the model structure in some way. Thus, as Howrey *et al.* [1974] note, a good deal of experimentation is involved before a satisfactory fitted model is achieved. These authors, rather charmingly, refer to the whole process as the application of "tender loving care," a concept which, while difficult to define rigorously, plays, as will be seen in Chapter 8, an important role in controversies concerning the evaluation of econometric model forecasts. Another area in which experimentation in model building is much used is in the determination of an appropriate time lag structure in specific equations. A number of alternatives may be tried, an appeal to economic theory perhaps being employed to suggest fruitful areas for the search. In our experience, however, although economic theory frequently postulates the existence of time lags in relationships, it is rarely sufficiently specific to be of much help in determining their algebraic nature. A final, and to our minds vital, point concerns the specification of an appropriate time series structure for the errors from the individual equations of econometric models. We have occasionally heard it suggested that, even here, economic theory can be used to suggest an appropriate structure. However, the arguments advanced in favor of such an assertion have always been less than convincing, and accordingly we cannot accept it as realistic. It is very common to see in applied econometric work the tacit assumption that residuals are white noise, even when the reported Durbin–Watson statistic renders such an assumption improbable. Thus, all too often, serial correlation in equation residuals is ignored—at least in the model building stage of the analysis. We could quote many recent examples, but doubtless the reader familiar with applied econometric literature will be well aware of the problem. On occasions, attempts are made to allow for the presence of autocorrelated errors, but the assumption of first-order autoregression is generally substituted for that of white noise in such circumstances (see, e.g., Fair [1970]). Doubtless this will go some way toward alleviating the problems caused by autocorrelated errors. Nevertheless, if some other structure is appropriate, the model remains misspecified, and any conclusions drawn from it will be more or less invalid, depending on the extent of the misspecification. In Chapter 3 a methodology whereby, on the evidence of the data, an appropriate time series model is chosen from a general class was discussed. This technique is only very rarely employed in applied econometric work and, to the best of our knowledge, has never been used in the construction of large macroeconomic forecasting models, although Hendry [1974] does experiment with a number of time series error structures in a small model. Thus it must be assumed, a priori, that the error structure of such models is misspecified. In the next section the consequences that can arise from misspecification of this kind are illustrated.

The progression from completed model to derived forecast (whether conditional or unconditional) is typically not straightforward. Perhaps the most readable discussion of what generally takes place is given by Evans *et al.* [1972]. These authors distinguish three separate steps that are frequently taken, in addition to mechanical solution of the structural model. First, in order to forecast the endogenous variables, it is necessary, as already noted, to form an assessment of the likely future values of exogenous variables. Almost invariably, the procedures employed here are neither quantifiable nor rigorous. Typically, exogenous variables are forecasted judgmentally, the investigator relying on his knowledge and experience, together with any external indications, or advice from co-workers, which might be available. Second, before the model is solved, adjustments are made to individual equations (or some simple transformation of these equations). These adjustments are considered as either the setting of a nonzero value for the error from an equation or, equivalently, the adjustment of the constant term in the equation. The adjustments made can be either subjective or mechanistic, or some combination of the two. For example, it might be desired to incorporate into the forecast, information about probable future exogenous developments, such as strikes. The adjustment necessary in such circumstances is generally in the nature of a "best guess" since typically no rigorous quantitative framework is available. It is often the case that residuals from many equations in a model are autocorrelated, and attempts are often made to take account of this in calculating forecasts. Green *et al.* [1972a] suggest fitting, by least squares, to the errors u_t either a first-order autoregressive process $u_t = \rho u_{t-1} + \epsilon_t$ or a second-order process $u_t = \rho_1 u_{t-1} + \rho_2 u_{t-2} + \epsilon_t$. Residuals over the forecast period are set, not equal to zero, but according to one of the formulas

$$u_{n+h} = \hat{\rho}^h u_n, \qquad u_{n+h} = \hat{\rho}^h \left(\frac{u_n + \hat{\rho} u_{n-1}}{2} \right) \qquad (6.3.7)$$

or

$$u_{n+h} = \hat{\rho}_1 u_{n+h-1} + \hat{\rho}_2 u_{n+h-2}, \qquad h = 1, 2, \ldots$$

For reasons that we find neither entirely clear nor convincing, Green *et al.* tend to favor the second form in (6.3.7). It should be added that, even if the assumed form of serial correlation is correct, forecasts obtained through mechanical adjustment procedures such as this can be far from optimal if autocorrelation is not *also* taken into account when the model is estimated. However, at least for forecasting a short distance ahead, the adjustment procedure may well be a good deal better than ignoring autocorrelated errors altogether. In addition to the examples just described, adjustments may also be made to reflect structural change not accounted for in the model and also any data revisions. For the uninitiated, it should be noted that a good deal of macroeconomic data is subject to continual revision, even several years after the event. This problem could be worthy of a separate book, and most certainly is too large to treat here. However, two questions should at least be posed. Suppose one makes the heroic assumption that the final published figure is the "correct" one, and that successive estimates converge to this value. Suppose, now, that in building a model, the latest available data are used. It follows that a severe errors in observation problem may arise—the variance of the errors being higher for recent observations than for distant ones. As far as we know, no attempt has been made to solve the estimation problems posed by such a phenomenon. The second difficulty involves deciding what it is that one should try to forecast. Should it be the first available figure, on the grounds that it is this value to which the policy maker will react? On the other hand, since a model attempts to represent actual behavior in the economy is it not geared toward forecasting the "true value" or the "true" change in the quantity of interest over the forecast period? These questions would seem to merit further study.

At the final stage of the forecasting procedure, the modified model is solved to obtain predictions of the endogenous variables. However, it may well be the case that the forecasts so obtained do not accord with the econometrician's a priori concept of the likely range of future values. In such situations the forecasts may again be modified judgmentally, perhaps through a modification of the forecasts of exogenous variables or of the adjustments to individual structural equations employed in the previous stages. The model must then be re-solved, leading to a further set of forecasts. Not all forecasts derived from econometric models are subject to the high degree of judgmental modification just described. For example, in computing predictions from his model, Fair [1970, 1974] does no more than insert judgmental forecasts of future exogenous variables.

Although, in the presence of severe judgmental modification, it is virtually impossible to specify analytically the forecast generating mechanism, except perhaps in a Bayesian framework (which would probably lead to a prohibitively cumbersome scheme in practice), this should not be taken as an argument against the insertion of judgment into the forecasting procedure. On the contrary, the forecaster ought to employ *all* information available to him at the time the forecast is made, irrespective of whether or not such information lends itself to incorporation in a formal quantitative framework. The quality of the forecasts will then reflect both the quality of the information employed and the efficiency of its use. However, given the prevalence of judgmental adjustments in the derivation of forecasts from econometric models, a difficulty arises when one attempts to evaluate the quality of such work. Should one simply judge the end-products—the forecasts—on their merits? This is certainly worthwhile, but suppose that good forecasts are obtained primarily through the judgmental skill of the investigator. Perhaps he could have done as well, or even better, without the model. Again, it is possible for an adequate model to produce unsatisfactory forecasts if the investigator's judgment is seriously faulty. Thus it would be desirable also to evaluate objectively the worth of the model as a forecasting tool. After all, if reasonable forecasts are obtained without significant aid from the model, one might argue on grounds of efficiency for discarding it or, in the hopes of advancing knowledge or obtaining superior forecasts in the future, for radically modifying it. Discussion of the quality of econometric models and forecasts will be postponed until the methodology of forecast evaluation has been examined in Chapter 8.

Given any econometric model, it is possible in principle, given distributional assumptions, to obtain interval as well as point predictors. In practice, this can be a formidable task when the system contains nonlinearities in the endogenous variables. A good discussion of the problem of interval prediction is given by Klein [1971a], and will not be discussed further here, other than to note that when judgmental adjustment is employed in the forecasting process the usual probability statements relating to confidence intervals are invalidated.

Attempts have been made to improve the forecasting ability of econometric models by including in the structural equations quantities whose presence is justified, not on grounds of economic theory, but rather because changes in them might reasonably be expected to herald changes in the variables of interest. Such quantities are referred to as "anticipations variables." The econometric model relying most heavily on these variables is that of Fair [1970]. Occasionally attempts are made to take an existing model and incorporate into its structure anticipations variables in an effort to determine whether or not forecasting performance can be improved. For example, Adams and Duggal [1974] examine the use of an index of consumer sentiment, investment anticipations based on surveys of businessmen, and housing starts within the framework of the Wharton econometric model. Generally speaking, they conclude that use of these anticipations variables produces some improvement in forecasting accuracy.

6.4 Danger of Spurious Regressions in Econometric Models

The reader may well already have gathered that any skepticism we feel with regard to the worth of econometric models, as generally constructed, as forecasting tools is based not on doubts as to their ability to represent adequately the structure of economic relationships in a functional sense, but rather on their frequently cursory and invariably insufficiently general treatment of the specification of lag and error structure. This should hardly be surprising since, after all, forecasting is preeminently a time-oriented exercise. It is surely only natural, therefore, to expect that inadequate attention to time series concepts would lead to unnecessarily poor forecasts. Of course, misspecification of lag or error structure could occur in a simultaneous equation econometric model in any number of ways. In order to obtain some insight into the potential problems involved, however, it will be convenient to examine again the single equation model. There appears to be no reason to suppose that any difficulties that arise from misspecification here will not also be present in the context of a large simultaneous equation econometric model.

It is by no means uncommon to find in published applied econometric work equations of an apparently high degree of goodness of fit, as measured by the coefficient of multiple correlation R^2 (6.2.6) or the "corrected" coefficient $\bar{R}^2$ (6.2.7), but with extremely low values for the Durbin–Watson d statistic (6.2.17). (The reader who doubts this assertion is referred to the equations of the various models given in Hickman [1972] or Renton [1975].) Perhaps the most severe case of this kind to come our way recently concerns the "St. Louis Model," reported by Andersen and Carlson [1974]. This model contains only five behavioral equations, and in three of them R^2 is bigger than d. It has already been noted that one of the problems raised by serially correlated errors is that the usual tests of significance are invalid. In this section, the possibility of obtaining a regression equation relating economic time series exhibiting typical behavior, with an apparently high degree of fit, when in fact the independent variables have no explanatory power whatever, is examined.

As a first step, one might ask whether economic variables, taken as a class, exhibit typical time series behavior of any kind. The available evidence, based on both autocorrelation analysis (Reid [1969], Newbold and Granger [1974]) and spectral analysis (Granger [1966]), suggests that some generalization is possible. It appears that, while levels of economic time series are generally nonstationary, stationarity can frequently be achieved by first differencing. Thus, in the terminology of Chapter 3, ARIMA(p, d, q) models with $d = 1$ are frequently appropriate. The simplest model in this class is the random walk $X_t - X_{t-1} = \epsilon_t$ and, indeed, this model has been found to represent well price series in speculative markets, an observation that dates back to Bachelier [1900]. A more general formulation is provided by the first-order moving average process for changes:

$$X_t - X_{t-1} = \epsilon_t + b_1 \epsilon_{t-1} \qquad (6.4.1)$$

In our experience, the model (6.4.1)—possibly with the addition of a constant term—provides a very good representation of a wide range of economic time series. We certainly do not advocate the adoption of this model on all occasions, preferring rather to go through the model building procedure described in Chapter 3. However, if some simple specific model is to be assumed on a priori grounds, we feel that the first-order integrated moving average process is a serious candidate for economic time series in general, and would certainly expect it typically to provide a better representation than the first-order autoregressive process $X_t - a_1 X_{t-1} = \epsilon_t$, with a_1 constrained to be less than unity, commonly assumed by earlier writers.

Consider now the regression model

$$Y_t = \mathbf{x}_t' \boldsymbol{\beta} + u_t \qquad (6.4.2)$$

where a constant term is included in the regression, so that $x_{1t} = 1$ for all t. Suppose that the null hypothesis

$$H_0: \quad \beta_2 = \beta_3 = \cdots = \beta_K = 0 \qquad (6.4.3)$$

is true, so that Y_t does not depend (linearly) on the "explanatory" variables $X_2, X_3, \ldots, X_K$ at all. It is reasonable to ask whether, given the time series properties of the individual series, high values of R^2 are likely to obtain, leading to rejection of the null hypothesis (6.4.3) through the conventional test based on the F statistic, if ordinary linear regression methods are applied to (6.4.2) and the message of the Durbin–Watson statistic is unheeded. This question is approached, largely through simulation, by Granger and Newbold [1974]. First, it should be noted that the usual F statistic

$$F = \frac{R^2}{1 - R^2} \frac{n - K}{K - 1}$$

is only distributed as Fisher's F with $K - 1$ and $n - K$ degrees of freedom under the null hypothesis if the error series u_t is white noise. But, if the null hypothesis is true, then $u_t = Y_t - \beta_1$, in which case the time series structure of

u_t is the same as that of Y_t. Thus, if Y_t represents the level of an economic variable, in which case its time series structure will be very far from that of white noise, the conventional test statistic cannot follow its assumed distribution under the null hypothesis, and hence the associated test is invalid. That is to say, high values of R^2 may well occur even if the null hypothesis is true, and one would observe a spurious regression.

The simplest case in which a spurious regression can arise involving IMA(1, 1) series would be the regression of a random walk on an independent random walk. To get some notion of the magnitude of the problems involved, suppose Y_t and X_t are *independent* first-order autoregressive processes $Y_t = aY_{t-1} + \epsilon_t$ and $X_t = a^*X_{t-1} + \eta_t$. Then it is well known that the sample correlation R between X_t and Y_t has variance

$$\text{var}(R) = n^{-1}(1 + aa^*)/(1 - aa^*)$$

(see, e.g., Kendall [1954]). It is instructive to consider the probability distribution of R. Since the whole density must lie in the region $(-1, 1)$, it follows that the distribution cannot have a single mode at zero if its variance is greater than $\frac{1}{3}$, this being the variance of a rectangular (uniform) distribution on $(-1, 1)$. For $n = 20$ and $a = a^*$, $\text{var}(R)$ is greater than $\frac{1}{3}$ if $a > 0.86$, and if $a = 0.9$, $E(R^2) = 0.47$. Thus it is clear that high values of R^2 can arise even from relating independent stationary series.

In order to assess the consequences of relating independent integrated processes, using conventional linear regression methods, Granger and Newbold conducted two simulation experiments. First, the equation $Y_t = \beta_1 + \beta_2X_t + u_t$ was fitted to generated series of 50 observations, with Y and X independent random walks. The ratio of $\hat{\beta}_2$, in absolute value, to its estimated standard error

$$S = \frac{|\hat{\beta}_2|}{\text{S}\hat{\text{E}}(\hat{\beta}_2)}$$

is generally employed to test the null hypothesis $\beta_2 = 0$. In this expression, $\text{S}\hat{\text{E}}(\hat{\beta}_2)$ is the estimated standard error of the coefficient estimate obtained from the usual linear regression procedure and is, of course, inappropriate when the error series is not white noise. The frequency distribution of this quantity over 100 simulations is given in Table 6.1. Using the traditional t test

Table 6.1 *Regressing two independent random walks*

S:	0–1	1–2	2–3	3–4	4–5	5–6	6–7	7–8
Frequency:	13	10	11	13	18	8	8	5

S:	8–9	9–10	10–11	11–12	12–13	13–14	14–15	15–16
Frequency:	3	3	1	5	0	1	0	1

at the 5% level, the null hypothesis of no relation would be rejected (wrongly) on about three-fourths of all occasions. Note, further, that for more than one-half of the simulation runs the estimate of β_2 is more than four times its standard error, and is more than six times its standard error on over one-fourth of the runs. Thus an apparently high degree of fit can very often be achieved simply by regressing independent random walks.

A second, more comprehensive, simulation involved regressing a series Y on m independent series $X_j, j = 2, 3, \ldots, m + 1$, with m taking values from one to five. The series involved all followed the same time series models, which were taken to be:

(i)   random walks;
(ii) white noises—i.e., changes in random walks;
(iii) IMA(1, 1) processes;
(iv) MA(1) processes—i.e., first differences of IMA(1, 1).

All the series used in a given run were independent of one another. Typical series were generated as follows. Set W_0 equal to 100 and

$$W_t = W_{t-1} + \eta_t, \qquad t = 1, 2, \ldots, 50$$

where η_t is a white noise process. Set

$$W_t^* = W_t + k\epsilon_t, \qquad t = 1, 2, \ldots, 50, \quad k = 0 \text{ or } 1$$

where ϵ_t is white noise independent of η_t. A value $k = 0$ gives a random walk for W_t^*, and if $k = 1$, W_t^* is IMA(1, 1). The white noise processes used were normally distributed, with zero means and unit variances. Table 6.2 summarizes the results obtained over 100 simulations, the null hypothesis tested being that the set of independent variables contributes nothing toward explanation of variation in the dependent variable. The results in this table ought to occasion a good deal of alarm among the fraternity of applied econometricians prone to report equations with high $\bar{R}^2$'s and dubious values for the Durbin–Watson statistic. Indeed, it is not unreasonable to suggest that a good many economic hypotheses, hitherto regarded as "empirically veri-fied," might warrant reexamination on statistical grounds. The message from the table is very clear. When random walks or integrated moving average processes are involved, the chances of "discovering" a spurious relationship using conventional test procedures are very high indeed, increasing with the number m of independent variables included in the regression. Although the values for $\bar{R}^2$ are not quite as high, on average, when regressions involve IMA(1, 1) processes, the results here are if anything more disquieting than those for random walks since the average values for the associated Durbin–Watson statistics are a good deal higher and might thus be expected to occasion less alarm among the unwary. The main conclusion to be drawn from the results concerning regressions on levels in Table 6.2 is that it will be the rule rather than the exception to find spurious relationships and hence

Table 6.2 *Regressions of a series on m independent "explanatory" series*

	Percent times H_0 rejected[a]	Average Durbin–Watson d	Average $\bar{R}^2$	Percent $\bar{R}^2 > 0.7$
		Random walks		
Levels $m = 1$	76	0.32	0.26	5
$m = 2$	78	0.46	0.34	8
$m = 3$	93	0.55	0.46	25
$m = 4$	95	0.74	0.55	34
$m = 5$	96	0.88	0.59	37
Changes $m = 1$	8	2.00	0.004	0
$m = 2$	4	1.99	0.001	0
$m = 3$	2	1.91	−0.007	0
$m = 4$	10	2.01	0.006	0
$m = 5$	6	1.99	0.012	0
		IMA(1, 1)		
Levels $m = 1$	64	0.73	0.20	3
$m = 2$	81	0.96	0.30	7
$m = 3$	82	1.09	0.37	11
$m = 4$	90	1.14	0.44	9
$m = 5$	90	1.26	0.45	19
Changes $m = 1$	8	2.58	0.003	0
$m = 2$	12	2.57	0.01	0
$m = 3$	7	2.53	0.005	0
$m = 4$	9	2.53	0.025	0
$m = 5$	13	2.54	0.027	0

[a] Overall F test, based on $\bar{R}^2$, at 5% level.

that a high value of R^2 or $\bar{R}^2$ is an indication of nothing at all if the associated value of d is low, except that the model is in some way misspecified. We do not advocate first differencing as a universal panacea for all problems in econometric work. For example, in the IMA(1, 1) case, the null hypothesis is still rejected twice as often as it should be when first differences are employed since the assumed error structure is still not the correct one. The optimal strategy, if at all possible, remains the selection of an appropriate time series specification for the errors from the general autoregressive integrated moving average class. However, in small samples or complex models, this may not be feasible, and in the presence of severe autocorrelation of the errors in such situations, first differencing might be expected to go a long way toward alleviating the problem and is certainly preferable to doing nothing at all. As an example of the effect of such a transformation on the conclusions that might be drawn from a regression analysis, consider some results of Sheppard [1971], who regressed U.K. consumption on autonomous expenditure and midyear money stock for both levels and changes, using annual data

Table 6.3 *Regression of U.K. consumption on autonomous expenditure and midyear money stock; annual data 1947–1962*

	$\bar{R}^2$	d
Levels	0.99	0.59
Changes	−0.03	2.21

over the period 1947–1962. The results are shown in Table 6.3, from which it can be seen that an apparently highly significant relationship (ignoring the message of the d statistic) disappears entirely when first differences are employed.

What can be said then of the many large econometric models containing equations with high values for R^2 associated with suspiciously low values for d? Do they contain relationships that are, in fact, spurious? Such a question could be answered only by reestimating the models with appropriate error structures, but it must be added that the potential for finding such relations inherent in the mode of construction is very high indeed, and this potential is further increased by the experimentation involved in the model building process. One piece of evidence in this context concerns the emphasis placed by model builders on the need to adjust models when estimated coefficients, while significant statistically, have the "wrong" signs. We doubt whether this phenomenon would arise nearly so often if appropriate error structures were employed. It would, of course, be foolish to expect models with spurious equations to forecast successfully. Indeed, if spurious equations are present, we would expect conditional forecasts, the ability to derive which is often claimed as the great strength of econometric models, to be most seriously affected.

The results so far presented in this section serve to highlight a phenomenon that has long been recognized by econometricians, that autocorrelated errors, if ignored, can seriously invalidate the usual tests of significance. Perhaps the best treatment in the econometric literature is given in Chapter 13 of Malinvaud [1966]. Very often, in applied work dealing with single equations, some corrective action is taken when serially correlated errors are detected. However, this is rather more rare in large simultaneous equations model estimation—a notable exception is Fair [1970]. However, as indicated in Section 6.2, the classical approach typically is to test the null hypothesis of white noise residuals against the alternative of first-order autoregression, no other possible formulation being contemplated in the subsequent estimation. We find it very difficult to justify the first-order autoregressive model as being generally appropriate for regression equation error structure. Indeed it would be difficult to make out an a priori case for *any* time series structure as being always the "correct" one. However, some sort of a case can be made for considering the IMA(1, 1) process as at least a reasonable alternative in many situations. It has already been seen that, if the regression equation (6.4.2) involves the levels of economic time series, the integrated moving average formulation for the errors might reasonably be expected to be appropriate under the null hypothesis (6.4.3). Even if the null hypothesis is false, one might argue that the residual u_t is simply the sum of a (possibly very large) number of factors that for one reason or another have been omitted from the list of explanatory variables. If these factors are (or behave like) levels of economic time series, then u_t is the sum of integrated moving average processes of first order, and hence is itself IMA(1, 1), provided the individual processes are independent or have specific simple inter-relationships. In fact the argument can be taken further since the sum of independent IMA(1, 1) processes and white noise processes remains IMA(1, 1). It is thus reasonable to conclude that the first-order integrated moving average process deserves consideration as an alternative specification of error structure. (Similar arguments would suggest the possibility of the first-order moving average process when the regression involves first differences.) One is thus led to ask how much is lost, both in testing for autocorrelated errors and in making inference upon correcting for serial correlation, when the alternative of first-order autoregression is assumed and the correct specification is IMA(1, 1). Intuitively the answer would seem to be that this procedure ought to be better than doing nothing at all, but will remain suboptimal since the assumed error structure is still incorrect. Newbold and Davies [1975] attempt to assess the importance of the misspecification through a simulation study of the simplest case.

Newbold and Davies generated the series

$$X_t = X_{t-1} + \epsilon_t + b\epsilon_{t-1}, \qquad X_0 = 100, \quad t = 1, 2, \ldots, 50$$

$$Y_t = Y_{t-1} + \eta_t + b^*\eta_{t-1}, \qquad Y_0 = 100, \quad t = 1, 2, \ldots, 50$$

where ϵ_t and η_t are independent $N(0, 1)$ white noise series, so that the two IMA(1, 1) processes X_t and Y_t are independent of one another. The regression equation

$$Y_t = \beta_1 + \beta_2 X_t + u_t \tag{6.4.4}$$

was fitted by ordinary least squares to 1000 pairs of series for various values of b and b^*. Results for the usual t test of significance of $\hat{\beta}_2$ and the Durbin–Watson test are shown in Table 6.4. The main conclusion from this table is that employment of the decision procedure "reject the null hypothesis of no relationship between the two series only if $|t|$ differs significantly from zero and d does not differ significantly from two" will generally not lead one astray (although, of course, neglect of the second condition will do so). The exception is for moderately large values of $-b^*$ with $-b$ not too large. In such situations, the t statistic is on average fairly high, while on average d is not far below two. It can be seen that, given this combination of circumstances, the null hypothesis of no relationship would be rejected about 20% of the time.

The relatively hopeful picture presented thus far is, however, only part of the story, for it remains to consider reestimation of the regression equation under the assumption of first-order serial correlation of the residuals. Suppose the errors from an ordinary least-squares fit of equation (6.4.4) are $\hat{u}_t = Y_t - \hat{\beta}_1 - \hat{\beta}_2 X_t$. The autoregressive parameter is then estimated by applying ordinary least squares to $\hat{u}_t = a\hat{u}_{t-1} + \epsilon_t$. Denote the estimate as $\hat{a}$ and transform (6.4.4) to the form

$$Y_t^* = \beta_1^* + \beta_2 X_t^* + \epsilon_t \qquad (6.4.5)$$

where $Y_t^* = Y_t - \hat{a}Y_{t-1}$; $X_t^* = X_t - \hat{a}X_{t-1}$, and $\beta_1^* = \beta_1(1 - \hat{a})$. The results of applying ordinary least squares to (6.4.5) over the same sets of generated data as already employed are shown in Table 6.5. The picture here is not nearly so encouraging for users of the simple first-order autoregression correction. The usual decision rule tends to reject the null hypothesis of no relationship rather too often for comfort. For example, over a wide range of values of $-b^*$, when $-b$ is not overly large, a relationship between the two series will be "found" on 20–30% of all occasions. The conclusion, then, for econometric model builders, is that one can often be led astray if only the first-order autoregressive process is contemplated as an alternative to white noise residuals, and that accordingly it may well be profitable to consider a wider class of time series models in assessing the error structure of particular equations.

A more efficient estimation procedure is an iterative variant of the method just examined. Suppose that the estimates of β_1 and β_2 derived from (6.4.5) are denoted $\hat{\hat{\beta}}_1$ and $\hat{\hat{\beta}}_2$. One can then reestimate the residuals as

$$\hat{\hat{u}}_t = Y_t - \hat{\hat{\beta}}_1 - \hat{\hat{\beta}}_2 X_t$$

This, in turn, leads to a new estimate of a, which can be employed in (6.4.5) to reestimate β_1 and β_2. The iterative cycle is continued until the resulting estimates converge. Table 6.6 shows the percentage times that the resulting estimates were significant at the 5% level when applying an ordinary Student's t test to the equation obtained at the last stage of the iterative cycle. The conclusion from these results remains that, for a wide range of wholly reasonable parameter values, the null hypothesis is rejected far too frequently.

In going from consideration of univariate time series to contemplation of complex systems involving hundreds of equations, we have taken a heroic leap forward. The results presented in this section indicate that, from the time series analysis viewpoint, not all bases have been touched on the way. Accordingly, in the next chapter, we examine the possibility of building

Table 6.4 Percentage of times t and d statistics are significant at 5% level for regression of an $IMA(1, 1)$ series on an independent $IMA(1, 1)$ series

	b = 0.0		b = −0.2		b = −0.4		b = −0.6		b = −0.8	
	t		t		t		t		t	
	Not Significant	Significant	Not Significant	Significant	Not Significant	Significant	Not Significant	Significant	Not Significant	Significant
b* = 0.0										
d { Not Significant	0	0	0	0	0	0	0	0	0	0
d { Inconclusive	0	0	0	0	0	0	0	0	0	0
d { Significant	33.1	66.9	37.7	62.3	36.4	63.6	42.5	57.5	60.8	39.2
Mean d	0.33		0.36		0.38		0.42		0.36	
Mean |t|	4.07		3.70		3.50		2.98		1.87	
b* = −0.2										
d { Not Significant	0	0.1	0	0.1	0.1	0	0	0	0	0
d { Inconclusive	0	0.2	0	0	0	0.1	0	0.1	0	0.1
d { Significant	33.4	66.3	36.0	63.9	39.5	60.3	45.7	54.2	61.5	38.4
Mean d	0.45		0.46		0.50		0.51		0.46	
Mean |t|	3.70		3.65		3.41		2.81		1.90	

$b^* = -0.4$ $\Big\{ d \Big\{$ Not Significant	0.3	0.7	0.5	1.0	0.5	1.1	0.5	1.3	0.4	0.3
Inconclusive	0.5	0.6	0.2	0.8	0.1	0.6	0.4	1.6	0.2	0.2
Significant	35.5	62.4	37.0	60.5	39.3	58.4	47.0	50.2	60.9	38.0
Mean d	0.68		0.72		0.71		0.73		0.65	
Mean $|t|$	3.43		3.44		3.32		2.59		1.81	
$b^* = -0.6$ $\Big\{ d \Big\{$ Not Significant	5.0	6.3	5.1	8.3	5.1	7.6	5.0	5.5	6.6	2.4
Inconclusive	2.3	2.6	2.1	2.3	2.1	3.1	2.3	1.1	2.6	1.0
Significant	37.8	46.0	38.2	44.0	38.4	43.7	46.2	39.1	61.9	25.5
Mean d	1.10		1.09		1.12		1.08		1.03	
Mean $|t|$	2.89		2.72		2.78		2.27		1.52	
$b^* = -0.8$ $\Big\{ d \Big\{$ Not Significant	36.8	20.7	39.8	19.2	41.3	20.3	37.4	18.6	44.2	8.6
Inconclusive	6.2	3.2	5.8	3.9	5.7	2.7	5.3	2.6	8.0	1.8
Significant	20.4	12.7	18.2	13.1	19.1	10.9	25.9	10.2	30.4	7.0
Mean d	1.66		1.67		1.67		1.64		1.60	
Mean $|t|$	1.85		1.84		1.74		1.59		1.21	

Table 6.5 *Percentage of times t and d statistics are significant at 5% level for regression of an* IMA(1, 1) *series on an independent* IMA(1, 1) *series, "allowing" for first-order serial correlation in residuals by partial differencing*

| | | $b = 0.0$ | | $b = -0.2$ | | $b = -0.4$ | | $b = -0.6$ | | $b = -0.8$ | |
| | | t | | t | | t | | t | | t | |
		Not Significant	Significant	Not Significant	Significant	Not Significant	Significant	Not Significant	Significant	Not Significant	Significant
$b^* = 0.0$	d Not Significant	62.9	11.9	59.8	10.0	55.1	7.2	51.7	4.5	59.4	3.4
	d Inconclusive	5.8	2.1	6.9	2.8	8.1	1.4	6.3	1.1	9.0	0.5
	Significant	11.6	5.7	14.9	5.6	19.7	8.5	27.5	8.9	24.3	3.4
	Mean d	1.77		1.74		1.66		1.59		1.65	
	Mean $|t|$	1.33		1.25		1.22		1.13		0.86	
$b^* = -0.2$	d Not Significant	75.0	20.0	75.9	16.2	70.8	13.4	69.5	8.1	76.0	5.5
	d Inconclusive	1.4	0.4	1.6	0.5	2.7	1.6	3.8	1.0	3.6	0.3
	Significant	2.3	0.9	3.1	2.7	6.3	5.2	10.3	7.3	11.1	3.5
	Mean d	2.01		2.00		1.92		1.86		1.90	
	Mean $|t|$	1.35		1.38		1.37		1.18		0.95	

$b^* = -0.4$ Not Significant	72.0	27.2	72.6	25.3	70.9	24.1	77.0	13.9	81.8	9.5
d Inconclusive	0.2	0.4	0.7	0.6	0.7	0.8	1.6	1.8	2.2	0.6
Significant	0.1	0.1	0.7	0.1	1.6	1.9	3.5	2.2	3.6	2.3
Mean d	2.17		2.12		2.08		2.03		2.08	
Mean $|t|$	1.55		1.60		1.56		1.27		1.00	
$b^* = -0.6$ Not Significant	66.6	33.4	70.1	29.9	68.7	31.0	76.2	22.1	87.2	10.5
d Inconclusive	0	0	0	0	0	0	0.1	0.6	0.7	0.2
Significant	0	0	0	0	0.1	0.2	0.9	0.1	0.8	0.6
Mean d	2.14		2.12		2.09		2.07		2.10	
Mean $|t|$	1.80		1.67		1.73		1.40		1.01	
$b^* = -0.8$ Not Significant	72.0	28.0	71.4	28.5	72.0	27.9	75.2	24.8	86.0	13.4
d Inconclusive	0	0	0.1	0	0	0	0	0	0.4	0
Significant	0	0	0	0	0.1	0	0	0	0.2	0
Mean d	2.00		1.99		2.00		2.00		1.99	
Mean $|t|$	1.58		1.56		1.50		1.39		1.05	

Table 6.6 *Percentage times t statistic is significant at 5% level for regression of an IMA(1, 1) series on an independent IMA(1, 1) series, "allowing" for first-order serial correlation in residuals through an iterative estimation technique employing partial differencing*

$b^* \backslash b$	0.0	−0.2	−0.4	−0.6	−0.8
0.0	9.6	8.2	5.2	4.3	4.9
−0.2	15.6	10.9	10.2	7.7	6.8
−0.4	21.1	19.8	16.2	9.9	7.7
−0.6	31.3	28.7	23.4	18.7	9.5
−0.8	28.6	28.2	26.8	21.2	11.3

models, describing the interrelationship of variables, employing consideration of their time series properties. It will not be possible, through such methods, to construct large systems of equations such as are currently in vogue in macroeconomic forecasting. However, the methodology derived ought to be of considerable assistance in that context.

CHAPTER SEVEN

MULTIPLE SERIES MODELING AND FORECASTING

Sagittarius   (November 22nd–December 22nd)
Precautions should be taken against running into unforeseen occurrences or events.

ASTROLOGER, NEWS CHRONICLE

7.1 Introduction

The first five chapters of this book were concerned with model building and forecasting for a single time series, without use of information provided by other series. In this chapter, a methodology for constructing multiple time series models is developed and illustrated, and the derivation of forecasts from such models considered. In fact, only the bivariate model building problem will be discussed in any great practical detail, though extensions to the more general case will be indicated.

An alternative methodology, whereby forecasts are calculated on the basis of a very wide information set, is the classical econometric approach, discussed in the previous chapter. The procedures employed here differ in two respects. First, the form of the model will be determined by the data alone, economic theory being employed only to suggest what variables might be relevant. Second, just as in the single series case the objective is to find a model that transforms a given series to white noise, the aim here is to build a model that transforms a vector of time series to a white noise vector.

7.2 Theoretical Models for Multiple Time Series

Before examining the practical model building problem, it is necessary to develop some multiple time series theory. Let $\mathbf{X}_t$ be a vector time series, with $\mathbf{X}'_t = (X_{1,t}, X_{2,t}, \ldots, X_{m,t})$, where each individual series is stationary and purely nondeterministic. The set of m series is called jointly covariance stationary if

$$\lambda_\tau^{(i,j)} = \mathrm{cov}(X_{i,t}, X_{j,t-\tau})$$

is independent of t for all i, j, and τ, and $\lambda_0^{(i,i)} < \infty$ for all i. The multivariate generalization of Wold's decomposition theorem then states that $\mathbf{X}_t$ can always be represented by

$$\mathbf{X}_t = \mathbf{c}(B)\boldsymbol{\eta}_t \tag{7.2.1}$$

where $\mathbf{c}(B)$ is an $m \times m$ matrix in the back-shift operator B, with typical element

$$c_{ij}(B) = \sum_{k=0}^{\infty} c_{ij,k} B^k$$

and $\boldsymbol{\eta}_t$ is a vector white noise process, with $\boldsymbol{\eta}_t' = (\eta_{1,t}, \eta_{2,t}, \ldots, \eta_{m,t})$ having the properties

$$E[\eta_{i,t}] = 0, \quad \text{all } i; \qquad E[\eta_{i,t}\eta_{j,t-\tau}] = 0, \quad \text{all } \tau, \quad i \neq j$$
$$= 0, \quad \tau \neq 0, \quad i = j$$

A proof of this theorem is given in Hannan [1970].

The typical equation in (7.2.1) is

$$X_{i,t} = \sum_k c_{i1,k}\eta_{1,t-k} + \sum_k c_{i2,k}\eta_{2,t-k} + \cdots + \sum_k c_{im,k}\eta_{m,t-k}$$

so that any $X_{i,t}$ is a weighted sum of current and past values of each of m uncorrelated white noise series. The advantage of the matrix notation is obvious.

If the matrix $\mathbf{c}(B)$ can in some way be approximated by the product of two matrices, $\mathbf{a}^{-1}(B)$ and $\mathbf{b}(B)$, each involving only finite-order polynomials in B, one is led to consider the class of linear models

$$\mathbf{a}(B)\mathbf{X}_t = \mathbf{b}(B)\boldsymbol{\eta}_t \tag{7.2.2}$$

where typical elements of $\mathbf{a}(B)$ and $\mathbf{b}(B)$ are

$$a_{ij}(B) = \sum_{k=0}^{p_{ij}} a_{ij,k} B^k \qquad \text{and} \qquad b_{ij}(B) = \sum_{k=0}^{q_{ij}} b_{ij,k} B^k$$

Equation (7.2.2) represents a multivariate mixed autoregressive moving average process, denoted ARMA(p, q) where $\mathbf{p} = \{p_{ij}\}$ and $\mathbf{q} = \{q_{ij}\}$ are $m \times m$ matrices. If $\mathbf{b}(B) = \mathbf{b}_0$, a matrix of degree zero in B, then the process is multivariate autoregressive; and if $\mathbf{a}(B) = \mathbf{a}_0$, the model is multivariate moving average. The process is stationary, in the sense that the model has no explosive solutions, if the roots of $|\mathbf{a}(z)| = 0$ all lie outside the unit circle, and is said to be invertible if the roots of $|\mathbf{b}(z)| = 0$ lie outside the unit circle. It will be assumed here and throughout this chapter that both conditions hold (possibly after the given series have been suitably differenced). It will also be assumed that $\mathbf{a}(z)$ is of full rank for every z in or on the unit circle, so that $\mathbf{a}(B)$ can be taken to possess an inverse. A similar assumption is made for $\mathbf{b}(z)$.

If a model of the form (7.2.2) is an adequate approximation to reality, with the largest elements of $\mathbf{p}$ and $\mathbf{q}$ of reasonable size, it seems plausible that such models could be identified and estimated, given sufficient data. Practical problems of this kind will be discussed in later sections of this chapter, but for the present it is assumed that data are generated by some multivariate ARMA model, and some consequences of this assumption will be explored.

To begin, a number of alternative representations are considered. It is straightforward to verify that an equivalent representation to (7.2.2) is the transfer function–noise form

$$X_{i,\,t} = \sum_{j \,\neq\, i} \frac{\omega_{ij}^*(B)}{\delta_{ij}^*(B)} X_{j,\,t} + \frac{\theta_i^*(B)}{\phi_i^*(B)} \eta_{i,\,t}, \qquad i = 1, 2, \ldots, m \qquad (7.2.3)$$

where all polynomials in B are of finite order.

For example, equations of the form (7.2.3) can be derived from (7.2.2) by considering the inverse of $\mathbf{b}(B)$, $\mathbf{b}^{-1}(B) = |\mathbf{b}(B)|^{-1}\mathbf{b}^*(B)$ where $\mathbf{b}^*(B)$ denotes the adjoint matrix (the transpose of the matrix of cofactors). Hence (7.2.2) can be written

$$\mathbf{b}^*(B)\mathbf{a}(B)\mathbf{X}_t = |\mathbf{b}(B)|\boldsymbol{\eta}_t$$

from which equations of the type (7.2.3) immediately follow. Models of this form will be fitted to actual data in later sections of this chapter.

Since, in (7.2.2), $\mathbf{a}(B)$ is taken to possess an inverse, the corresponding MA(∞) form is (7.2.1), with $\mathbf{c}(B) = \mathbf{a}^{-1}(B)\mathbf{b}(B)$. As in the single series case, this form is particularly useful in forecasting.

A further form of the model, of some interest, follows by noting that

$$\mathbf{a}^{-1}(B) = \mathbf{a}^*(B)/|\mathbf{a}(B)|$$

where $\mathbf{a}^*(B)$ is the adjoint matrix of $\mathbf{a}(B)$. It follows that (7.2.2) can be written

$$|\mathbf{a}(B)|\mathbf{X}_t = \mathbf{a}^*(B)\mathbf{b}(B)\boldsymbol{\eta}_t$$

and the ith equation is simply

$$|\mathbf{a}(B)|X_{i,\,t} = \boldsymbol{\alpha}_i'(B)\boldsymbol{\eta}_t \qquad (7.2.4)$$

where $\boldsymbol{\alpha}_i'(B)$ is the ith row of $\mathbf{a}^*(B)\mathbf{b}(B)$. Thus each $X_{i,\,t}$ is now explained in terms of its own past plus an error series in moving average form, although the error series are interrelated from one explanatory equation to another. Now, it appears from (7.2.4) that the autoregressive parts of each of the ARMA single series models, appropriate to the $X_{i,\,t}$, are identical. In fact this is illusory since $|\mathbf{a}(B)|$ will typically have common factors, or near common factors, with the moving average operator that follows from the right-hand side of (7.2.4). However, this equation does provide a further illustration of why the single series mixed ARMA model might be expected to occur frequently.

In certain circumstances, often assumed by econometricians to arise, the variables $\mathbf{X}_t$ may be partitioned into two distinct groups. Consider the model (7.2.2), written as

$$
\begin{bmatrix}
\mathbf{a}^{11}(B) & \mathbf{a}^{12}(B) \\
\mathbf{a}^{21}(B) & \mathbf{a}^{22}(B)
\end{bmatrix}
\begin{bmatrix}
\mathbf{X}_t^{(1)} \\
\mathbf{X}_t^{(2)}
\end{bmatrix}
=
\begin{bmatrix}
\mathbf{b}^{11}(B) & \mathbf{b}^{12}(B) \\
\mathbf{b}^{21}(B) & \mathbf{b}^{22}(B)
\end{bmatrix}
\begin{bmatrix}
\boldsymbol{\eta}_t^{(1)} \\
\boldsymbol{\eta}_t^{(2)}
\end{bmatrix}
$$

where $\mathbf{X}_t' = (\mathbf{X}_t^{(1)\prime}, \mathbf{X}_t^{(2)\prime})$, $\boldsymbol{\eta}_t' = (\boldsymbol{\eta}_t^{(1)\prime}, \boldsymbol{\eta}_t^{(2)\prime})$. If, in this equation, $\mathbf{a}^{21}(B)$, $\mathbf{b}^{12}(B)$, and $\mathbf{b}^{21}(B)$ are all identically zero, the equations decompose into two blocks

$$
\mathbf{a}^{11}(B)\mathbf{X}_t^{(1)} + \mathbf{a}^{12}(B)\mathbf{X}_t^{(2)} = \mathbf{b}^{11}(B)\boldsymbol{\eta}_t^{(1)} \tag{7.2.5}
$$

and

$$
\mathbf{a}^{22}(B)\mathbf{X}_t^{(2)} = \mathbf{b}^{22}(B)\boldsymbol{\eta}_t^{(2)} \tag{7.2.6}
$$

where $\mathbf{X}_t^{(1)}$ and $\boldsymbol{\eta}_t^{(1)}$ are vectors of size m_1; $\mathbf{X}_t^{(2)}$ and $\boldsymbol{\eta}_t^{(2)}$ are vectors of size m_2; and $m_1 + m_2 = m$. In fact the decomposition arises even if $\mathbf{b}^{12}(B) \neq 0$, for then the additional term $\mathbf{b}^{12}(B)\boldsymbol{\eta}_t^{(2)}$ in (7.2.5) is equal, by (7.2.6), to

$$
\frac{\mathbf{b}^{12}(B)(\mathbf{b}^{22})^*(B)\mathbf{a}^{22}(B)\mathbf{X}_t^{(2)}}{|\mathbf{b}^{22}(B)|}
$$

where $(\mathbf{b}^{22})^*(B)$ is the adjoint matrix. Substituting for $\mathbf{b}^{12}(B)\boldsymbol{\eta}_t^{(2)}$ in the augmented (7.2.5), multiplying through by $|\mathbf{b}^{22}(B)|$ and collecting terms yields an equation in the same form as (7.2.5) above. For such a decomposition to occur, it is necessary that the components of $\mathbf{X}_t^{(2)}$ cause those of $\mathbf{X}_t^{(1)}$, but not vice versa. The concept of causality is discussed at some length in Section 7.4, and so will not be defined here. The variables $\mathbf{X}_t^{(1)}$ are then termed endogenous and $\mathbf{X}_t^{(2)}$ exogenous, corresponding to the definitions introduced in Section 6.3. The set of equations (7.2.5) then constitutes the structural equations of the model, while (7.2.6) determines the quantities $\mathbf{X}_t^{(2)}$ exclusively in terms of previous values of this vector. The structural form is frequently employed by econometricians, economic theory being employed to suggest some specific values in $\mathbf{a}^{11}(B)$ and $\mathbf{a}^{12}(B)$ that may well be useful both in identifying model structure and ensuring that the achieved model is estimable, a problem that will be discussed in the following section. The use of structural equations for forecasting purposes was discussed in Section 6.3. However, their primary use is in the analysis of the structure of an economic system.

If $\mathbf{a}^{11}(z)$ is of full rank for all $|z| \leqslant 1$, then (7.2.5) can be transformed to

$$
\mathbf{X}_t^{(1)} = -(\mathbf{a}^{11}(0))^{-1}(\mathbf{a}^{11}(B) - \mathbf{a}^{11}(0))\mathbf{X}_t^{(1)} - (\mathbf{a}^{11}(0))^{-1}\mathbf{a}^{12}(B)\mathbf{X}_t^{(2)}
$$
$$
+ (\mathbf{a}^{11}(0))^{-1}\mathbf{b}^{11}(B)\boldsymbol{\eta}_t^{(1)}
$$

so that each member of $\mathbf{X}_t^{(1)}$ is expressed as a linear sum of past $\mathbf{X}_t^{(1)}$ values, current and past $\mathbf{X}_t^{(2)}$ values, and a moving average error term. This corresponds to the reduced form of Section 6.3, and is clearly useful for forecasting $\mathbf{X}_t^{(1)}$ conditioned on $\mathbf{X}_t^{(2)}$ being given.

One further form is obtained by multiplying through (7.2.5) by $(\mathbf{a}^{11}(B))^{-1}$, so that each endogenous variable is expressed as the linear combination of current and past values of the exogenous variables and an ARMA error term. The resulting system of transfer function–noise equations is called the final form of the model. The uses of the various forms of the basic model (7.2.2) have been discussed by Zellner and Palm [1974]. Although theoretically equivalent, some forms are easier to estimate, some convenient for the introduction of prior information, and some for forecasting. Various situations may call for one form rather than another, as will be seen in later sections.

One final point concerning structural forms should be added. Econometricians will typically analyze a system of equations like (7.2.5), ignoring (7.2.6). However, it follows from the derivation, and is otherwise obvious since in general m equations are required to determine m variables, that such a system is incomplete, requiring addition of the mechanism by which the exogenous variables are generated. This is the reason that econometric models can typically be used directly to calculate only conditional rather than unconditional forecasts. In the computation of unconditional forecasts, the usual econometric approach is to first forecast the exogenous variables judgmentally, rather than through a system of equations like (7.2.6).

7.3 Identification

In the past few years, the term "identification" has been used by time series analysts and econometricians in rather different ways. Since we would contend that econometricians and time series analysts, working with economic time series data, should be identical groups, this problem of terminology is particularly unfortunate. To the time series analyst, the identification problem is to find, based on the evidence of the data, appropriate orders for the lags used in his model. In the multivariate context, this involves the determination of integer values for the elements in the matrices $\mathbf{p}$ and $\mathbf{q}$ of the ARMA($\mathbf{p}$, $\mathbf{q}$) model (7.2.2). In this section only, this will be denoted as TS-identification. The econometrician might regard the TS-identification problem as part of the overall problem of model specification. The econometric identification problem, denoted E-identification, relates to finding a *unique* model, of a structural form, compatible with the joint density function of the given data. Econometricians typically solve this problem by imposing constraints, suggested by economic theory, on the parameters of the model. Time series analysts have, on occasion, referred to E-identification as the problem of "model multiplicity."

It is straightforward to see that an E-identification problem exists in multiple time series analysis. Denote

$$\lambda_\tau^{(ij)} = \text{cov}(X_{i,t}, X_{j,t-\tau}), \qquad \Lambda_\tau = \{\lambda_\tau^{(ij)}\} = E[\mathbf{X}_t \mathbf{X}'_{t-\tau}]$$

and

$$\Lambda(z) = \sum_{\tau=-\infty}^{\infty} \Lambda_\tau z^\tau \tag{7.3.1}$$

Thus $\Lambda(z)$ is the lag covariance generating function and, putting $z = e^{-i\omega}$, it will also be proportional to the spectral/cross-spectral matrix of the component series of $\mathbf{X}_t$. If $\mathbf{X}_t$ is generated by the multivariate ARMA model (7.2.2) with

$$E[\boldsymbol{\eta}_t \boldsymbol{\eta}'_s] = \mathbf{0}, \qquad t \neq s$$

$$= \boldsymbol{\Sigma}, \qquad t = s$$

then by consideration of multivariate filters, generalizing the discussion of Section 1.9, it is found that

$$\Lambda(z) = \mathbf{c}(z)\boldsymbol{\Sigma}\mathbf{c}^*(z) \tag{7.3.2}$$

where $\mathbf{c}(z) = \mathbf{a}^{-1}(z)\mathbf{b}(z)$ and $\mathbf{c}^*(z) = \mathbf{c}'(\bar{z})$ i.e., the transposed matrix with conjugation. Since $\boldsymbol{\Sigma}$ is necessarily a positive definite matrix, it can always be decomposed as $\boldsymbol{\Sigma} = \boldsymbol{\Gamma}\boldsymbol{\Gamma}'$ and (7.3.2) may be written

$$\Lambda(z) = \mathbf{C}(z)\mathbf{C}^*(z) \tag{7.3.3}$$

where $\mathbf{C}(z) = \mathbf{c}(z)\boldsymbol{\Gamma}$.

Given data, it is reasonable to assume that an estimate of $\Lambda(z)$ of (7.3.1) is available, and any model used will be based on this estimate. However, clearly if $\mathbf{C}(z)$ satisfies (7.3.3), then so will $\mathbf{C}(z)\boldsymbol{\alpha}(z)\mathbf{J}\boldsymbol{\beta}(z)$ where $\boldsymbol{\alpha}(z)$ and $\boldsymbol{\beta}(z)$ are diagonal matrices whose elements, respectively, are $\alpha_j(z)/\alpha_j(z^{-1})$ and $\beta_i(z^{-1})/\beta_i(z)$ and $\mathbf{J}$ is any matrix satisfying $\mathbf{J}\mathbf{J}' = \mathbf{I}$. Even if $\mathbf{C}(z)$ were assumed to be known, there would remain the problem of finding matrices $\mathbf{a}(z)$, $\mathbf{b}(z)$, and $\boldsymbol{\Gamma}$ such that $\mathbf{C}(z) = \mathbf{a}^{-1}(z)\mathbf{b}(z)\boldsymbol{\Gamma}$ and clearly there is no unique way of doing this, without imposing restrictions on the model.

If one does not have a unique model, and it is our contention that this is the usual situation, is this a source of real concern? In some ways, the extra freedom one has in formulating the model is an advantage since in certain circumstances the model is better phrased in one form, while in other cases a different form might be preferable. However, the nonuniqueness of the models can be worrying when carrying out TS-identification and estimation. The problem can be partially resolved by adding necessary constraints to the model, and completely resolved if sufficient constraints are used.

The E-identification problem will first be considered for a system of structural equations, as defined in (7.2.5), but now denoted

$$\mathbf{a}(B)\mathbf{X}_t^{(1)} + \mathbf{d}(B)\mathbf{X}_t^{(2)} = \mathbf{b}(B)\boldsymbol{\eta}_t \tag{7.3.4}$$

where $\mathbf{a}(B)$ and $\mathbf{b}(B)$ are $m_1 \times m_1$ matrices and $\mathbf{d}(B)$ is an $m_1 \times m_2$ matrix. All elements in these matrices are polynomials in B, with maximum orders p,

q, and r, respectively. From a theoretical viewpoint, Hannan [1971] has specified necessary and sufficient conditions that the equations in (7.3.4) are uniquely determined from given autocovariance sequences for each $X_{j,t}^{(1)}$ and $X_{j,t}^{(2)}$ and the cross-covariance sequences between every pair of series. These conditions may be written as follows:

(i) either $\mathbf{a}(0) = \mathbf{b}(0)$ or $\boldsymbol{\Sigma} = \mathbf{I}$, where $\mathbf{I}$ is the identity matrix of order m_1 and $\boldsymbol{\Sigma}$ is the variance–covariance matrix of $\boldsymbol{\eta}_t$ in (7.3.4).

(ii) The roots of $|\mathbf{a}(z)| = 0$ and $|\mathbf{b}(z)| = 0$, where $|\mathbf{a}|$ is the determinant of $\mathbf{a}$, all lie outside the unit circle. These are the generalized stationarity and invertibility conditions for the model.

(iii) The rank of $[\mathbf{a}(z) : \mathbf{d}(z) : \mathbf{b}(z)]$ is equal to m_1, for all z.

(iv) The rank of $[\mathbf{a}_p : \mathbf{d}_r : \mathbf{b}_q]$ is equal to m_1, where, for example, $\mathbf{a}_p$ is defined from

$$\mathbf{a}(z) = \sum_{j=0}^{p} \mathbf{a}_j z^j$$

(v) Consider the composite matrix

$$\mathbf{V} = [\mathbf{a}_0 : \mathbf{a}_1 : \cdots : \mathbf{a}_p : \mathbf{d}_0 : \mathbf{d}_1 : \cdots : \mathbf{d}_r : \mathbf{b}_0 : \mathbf{b}_1 : \cdots : \mathbf{b}_q]$$

and define $\mathbf{V}_i$ to be the submatrix consisting of all columns of $\mathbf{V}$ with zeros in the ith row. Then, if there are at least $m_1 - 1$ zeros in each row of $\mathbf{V}$, a sufficient condition is that the rank of $\mathbf{V}_i$ is $m_1 - 1$ for all $i = 1, \ldots, m_1$.

Clearly, condition (v) is an extension of the classical condition for E-identification in econometric models described in Section 6.3. It should be added that, as in any regression equation, a normalization condition is also required, so that the coefficient on $X_{j,t}^{(1)}$ is unity in the jth equation, and then $\mathbf{a}(0)$ has unit values along its diagonal. These conditions are further explained and interpreted by Preston and Wall [1973].

Clearly the checking of these conditions would be frequently very complicated in practice. Moreover, it would be necessary to perform a successful TS-identification before E-identification is attempted, as the values of p, q, and r are taken to be known, and it is important in E-identification that these values be correctly specified.

The reason for considering E-identification conditions in terms of the structural form is that econometricians find it easier to impose the restrictions on the model implied by economic theory on this form, rather than the reduced or final forms. Theory tells one that certain coefficients in the model will take the value zero, hence hopefully leading to the satisfaction of condition (v). In fact, more general restrictions on the coefficients of an equation could be handled, but these will not be discussed here.

Rather simpler conditions arise if a more simple attitude is taken to the form of the restrictions that economic theory is likely to provide. Hatanaka [1974b] considers the situation in which the only a priori restrictions available for identification are in the form of whether or not a certain variable is excluded from an equation, and that $\mathbf{a}(B)$ and $\mathbf{d}(B)$ are the only parameters to be E-identified. No assumption is made about the orders of the polynomials comprising $\mathbf{a}(B)$ or $\mathbf{d}(B)$, and nothing is assumed about the structure of the right-hand side of (7.3.4) apart from its stationarity. (In particular, the component polynomials of $\mathbf{b}(B)$ could be of infinite order.) A variable is said to be excluded from an equation if and only if it enters neither in the current nor in the lagged values. Expressing the problem in this way suppresses the time series aspect and leads to the classical E-identification order and rank conditions:

Order condition: In a model of m_1 linear equations, to be E-identified an equation must exclude at least $m_1 - 1$ of the m variables appearing in the model.

Rank condition: In a linear model of m_1 equations, an equation is E-identified if and only if the matrix of coefficients with which those variables, excluded from the equation in question, appear in other equations, has rank $m_1 - 1$.

In these conditions, the "variables" are $X_{j,t}^{(1)}$ and $X_{j,t}^{(2)}$ and the "coefficients" are the components of $\mathbf{a}(B)$ and $\mathbf{d}(B)$, such as $a_{ij}(B)$, and so are functions of the operator B. The rank condition is checked by replacing B by z and is required to hold for all $|z| \leqslant 1$. The order condition is necessary, and the rank condition both necessary and sufficient. Framing the identification problem in this way allows the econometrician to specify, from theory, enough restrictions for the structural form to be identified, but leaves open the time series identification question.

Hatanaka's solution to E-identification depends entirely on the availability of a sufficient number of exogenous variables. However, if one is forced to analyze data without any assumption about some variables being exogenous or some coefficients zero, then one is faced with the fact that there may not be a unique model producing the set of auto- and cross covariances derived from the data.

In a special case, a necessary and sufficient condition for the existence of a unique model does exist. Consider the model (7.2.2), without exogenous variables, where

$$\mathbf{a}(B) = \sum_{j=0}^{p} \mathbf{a}_j B^j, \qquad \mathbf{b}(B) = \sum_{j=0}^{q} \mathbf{b}_j B^j$$

with

$$E[\boldsymbol{\eta}_t \boldsymbol{\eta}_s] = 0, \qquad t \neq s$$

$$= \Sigma, \qquad t = s$$

Then, if $\mathbf{a}_0 = \mathbf{b}_0 = \mathbf{I}$, and if the only vector λ such that $\lambda'(\mathbf{a}_p - \mathbf{b}_q) = \mathbf{0}$ is $\lambda = \mathbf{0}$, then the model is uniquely determined, or E-identified. This result has been proved by Hannan [1969]. The last condition necessarily holds if either $\mathbf{a}_p$ or

$\mathbf{b}_q$ is a nonsingular matrix. It should be noted, however, that a correct TS-identification is required before this condition can be checked.

From the construction of the model it is obvious that the condition $\mathbf{b}_0 = \mathbf{I}$ can be replaced by the condition $\mathbf{\Sigma} = \mathbf{I}$. The condition $\mathbf{a}_0 = \mathbf{I}$ gives a very convenient model for forecasting purposes, but it should be noted that it cannot generally be coupled with the conditions $\mathbf{b}_0 = \mathbf{I}$ *and* $\mathbf{\Sigma}$ a diagonal matrix. If all three of these conditions held, one would have a model without any instantaneous causality, but particularly in economics, when using data observed at intervals of time that are not very short, it is necessary to have either $\mathbf{a}_0 \neq \mathbf{I}$ or $\mathbf{\Sigma}$ not diagonal. Since $\mathbf{\Sigma}$ is a positive definite matrix, there exists a nonsingular triangular matrix $\mathbf{P}$, with ones along the main diagonal, such that $\mathbf{P}\mathbf{\Sigma}\mathbf{P}'$ is diagonal. So the general model can have either $\mathbf{a}_0 = \mathbf{I}$, with unrestricted $\mathbf{\Sigma}$, or $\mathbf{a}_0 = \mathbf{P}$ with $\mathbf{\Sigma}$ diagonal. When $\mathbf{a}_0$ is triangular, the model is called recursive, and is still in a useful form for forecasting purposes. Clearly, there is no unique recursive model, but it is often convenient when considering model theory and estimation to have error series $\eta_{j,\,t}$ that are mutually stochastically uncorrelated. The recursive form of the transfer function–noise model (7.2.3) is obtained by setting $\omega_{ij}^*(0) = 0$ for all $j > i$. The use of recursive models in econometrics has been forcefully argued over a number of years by Wold [1954b, 1964].

To illustrate the various forms, consider the following three bivariate models:

(A) $X_{1,\,t} = 0.7X_{1,\,t-1} + 0.3X_{2,\,t-1} + \eta_{1,\,t} + 0.5\eta_{2,\,t-1}$

$X_{2,\,t} = 0.4X_{1,\,t-1} + 0.4X_{2,\,t-1} + \eta_{2,\,t} + 0.6\eta_{2,\,t-1}$

where $\boldsymbol{\eta}_t$ is a white noise vector, with

$$\mathbf{\Sigma}_\eta = \begin{bmatrix} 1.09 & 0.7 \\ 0.7 & 1.16 \end{bmatrix}$$

(B) $X_{1,\,t} = 0.7X_{1,\,t-1} + 0.3X_{2,\,t-1} + e_{1,\,t} + 0.3e_{2,\,t} + 0.2e_{1,\,t-1} + 0.5e_{2,\,t-1}$

$X_{2,\,t} = 0.4X_{1,\,t-1} + 0.4X_{2,\,t-1} + e_{2,\,t} + 0.4e_{1,\,t} + 0.24e_{1,\,t-1} + 0.6e_{2,\,t-1}$

where $\mathbf{e}_t$ is a white noise vector, with

$$\mathbf{\Sigma}_e = \begin{bmatrix} 1 & 0 \\ 0 & 1 \end{bmatrix}$$

(C) $X_{1,\,t} = 0.7X_{1,\,t-1} + 0.3X_{2,\,t-1} + u_{1,\,t} + 0.321u_{1,\,t-1} + 0.5u_{2,\,t-1}$

$X_{2,\,t} = 0.642X_{1,\,t} + 0.207X_{2,\,t-1} - 0.049X_{1,\,t-1}$

$\qquad + u_{2,\,t} + 0.279u_{2,\,t-1} + 0.179u_{1,\,t-1}$

where $\mathbf{u}_t$ is a white noise vector, with

$$\Sigma_{\mathbf{u}} = \begin{bmatrix} 1.09 & 0 \\ 0 & 0.71 \end{bmatrix}$$

In (A), $\mathbf{a}_0 = \mathbf{I} = \mathbf{b}_0$ and the series $\eta_{1,t}$ and $\eta_{2,t}$ are correlated. In (B), $\mathbf{a}_0 = \mathbf{I}$, and $\Sigma_{\mathbf{e}} = \mathbf{I}$, but $\mathbf{b}_0$ is a full 2×2 matrix. In (C), the residual series $u_{1,t}$ and $u_{2,t}$ are uncorrelated, $\mathbf{b}_0 = \mathbf{I}$, and $\mathbf{a}_0$ is triangular. All three models are identical, as can be seen by noting that

$$\eta_{1,t} = e_{1,t} + 0.3 e_{2,t}, \qquad \eta_{2,t} = e_{2,t} + 0.4 e_{1,t}$$

and $\mathbf{P}\boldsymbol{\eta}_t = \mathbf{u}_t$ where

$$\mathbf{P} = \begin{bmatrix} 1 & 0 \\ -0.642 & 1 \end{bmatrix} \qquad \text{and} \qquad \mathbf{P}\Sigma_{\eta}\mathbf{P}' = \Sigma_{\mathbf{u}}$$

Thus (B) is derived from (A) by substitution and (C) from (A) by premultiplying throughout by $\mathbf{P}$ and substitution. It should also be noted that the residual series $\boldsymbol{\eta}_t$, $\mathbf{e}_t$, and $\mathbf{u}_t$ are not directly observable, but can only be estimated by inverting the model and substituting observations $\mathbf{x}_t$ of $\mathbf{X}_t$.

When building models, one of the $\mathbf{a}_0$, $\mathbf{b}_0$, Σ combinations of restrictions used in these examples should be specified before the analysis is attempted.

7.4 Causality and Feedback

Many of the prior restrictions placed on the models to achieve E-identification relate to the question of whether there is unidirectional causality or feedback between pairs of variables. To give operational meaning to these terms, acceptable definitions need to be derived, and in particular a testable definition of causality obtained. This is attempted in the present section.

Virtually any statistics textbook tells readers that a correlation between a pair of random variables cannot be interpreted as causality. None of the books then go on to tell what *can* be interpreted as causality. Even the stated lesson is not completely correct. If X and Y are the only two random variables in a universe and one *knows* that X cannot cause Y, but if the possibility of Y causing X is still open to question, then an observed significant correlation could be interpreted causally. The assumption that X does not cause Y gives sufficient structure to the situation for a causal interpretation to be given. A method of giving structure to a group of economic variables is to apply the following two rules, that seem to be fairly generally acceptable:

(i) The future cannot cause the past. Strict causality can occur only with the past causing the present or future.

(ii) It is sensible to discuss causality only for a group of stochastic processes. It is not possible to detect causality between two deterministic processes.

Given these rules, a possible definition of causality is as follows: Let $P(A|B)$ denote the conditional distribution function of A given B, let Ω_t represent all the information in the universe at time t, and ask, Does the series Y_t cause the series X_t? Then if

$$P(X_{t+1}|\Omega_t) = P(X_{t+1}|\Omega_t - Y_t) \qquad (7.4.1)$$

where $\Omega_t - Y_t$ is all the information in the universe apart from Y_t, Y_t does not cause X_t. If (7.4.1) does not hold, then Y_t does cause X_t. It is doubtful that philosophers would completely accept this definition, and possibly *cause* is too strong a term, or one too emotionally laden, to be used. A better term might be *temporally related*, but since cause is such a simple term we shall continue to use it. The above rules postulate that instantaneous causality is impossible, so that there must be a time delay between a cause and effect. Unfortunately, data limitations may make instantaneous causality (IC) appear likely. If the true delay between cause and effect is one day but if our stochastic processes are only observed monthly, IC will seem to occur. A definition of IC between Y_t and X_t would be: if

$$P(X_{t+1}|\Omega_{t+1} - X_{t+1}) \neq P(X_{t+1}|\Omega_{t+1} - X_{t+1} - Y_{t+1})$$

then there is IC between Y_{t+1} and X_{t+1}.

Feedback will occur if there is a pure (non-IC) causality between Y_t and X_t and also between X_t and Y_t. These need not both occur, as some simple examples can show. However, it is not possible, in general, to differentiate between instantaneous causality, in either direction, and between instantaneous feedback. One is back to the simple correlation idea mentioned at the beginning of this section. Differentiation is possible only if one adds some extra structure, such as "I know X_t cannot cause Y_t."

The definition as it stands is far too general to be testable. It is possible to reach a testable definition only by imposing considerable simplification and particularization to this definition. It must be recognized that, in so doing, the definition will become less intuitively acceptable and more error-prone.

Rather than dealing with all of the information in the universe, a plausible set of observed series, or information set I_t, will have to be used. The selection of this set will presumably depend on some underlying theory, on one's intuition, and on what data are available. It will always be possible to obtain spurious causality between two variables because a third variable, causal to both, has been left out of I_t. Thus, one should properly speak of "causality with respect to the particular I_t used."

It is also impractical to hope to deal with conditional distribution functions when given only a finite amount of data, without making a specific assumption about the form of multivariate distribution involved. However it is rare to have precise information about this distribution and an assumption of normality, for example, is not generally acceptable. An alternative route is to use a summary statistic instead of the whole conditional distribution, an example being the conditional mean. If this route is taken, then the above definition of pure causality can be expressed in terms of predictions. Let $f_t(I)$ represent the optimal prediction of X_{t+1} using the information in I_t, with consequent error $e_t(I) = X_{t+1} - f_t(I)$ and error variance $\sigma_t^2(I) = \text{var}(e_t(I))$. If $I_t - Y_t$ represents the information set apart from the series Y_t, then it can be said that Y_t causes X_t with respect to I_t if $\sigma_t^2(I - Y) > \sigma_t^2(I)$. In the stationary case, the variances will be independent of time and can be estimated. There is, of course, no overpowering reason why the cost function used should be a squared error one, so that the criterion for optimality of a forecast need not be least squares. Other metrics could be used and different conclusions about causation may be reached when one metric is used rather than another. One further problem still remains, and that is the form of the prediction used. The completely general optimal prediction could well be a complicated nonlinear function of the components of I_t. Unless a specific theory is available to suggest the form of nonlinearity involved, this representation is too general to be usable. One is forced to consider a specific form for the prediction, and that most used, due to simplicity, is the linear form. Thus, for purely pragmatic reasons, the "optimal prediction" in the most recent definition should be replaced by "optimal linear prediction." If one makes all of these simplifications, one ends up with "linear, least-squares causality with respect to the information set I_t." This is a long way from the rather general definition started with in (7.4.1), but it does provide a testable definition. It must be emphasized that true causality may be missed, or spurious causality observed, because of these simplifications, but this is a common situation in all linear statistics, such as correlation and linear regression techniques. If anyone prefers a more general formulation of the definition, they can always try to use some other cost function or a particular nonlinear prediction, if they feel that they can handle the statistical problems involved, and if they have sufficient data.

Using just the linear definition, one can derive alternative but equivalent definitions. For example, if I_t consists of just the two stationary series X_t and Y_t, then these can always be considered to have been generated by the model

$$a(B)X_t + b(B)Y_t = \epsilon_t, \qquad c(B)Y_t + d(B)X_t = \eta_t$$

where a, b, c, d are polynomials (possibly of infinite order) in the backward operator B, with $a(0) = c(0) = 1$, and where ϵ_t, η_t are a pair of uncorrelated white noise series. Then the linear definition of causality implies that Y_t does not cause X_t if $b(B) \equiv 0$. If $b(0) \neq 0$, then one has instantaneous causality.

Some relevant references, both discussing causality and applying the above definition, are Suppes [1970], Granger [1969b], Sims [1972], Haugh [1972], Pierce [1974], and Pierce and Haugh [1975].

7.5 Properties of Optimal Multiseries Forecasts

Suppose that $\mathbf{X}'_t = (X_{1,\,t}, X_{2,\,t}, \ldots, X_{m,\,t})$ is a vector time series generated by the model

$$\mathbf{a}(B)\mathbf{X}_t = \mathbf{b}(B)\boldsymbol{\eta}_t \tag{7.5.1}$$

where $\boldsymbol{\eta}_t$ is a zero-mean white noise vector, so that

$$E[\boldsymbol{\eta}_t] = \mathbf{0} \quad \text{and} \quad \begin{aligned} E[\boldsymbol{\eta}_t\boldsymbol{\eta}'_s] &= \mathbf{0}, & t \neq s \\ &= \Sigma, & t = s \end{aligned}$$

The completely general situation will be considered, so that the matrices $\mathbf{a}_0$, $\mathbf{b}_0$, and Σ are not constrained. As explained in Section 7.3, this situation is more general than is necessary, but clearly includes the constrained, or E-identified, models as special cases. The only conditions assumed to hold are that the roots of the equations $|\mathbf{a}(z)| = 0$, $|\mathbf{b}(z)| = 0$ all lie in the region $|z| > 1$. These conditions ensure stationarity and invertibility, respectively.

Throughout this section, the assumption will be made that the model (7.5.1) is known exactly. Given this assumption, the optimal h-steps ahead forecast of $\mathbf{X}_t$ will be derived and its properties examined, thus generalizing the results of Section 4.3. A comparison of the two sections will show that the multiseries results are virtually identical to those for single series, apart from the use of vectors and matrices.

Denote the $MA(\infty)$ model, equivalent to (7.5.1), by

$$\mathbf{X}_t = \mathbf{c}(B)\boldsymbol{\eta}_t \tag{7.5.2}$$

where

$$\mathbf{a}(B)\mathbf{c}(B) = \mathbf{b}(B) \tag{7.5.3}$$

Taking

$$\mathbf{a}(B) = \sum_{j=0}^{p} \mathbf{a}_j B^j, \quad \mathbf{b}(B) = \sum_{j=0}^{q} \mathbf{b}_j B^j, \quad \mathbf{c}(B) = \sum_{j=0}^{\infty} \mathbf{c}_j B^j$$

then, by equating coefficients of B^j in (7.5.3), one obtains the relationships

$$\sum_{k=0}^{j} \mathbf{a}_k \mathbf{c}_{j-k} = \mathbf{b}_j, \quad j = 0, 1, 2, \ldots \tag{7.5.4}$$

Now, let $\mathbf{f}_{n,\,h}$ be a vector of forecasts of $\mathbf{X}_{n+h}$ made at time n, so that the ith component of $\mathbf{f}_{n,\,h}$ is the h-step linear forecast of $X_{i,\,n+h}$. This forecast will be based on the information set

$$I_n = \{\mathbf{X}_{n-j}; j \geq 0\}$$

and is thus of the form

$$\mathbf{f}_{n,\,h} = \sum_{j=0}^{\infty} \boldsymbol{\lambda}_{j,\,h}\mathbf{X}_{n-j} \tag{7.5.5}$$

where $\boldsymbol{\lambda}_{j,\,h}$ is an $m \times m$ matrix. Thus one can write

$$\mathbf{f}_{n,\,h} = \boldsymbol{\lambda}_h(B)\mathbf{X}_n$$

Substituting from (7.5.2), this becomes

$$\mathbf{f}_{n,\,h} = \phi_h(B)\eta_n \tag{7.5.6}$$

where

$$\phi_h(B) = \lambda_h(B)\mathbf{c}(B) \tag{7.5.7}$$

To find the optimal forecast, it is necessary to determine the matrix $\lambda_h(B)$, or equivalently $\phi_h(B)$. It is also necessary, of course, to have a criterion of optimality. To derive such a criterion, the following definition is first required. If $\mathbf{V}_1$ and $\mathbf{V}_2$ are two real positive definite $m \times m$ matrices, then $\mathbf{V}_1$ will be said to be smaller than $\mathbf{V}_2$, provided that, for every nonzero $m \times 1$ vector $\mathbf{d}$,

$$\mathbf{d}'\mathbf{V}_1\mathbf{d} < \mathbf{d}'\mathbf{V}_2\mathbf{d}$$

Such a relationship will be denoted $\mathbf{V}_1 < \mathbf{V}_2$. If $\{\mathbf{V}(s)\}$ represents a set of positive definite $m \times m$ matrices, a particular one, $\mathbf{V}_0$, will be called the smallest if $\mathbf{V}_0 < \mathbf{V}_j$ for every $\mathbf{V}_j \neq \mathbf{V}_0$ contained in $\{\mathbf{V}(s)\}$.

Denote the h-step error series by

$$\mathbf{e}_{n,\,h} = \mathbf{X}_{n+h} - \mathbf{f}_{n,\,h} \tag{7.5.8}$$

and let this vector have covariance matrix

$$\mathbf{V}(h) = E[\mathbf{e}_{n,\,h}\mathbf{e}'_{n,\,h}] \tag{7.5.9}$$

since $E[\mathbf{e}_{n,\,h}] = \mathbf{0}$. Then a forecast vector $\mathbf{f}_{n,\,h}$ will be said to be (linearly) optimal if the corresponding $\mathbf{V}(h)$ is the smallest of all the possible error covariance matrices. The optimality criterion, stated in this form, is both more convenient and more general than the more frequently used ones of minimizing the determinant or trace of $\mathbf{V}(h)$.

Substituting for $\mathbf{X}_{n+h}$ and $\mathbf{f}_{n,\,h}$ from (7.5.2) and (7.5.6) into (7.5.8) gives

$$\mathbf{e}_{n,\,h} = [\mathbf{c}(B) - B^h\phi_h(B)]\eta_{n+h} = \sum_{j=0}^{h-1} \mathbf{c}_j\eta_{n+h-j} + \sum_{j=0}^{\infty} (\mathbf{c}_{j+h} - \phi_{h,\,j})\eta_{n-j} \tag{7.5.10}$$

so that, from (7.5.9),

$$\mathbf{V}(h) = \sum_{j=0}^{h-1} \mathbf{c}_j\Sigma\mathbf{c}'_j + \sum_{j=0}^{\infty} (\mathbf{c}_{j+h} - \phi_{h,\,j})\Sigma(\mathbf{c}_{j+h} - \phi_{h,\,j})'$$

As $\phi_h(B)$ varies, this set of positive definite matrices clearly has smallest value

$$\mathbf{V}(h) = \sum_{j=0}^{h-1} \mathbf{c}_j\Sigma\mathbf{c}'_j \tag{7.5.11}$$

and this is achieved by taking

$$\phi_{h,\,j} = \mathbf{c}_{j+h} \tag{7.5.12}$$

Thus, the optimal forecasts are given by

$$\mathbf{f}_{n,h} = \sum_{j=0}^{\infty} \mathbf{c}_{j+h}\boldsymbol{\eta}_{n-j} \tag{7.5.13}$$

with corresponding errors

$$\mathbf{e}_{n,h} = \sum_{j=0}^{h-1} \mathbf{c}_j\boldsymbol{\eta}_{n+h-j} \tag{7.5.14}$$

In particular, the one-step errors are $\mathbf{e}_{n,1} = \mathbf{c}_0\boldsymbol{\eta}_{n+1}$ and so comprise a white noise vector. It is also clear, from (7.5.11), that $\mathbf{V}(h_1) \geqslant \mathbf{V}(h_2)$ for $h_1 > h_2$, so that, as is to be expected, the further ahead one forecasts, the less well one does, on average.

Rearranging (7.5.13) gives

$$\mathbf{f}_{n,h} = \mathbf{c}_h\boldsymbol{\eta}_n + \sum_{j=0}^{\infty} \mathbf{c}_{j+h+1}\boldsymbol{\eta}_{n-1-j}$$

so that

$$\mathbf{f}_{n,h} = \mathbf{f}_{n-1,h+1} + \mathbf{c}_h\mathbf{c}_0^{-1}(\mathbf{X}_n - \mathbf{f}_{n-1,1}) \tag{7.5.15}$$

which is the generalization of the updating formula (4.3.24).

Now consider

$$\sum_{j=0}^{p} \mathbf{a}_j\mathbf{f}_{n,h-j} = \sum_{i=0}^{p}\sum_{j=0}^{p} \mathbf{a}_j\mathbf{c}_{i+h-j}\boldsymbol{\eta}_{n-i} = \sum_{i=0}^{\infty} \mathbf{b}_{i+h}\boldsymbol{\eta}_{n-i} \tag{7.5.16}$$

from (7.5.4), taking $\mathbf{b}_i \equiv \mathbf{0}$ for all $i > q$. This generalizes the single series equation (4.4.15) and is used in exactly the same way to form the sequence of optimal forecasts, as h increases with n fixed. Thus, for example, it is easily shown that the optimal one-step forecast for the generalized AR(p) process, with $\mathbf{b}(B) = \mathbf{b}_0$, is given by

$$\mathbf{f}_{n,1} = -\mathbf{a}_0^{-1}\sum_{j=1}^{p} \mathbf{a}_j\mathbf{X}_{n-j+1} \tag{7.5.17}$$

Similarly, the optimal h-step forecast of the AR(1) process $\mathbf{a}_0\mathbf{X}_t + \mathbf{a}_1\mathbf{X}_{t-1} = \boldsymbol{\eta}_t$ is given by

$$\mathbf{f}_{n,h} = \left(-\mathbf{a}_0^{-1}\mathbf{a}_1\right)^h\mathbf{X}_n \tag{7.5.18}$$

The frequency domain approach to the forecasting problem can also be generalized to the multiseries case, but will not be considered here, as it provides few, if any, extra results of practical importance. Details are given in Hannan [1970, Chapter 3].

Throughout this section the completely unrealistic assumption that the generating model is known exactly has been made. In the following sections of this chapter some practical multivariate time series model building techniques are described. To produce forecasts, the models thus derived are taken to be true models, and the methods of this section can then be employed.

7.6 Relating Bivariate Series: The Cross Correlogram

So far, the theoretical discussion of multiple time series has been completely general. It is now time to descend from these lofty heights in order to
consider what can be achieved in practice, given only data on a set of time
series. The bulk of the discussion will be concerned with the situation where
only a pair of observed time series are available. There are three reasons for
this. First, there is little point in considering the general case without first
assessing the possibilities of the simpler bivariate situation. Second, virtually
all the important points of principle can be examined in the bivariate
framework, and, finally, the methodological difficulties are such that the
construction of a general m-variable time series model, from series of length
likely to occur in practice, is prohibitively difficult unless m is very small.

The question of which form of the multivariate time series model is most
suitable for practical analysis is an open one. Our own preference is for the
recursive form of the transfer function–noise model (7.2.3). In the bivariate
case this can be written as

$$Y_{1,t} = V_1^*(B)Y_{2,t} + \psi_1^*(B)\eta_{1,t}^*, \qquad Y_{2,t} = V_2^*(B)Y_{1,t} + \psi_2^*(B)\eta_{2,t}^*, \quad (7.6.1)$$

where $Y_{1,t}$ and $Y_{2,t}$ are zero-mean stationary series, which could be obtained
from given series $X_{1,t}, X_{2,t}$ by suitable differencing and, if necessary, subtracting off nonzero means from the differenced series. In (7.6.1) it is assumed
that

$$V_j^*(B) = \frac{\omega_j^*(B)}{\delta_j^*(B)}, \quad j = 1, 2, \qquad \psi_j^*(B) = \frac{\theta_j^*(B)}{\phi_j^*(B)}, \quad j = 1, 2 \quad (7.6.2)$$

where the polynomials in B on the right-hand side of each equation in (7.6.2)
are of finite orders, with $\omega_2^*(0) \equiv 0$ and where the roots of the equations in z

$$\delta_j^*(z) = 0; \quad \theta_j^*(z) = 0; \quad \phi_j^*(z) = 0; \qquad j = 1, 2$$

all lie outside the unit circle $|z| = 1$. The error series $\eta_{1,t}^*$ and $\eta_{2,t}^*$ in (7.6.1) are
taken to be mutually stochastically uncorrelated white noise processes, i.e.,

$$E[\eta_{j,t}^*, \eta_{j,t-\tau}^*] = 0; \qquad j = 1, 2, \quad \text{all } \tau \neq 0$$

$$E[\eta_{1,t}^*, \eta_{2,s}^*] = 0; \qquad \text{all } t, s \qquad\qquad (7.6.3)$$

Now, the time series identification problem is to find, based on the
evidence of data, appropriate forms for the polynomials in B in (7.6.1). In
Chapter 3, it was seen that, for the single series case, useful information was
provided by the sample correlogram. In the bivariate situation, it is natural to
think in terms of the cross correlogram, though, for reasons that will become
clear later, the following discussion is carried out in terms of cross
covariances. Denote

$$\lambda_k^{(1,2)} = \text{cov}(Y_{1,t}, Y_{2,t-k}) \qquad \text{and} \qquad \lambda_k^{(j,j)} = \text{cov}(Y_{j,t}, Y_{j,t-k}), \qquad j = 1, 2$$

and let the corresponding generating functions be

$$\lambda^{(i,j)}(z) = \sum_{\text{all } k} \lambda_k^{(i,j)} z^k$$

Given (7.6.3), it follows from (7.6.1) that

$$E[(Y_{1,t} - V_1^*(B)Y_{2,t})(Y_{2,t-k} - V_2^*(B)Y_{1,t-k})] = 0$$

$$k = \ldots, -1, 0, 1 \ldots$$

and hence

$$\lambda_k^{(1,2)} - \sum_{i=0}^{\infty} V_{1,i}^* \lambda_{k-i}^{(2,2)} - \sum_{j=1}^{\infty} V_{2,j}^* \lambda_{k+j}^{(1,1)} + \sum_{i=0}^{\infty}\sum_{j=1}^{\infty} V_{1,i}^* V_{2,j}^* \lambda_{i-j-k}^{(1,2)} = 0$$

$$k = \ldots, -1, 0, 1 \ldots$$

since, by definition, $V_{2,0}^* = 0$. Multiplying each of these equations by z^k and summing over k gives, in terms of generating functions,

$$\lambda^{(1,2)}(z) + V_1^*(z)V_2^*(z^{-1})\lambda^{(1,2)}(z^{-1}) = V_1^*(z)\lambda^{(2,2)}(z) + V_2^*(z^{-1})\lambda^{(1,1)}(z)$$

$$(7.6.4)$$

Finally, substituting z^{-1} for z in (7.6.4), and solving the resulting equation, together with (7.6.4), yields

$$\lambda^{(1,2)}(z)$$

$$= \frac{V_2^*(z^{-1})(1 - V_1^*(z)V_2^*(z))\lambda^{(1,1)}(z) + V_1^*(z)(1 - V_1^*(z^{-1})V_2^*(z^{-1}))\lambda^{(2,2)}(z)}{1 - |V_1^*(z)V_2^*(z)|^2}$$

$$(7.6.5)$$

The difficulties involved in working with cross covariances of raw data are well illustrated by the special case where $Y_{2,t}$ causes $Y_{1,t}$ but there is no feedback, so that $V_2^*(z) \equiv 0$. Then (7.6.5) simplifies to

$$\lambda^{(1,2)}(z) = V_1^*(z)\lambda^{(2,2)}(z) \qquad (7.6.6)$$

Now, in this expression, $V_1^*(z)$ does not contain negative powers of z while, in general, $\lambda^{(2,2)}(z)$ will contain both negative and positive powers of z. Hence, even when there is no feedback, $\lambda^{(1,2)}(z)$ will contain both negative and positive powers of z, and hence $\lambda_k^{(i,j)}$ can be nonzero for both negative and positive k, thus making the cross correlogram difficult to interpret even in this simple situation.

The simplicity of the one-way causality case leads one to ask how this situation might be detected in practice. Suppose that $Y_{2,t}$ were white noise. Then $\lambda^{(2,2)}(z) = \lambda_0^{(2,2)}$ and it follows immediately from (7.6.6) that

$$\lambda_{-k}^{(1,2)} = 0, \quad k = 1, 2, \ldots \qquad \text{and} \qquad \lambda_k^{(1,2)} = V_{1,k}^* \lambda_0^{(2,2)}, \quad k = 0, 1, 2, \ldots$$

so that the plot of $\lambda_k^{(1,\,2)}$ against k should take a distinctive form. This observation suggests that, rather than look at the cross correlations between $Y_{1,\,t}$ and $Y_{2,\,t}$, one should first build univariate models

$$a_j(B)Y_{j,\,t} = b_j(B)\epsilon_{j,\,t}, \qquad j = 1, 2 \tag{7.6.7}$$

and examine the cross correlations between the individual white noise residual (or innovation) series $\epsilon_{1,\,t}$ and $\epsilon_{2,\,t}$.

The difficulties involved in examining the cross correlogram of raw data were described and illustrated in a particular extreme case by Box and Newbold [1971], who considered the consequences of cross correlating two *independent* random walk series $(1 - B)X_{j,\,t} = \epsilon_{j,\,t}, j = 1, 2$. It was noted that the resulting sample cross correlograms tended to contain quite large values, which is simply the spurious regression problem discussed in Section 6.4, and also to be very smooth. An example, based on series of 100 observations, is shown in Fig. 7.1a. For comparison, the cross correlogram between the differenced (white noise) series is shown in Fig. 7.1b. Box and Newbold showed that the smoothness resulted from the fact that the sample cross covariances behave approximately as an ARIMA(0, 2, 0) time series, i.e.,

$$\hat{\lambda}_k^{(1,\,2)} \approx 2\hat{\lambda}_{k-1}^{(1,\,2)} - \hat{\lambda}_{k-2}^{(1,\,2)} + u_k$$

where u_k is white noise. Thus, not only does one have to contend with the spurious regression phenomenon of the size of the cross correlations, but also the *shape* of the cross correlogram is impossible to interpret sensibly.

As a first step in bivariate model building, then, it would seem essential to undertake some prewhitening of the data. One could build single series models of the form (7.6.7) and calculate the cross correlogram of the innovation series $\epsilon_{1,\,t}$ and $\epsilon_{2,\,t}$. A useful intermediate step, particularly when there is feedback, might then be to build a bivariate model, of the form (7.6.1), linking the individual innovation series. This can be written as

$$\epsilon_{1,\,t} = V_1(B)\epsilon_{2,\,t} + \psi_1(B)\eta_{1,\,t}, \qquad \epsilon_{2,\,t} = V_2(B)\epsilon_{1,\,t} + \psi_2(B)\eta_{2,\,t} \tag{7.6.8}$$

where

$$V_j(B) = \omega_j(B)/\delta_j(B), \qquad \psi_j(B) = \theta_j(B)/\phi_j(B), \qquad j = 1, 2$$

and $\omega_2(0) \equiv 0$. The error terms $\eta_{1,\,t}$ and $\eta_{2,\,t}$ in (7.6.8) are mutually stochastically uncorrelated white noise processes.

We often find it convenient, before fitting (7.6.8), to standardize the series $\epsilon_{1,\,t}$ and $\epsilon_{2,\,t}$, so that each has unit variance. In this case, it follows from (7.6.5) that the cross covariance generating function for the individual innovation series is

$$\lambda_\epsilon^{(1,\,2)}(z) = \frac{V_2(z^{-1})(1 - V_1(z)V_2(z)) + V_1(z)(1 - V_1(z^{-1})V_2(z^{-1}))}{1 - |V_1(z)V_2(z)|^2}$$

$$= \frac{V_1(z)(1 - |V_2(z)|^2) + V_2(z^{-1})(1 - |V_1(z)|^2)}{1 - |V_1(z)V_2(z)|^2}$$

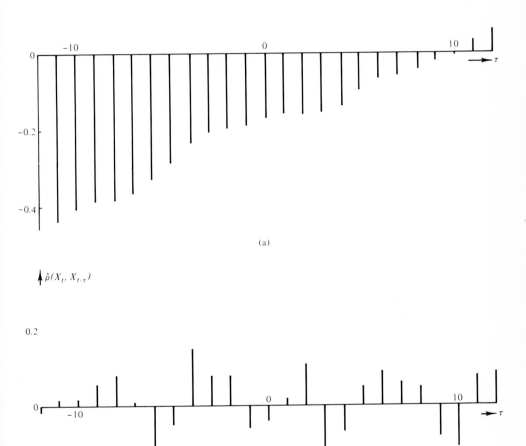

(a)

(b)

FIG. 7.1   (a) Sample cross correlations for series of length 100 generated from independent random walks. (b) Sample cross correlations of first differences of data used in (a).

which is a rather simpler form than (7.6.5). If there is one-way causality between the ϵ's, which implies one-way causality between the Y's, it follows that the cross-covariance function will be one-sided, and hence this situation should not be too difficult to detect in practice.

The fact that $\epsilon_{1,t}$ and $\epsilon_{2,t}$ are both white noise series implies certain constraints on the model (7.6.8). Solving these equations for the $\epsilon_{j,t}$ in terms of the $\eta_{j,t}$ yields

$$(1 - V_1(B)V_2(B))\epsilon_{1,t} = \psi_1(B)\eta_{1,t} + V_1(B)\psi_2(B)\eta_{2,t}$$

$$(1 - V_1(B)V_2(B))\epsilon_{2,t} = V_2(B)\psi_1(B)\eta_{1,t} + \psi_2(B)\eta_{2,t}$$

Taking autocovariance generating functions on each side of these equations implies since the $\epsilon_{j,t}$ have unit variance and $\eta_{1,t}$ and $\eta_{2,t}$ are mutually stochastically uncorrelated

$$|1 - V_1(z)V_2(z)|^2 = \sigma_1^2|\psi_1(z)|^2 + \sigma_2^2|V_1(z)|^2|\psi_2(z)|^2$$

$$|1 - V_1(z)V_2(z)|^2 = \sigma_1^2|V_2(z)|^2|\psi_1(z)|^2 + \sigma_2^2|\psi_2(z)|^2$$

where $\sigma_j^2 = E[\eta_{j,t}^2], j = 1, 2$. Hence

$$|\psi_j(z)|^2\sigma_j^2 = \frac{|1 - V_1(z)V_2(z)|^2(1 - |V_j(z)|^2)}{1 - |V_1(z)V_2(z)|^2} , \qquad j = 1, 2 \qquad (7.6.9)$$

Thus, in theory, the functions $\psi_j(z)$ are determined up to a factor z^k, given $V_j(z)$, $j = 1, 2$. This factor is of no consequence since it merely represents a lagging of a nondirectly observed white noise series. Equation (7.6.9) is only of practical importance in situations where $V_1(z)$ and $V_2(z)$ are relatively simple functions. Otherwise it becomes too difficult to compute. However, it is important when constructing models to simulate a pair of white noise series since it implies restrictions on the way such series can be related. In general it follows that the manner in which any two series, obeying *specific* single series models, can be related is constrained by the form of those single series models.

In the case of one-way causality, where $V_2(z) \equiv 0$, a remarkable simplification follows. In this case (7.6.9) becomes

$$|\psi_1(z)|^2\sigma_1^2 = 1 - |V_1(z)|^2 \qquad (7.6.10)$$

Writing the model as

$$\epsilon_{1,t} = \frac{\omega_1(B)}{\delta_1(B)}\epsilon_{2,t} + \frac{\theta_1(B)}{\phi_1(B)}\eta_{1,t}, \qquad \epsilon_{2,t} = \eta_{2,t} \qquad (7.6.11)$$

it can be shown that $\delta_1(B)$ and $\phi_1(B)$ are necessarily identical. This can be seen by considering the poles on each side of (7.6.10) or directly, as follows. Multiplying through the first equation of (7.6.11) by $\delta_1(B)$ yields

$$\delta_1(B)\epsilon_{1,t} = \omega_1(B)\eta_{2,t} + \frac{\delta_1(B)\theta_1(B)}{\phi_1(B)}\eta_{1,t}$$

Now, the left-hand side of this expression is finite order moving average, and the right-hand side is the sum of two uncorrelated components, one of which is finite order moving average. It follows from the theorem of Section 1.8 that this can only be so if the second term on the right-hand side is also finite order moving average. Hence $\phi_1(B)$ must be a factor of $\delta_1(B)\theta_1(B)$. However, it is assumed in constructing (7.6.11) that $\phi_1(B)$ and $\theta_1(B)$ have no common factors. Hence $\phi_1(B)$ is a factor of $\delta_1(B)$. Multiplying through the first equation of (7.6.11) by $\phi_1(B)$ yields

$$\phi_1(B)\epsilon_{1,\,t} = \frac{\phi_1(B)\omega_1(B)}{\delta_1(B)}\,\eta_{2,\,t} + \theta_1(B)\eta_{1,\,t}$$

An argument exactly analogous to the above leads to the conclusion that $\delta_1(B)$ is a factor of $\phi_1(B)$. Hence the two polynomials are identical, and (7.6.11) can be written as

$$\delta_1(B)\epsilon_{1,\,t} = \omega_1(B)\epsilon_{2,\,t} + \theta_1(B)\eta_{1,\,t} \tag{7.6.12}$$

Of course, using (7.6.10), $\theta_1(B)$ is known exactly, although this is typically of little use in practice. It does, however, follow from the theorem of Section 1.8 that in (7.6.12) the order of $\theta_1(B)$ must be equal to the larger of the orders of $\delta_1(B)$ amd $\omega_1(B)$, if these are different, and less than or equal to their common order if both are of the same order.

Unfortunately, if there is feedback in the system, no such similar simplifications are available. To illustrate the possible use of (7.6.9), consider the case

$$\epsilon_{1,\,t} = \frac{(\omega_{1,\,0} + \omega_{1,\,1}B)}{(1 - \delta_{1,\,1}B)}\,\epsilon_{2,\,t} + \psi_1(B)\eta_{1,\,t}$$

$$\epsilon_{2,\,t} = \omega_{2,\,1}B\epsilon_{1,\,t} + \psi_2(B)\eta_{2,\,t} \tag{7.6.13}$$

First, expressing the infinite order polynomials in z in (7.6.9) by corresponding ratios of finite order polynomials, one has

$$\left|\frac{\theta_j(z)}{\phi_j(z)}\right|^2 \sigma_j^2 = \omega(z)\frac{(|\delta_j(z)|^2 - |\omega_j(z)|^2)}{|\delta_j(z)|^2}\,, \qquad j = 1, 2 \tag{7.6.14}$$

where

$$\omega(z) = \frac{|\delta_1(z)\delta_2(z) - \omega_1(z)\omega_2(z)|^2}{|\delta_1(z)\delta_2(z)|^2 - |\omega_1(z)\omega_2(z)|^2} \tag{7.6.15}$$

For the model (7.6.13), one has

$$|\delta_1(z)|^2 - |\omega_1(z)|^2 = 1 + \delta_{1,\,1}^2 - \omega_{1,\,0}^2 - \omega_{1,\,1}^2 - (\delta_{1,\,1} + \omega_{1,\,0}\omega_{1,\,1})(z + z^{-1})$$

and one requires a function $u(z) = u_0 + u_1 z$ such that

$$|\delta_1(z)|^2 - |\omega_1(z)|^2 = |u(z)|^2$$

Since

$$|u(z)|^2 = u_0^2 + u_1^2 + u_0u_1(z + z^{-1})$$

equating powers of z gives

$$u_0^2 + u_1^2 = 1 + \delta_{1,1}^2 - \omega_{1,0}^2 - \omega_{1,1}^2, \qquad u_0u_1 = -(\delta_{1,1} + \omega_{1,0}\omega_{1,1})$$

Adding and subtracting twice the second of these equations to the first gives

$$(u_0 + u_1)^2 = (1 - \delta_{1,1})^2 - (\omega_{1,0} + \omega_{1,1})^2$$

$$(u_0 - u_1)^2 = (1 + \delta_{1,1})^2 - (\omega_{1,0} - \omega_{1,1})^2$$

so that a solution for $u(z)$ is

$$u_0 = \tfrac{1}{2}[M + N], \qquad u_1 = \tfrac{1}{2}[M - N] \tag{7.6.16}$$

where

$$M = \left[(1 - \delta_{1,1})^2 - (\omega_{1,0} + \omega_{1,1})^2\right]^{1/2}$$

$$N = \left[(1 + \delta_{1,1})^2 - (\omega_{1,0} - \omega_{1,1})^2\right]^{1/2} \tag{7.6.17}$$

Since the numerator of (7.6.15) is

$$|\delta_1(z)\delta_2(z) - \omega_1(z)\omega_2(z)|^2 = |1 - (\delta_{1,1} + \omega_{1,0}\omega_{2,1})z - \omega_{1,1}\omega_{2,1}z^2|^2$$

it follows that, up to a constant, a possible choice of $\theta_1(z)$ is

$$\left(1 - (\delta_{1,1} + \omega_{1,0}\omega_{2,1})z - \omega_{1,1}\omega_{2,1}z^2\right)(u_0 + u_1z)$$

Hence, normalizing, one has

$$\theta_1(z) = \left(1 - (\delta_{1,1} + \omega_{1,0}\omega_{2,1})z - \omega_{1,1}\omega_{2,1}z^2\right)(1 + (u_1/u_0)z) \tag{7.6.18}$$

where u_0 and u_1 are given by (7.6.16) and (7.6.17).

Now, the denominator of (7.6.15) is

$$|\delta_1(z)\delta_2(z)|^2 - |\omega_1(z)\omega_2(z)|^2 = 1 + \delta_{1,1}^2 - \omega_{2,1}^2(\omega_{1,0}^2 + \omega_{1,1}^2)$$
$$- (\delta_{1,1} + \omega_{2,1}^2\omega_{1,0}\omega_{1,1})(z + z^{-1})$$

and one requires a function $q(z) = q_0 + q_1z$ such that

$$|\delta_1(z)\delta_2(z)|^2 - |\omega_1(z)\omega_2(z)|^2 = |q(z)|^2$$

Using an argument identical to that employed in the derivation of (7.6.16), it follows that a solution is given by

$$q_0 = \tfrac{1}{2}[M' + N'], \qquad q_1 = \tfrac{1}{2}[M' - N'] \tag{7.6.19}$$

where

$$M' = \left[(1 - \delta_{1,1})^2 - \omega_{2,1}^2(\omega_{1,0} + \omega_{1,1})^2\right]^{1/2},$$

$$N' = \left[(1 + \delta_{1,1})^2 - \omega_{2,1}^2(\omega_{1,0} - \omega_{1,1})^2\right]^{1/2} \tag{7.6.20}$$

Since the denominator of (7.6.14) for $j = 1$ is $|\delta_1(z)|^2 = |1 - \delta_{1,1}z|^2$, it follows that, normalizing so that the leading term is unity, one has

$$\phi_1(z) = (1 - \delta_{1,1}z)\left(1 + \frac{q_1}{q_0}z\right) \tag{7.6.21}$$

where q_0 and q_1 are given by (7.6.19) and (7.6.20).

Now $|\delta_2(z)|^2 - |\omega_2(z)|^2 = 1 - \omega_{2,1}^2$, and hence, on normalizing, one has

$$\theta_2(z) = 1 - (\delta_{1,1} + \omega_{1,0}\omega_{2,1})z - \omega_{1,1}\omega_{2,1}z^2 \tag{7.6.22}$$

Finally, since $\delta_2(z) = 1$, it follows from (7.6.14) that

$$\phi_2(z) = 1 + (q_1/q_0)z \tag{7.6.23}$$

where q_0 and q_1 are given by (7.6.19) and (7.6.20).

Hence, for this simple example, it has been possible to determine analytically the ARMA error structures in (7.6.13). The example will be recalled when discussing the problem of identifying bivariate feedback models, but, if nothing else, it will have served to demonstrate the futility of attempts to employ (7.6.9) to identify error structure when the identified transfer functions contain more than a very few terms.

7.7 Building Bivariate Models: Unidirectional Causality

The simplest possible situation in multivariate time series analysis is where one has a pair of zero mean stationary series $Y_{1,t}$ and $Y_{2,t}$, which again may be derived from given series $X_{1,t}$ and $X_{2,t}$ by differencing, where $Y_{2,t}$ causes $Y_{1,t}$ but $Y_{1,t}$ does not cause $Y_{2,t}$. In the case of a linear relationship, a complete specification is given by the model

$$Y_{1,t} = \frac{\omega_1^*(B)}{\delta_1^*(B)} Y_{2,t} + \frac{\theta_1^*(B)}{\phi_1^*(B)} \eta_{1,t}^* \tag{7.7.1}$$

and

$$a_2(B)Y_{2,t} = b_2(B)\epsilon_{2,t} \tag{7.7.2}$$

where $\eta_{1,t}^*$ and $\epsilon_{2,t}$ are mutually stochastically uncorrelated white noise processes. Now, if one knows a priori that one's data exhibit unidirectional causality rather than feedback, it is not essential, as will be seen in the next section (and, indeed, as one might guess from (7.6.6)) to prewhiten the output series $Y_{1,t}$. However, one is frequently in the position of suspecting possible feedback in the system, in which case the best strategy is to first build the univariate models (7.7.2) and

$$a_1(B)Y_{1,t} = b_1(B)\epsilon_{1,t} \tag{7.7.3}$$

To illustrate the methods of this section, 144 monthly observations on numbers unemployed and index of industrial production in the United States, both seasonally adjusted, were analyzed. Denoting these series by $X_{1,t}$ and

$X_{2,t}$, respectively, the fitted univariate models were:

$$(1 + 0.06B - 0.21B^2 - 0.18B^3 - 0.19B^4)(1 - B)X_{1,t} = \epsilon_{1,t};$$
$$\quad [0.08] \qquad [0.08] \qquad\quad [0.08] \qquad\quad [0.08]$$

$$\hat{\sigma}^2_{\epsilon_1} = 13617 \qquad (7.7.4)$$

and

$$(1 - 0.26B)(1 - B)X_{2,t} = 0.21 + \epsilon_{2,t}, \qquad \hat{\sigma}^2_{\epsilon_2} = 0.49 \qquad (7.7.5)$$
$$\quad [0.08] \qquad\qquad\qquad\quad [0.06]$$

To assess the possible interrelationship of these series, the cross correlogram between the residuals from the fitted equations (7.7.4) and (7.7.5) is calculated. The resulting values are shown in Table 7.1, where

$$\hat{\lambda}_k = \widehat{\text{corr}}(\epsilon_{1,t}, \epsilon_{2,t-k})$$

To judge the magnitudes of these quantities, it should be noted that, under the hypothesis of no relationship between the series, these sample cross correlations would be asymptotically distributed as independent normal with zero means and standard deviations $n^{-1/2}$, where n is the number of observations employed in their calculation (see Haugh [1972] and Haugh and Box [1974]). In the present example, $n = 139$ and so $n^{-1/2} \approx 0.085$. Based on this criterion, the values in the second half of Table 7.1 do not seem unduly large, suggesting that industrial production "causes" unemployment but that there is no feedback (or, at least, if there is feedback, it is very weak).

Now, as an intermediate stage, one can first construct a model linking the residuals from the univariate models. In fact, for ease of interpretation, we prefer to divide these residuals by their respective standard deviations and fit a model to the standardized series. It follows from the results of the previous section that such a model will be of the form

$$\epsilon_{1,t} = \frac{\omega_1(B)}{\delta_1(B)} \epsilon_{2,t} + \frac{\theta_1(B)}{\delta_1(B)} \eta_{1,t} \qquad (7.7.6)$$

Now, write $\omega_1(B)/\delta_1(B) = V_1(B) = V_{1,0} + V_{1,1}B + V_{1,2}B^2 + \cdots$ Hence,

Table 7.1 *Cross correlations between residuals from Eqs. (7.7.4) and (7.7.5)*

k:	0	1	2	3	4	5	6	7
$\hat{\lambda}_k$:	-0.16	-0.22	-0.22	-0.06	-0.13	-0.02	-0.04	0.05
k:	8	9	10	11	12	13	14	15
$\hat{\lambda}_k$:	0.01	-0.06	-0.03	-0.04	0.13	0.01	0.12	0.07

k:		-1	-2	-3	-4	-5	-6	-7
$\hat{\lambda}_k$:		-0.08	-0.05	-0.07	-0.09	0.04	-0.08	-0.11
k:	-8	-9	-10	-11	-12	-13	-14	-15
$\hat{\lambda}_k$:	-0.07	-0.08	0.02	0.05	0.08	-0.10	0.05	-0.01

multiplying through (7.7.6) by $\epsilon_{2, t-k}$, $k \geq 0$, and taking expectations, it follows that, since $\epsilon_{2, t}$ and $\eta_{1, t}$ are mutually stochastically uncorrelated, $E[\epsilon_{1, t}\epsilon_{2, t-k}] = V_{1, k}$, $k = 0, 1, 2, \ldots$, since the $\epsilon_{j, t}$ have unit variance. Hence the quantities in the top half of Table 7.1 are estimates of the $V_{1, k}$.

Now, the first three $\hat{V}_{1, k}$ are quite large, and the remainder fairly small, suggesting $V_1(B) = \omega_{1, 0} + \omega_{1, 1}B + \omega_{1, 2}B^2$. Thus, in (7.7.6), one can write

$$\epsilon_{1, t} = (\omega_{1, 0} + \omega_{1, 1}B + \omega_{1, 2}B^2)\epsilon_{2, t} + \theta_1(B)\eta_{1, t}$$

and it follows from the results of the previous section that

$$\theta_1(B) = 1 - \theta_{1, 1}B - \theta_{1, 2}B^2$$

which completes the identification of (7.7.6).

Multiplying through (7.7.6) by $\delta_1(B)$ gives

$$\delta_1(B)\epsilon_{1, t} = \omega_1(B)\epsilon_{2, t} + \theta_1(B)\eta_{1, t}$$

This is now in a convenient form for estimation, which involves minimizing the sum of the squared $\eta_{1, t}$ using a nonlinear regression algorithm. Details are analogous to those for the estimation of (7.7.1), which will be discussed later in this section. The estimated model for the unemployment–industrial production data was

$$\epsilon_{1, t} = (-\underset{[0.08]}{0.16} - \underset{[0.08]}{0.26}B - \underset{[0.08]}{0.25}B^2)\epsilon_{2, t} + (1 - \underset{[0.09]}{0.21}B - \underset{[0.09]}{0.19}B^2)\eta_{1, t} \quad (7.7.7)$$

In general, the procedure for identifying the appropriate form of the transfer function in (7.7.6) is in two stages. First, one obtains estimates of $V_1(B) = V_{1, 0} + V_{1, 1}B + V_{1, 2}B^2 + \cdots$ from the cross correlogram of the univariate model residuals. Second, one examines the estimated $\hat{V}_{1, k}$ and tries to approximate these by a ratio of finite order polynomials in B, i.e., $V_1(B) = \omega_1(B)/\delta_1(B)$. Thus, for example, if

$$\hat{V}_{1, k} \approx \delta_{1, 1}\hat{V}_{1, k-1}, \qquad k = 1, 2, 3, \ldots \quad (7.7.8)$$

with $\hat{V}_{1, 0} = \omega_{1, 0}$, then the model $V_1(B) = \omega_{1, 0}/(1 - \delta_{1, 1}B)$ would be appropriate. However, if (7.7.8) held only for $k = 2, 3, \ldots$, the appropriate form would be

$$V_1(B) = (\omega_{1, 0} + \omega_{1, 1}B)/(1 - \delta_{1, 1}B)$$

In the same way, second order decay in the $\hat{V}_{1, k}$, which might be exponential or sinusoidal, of the form

$$\hat{V}_{1, k} \approx \delta_{1, 1}\hat{V}_{1, k-1} + \delta_{1, 2}\hat{V}_{1, k-2}$$

would be represented by taking $\delta_1(B) = 1 - \delta_{1, 1}B - \delta_{1, 2}B^2$.

The next stage in the model building process is to amalgamate the univariate models (7.7.2) and (7.7.3) with the model (7.7.6) linking the innovation series $\epsilon_{1,t}$ and $\epsilon_{2,t}$. In general, this yields

$$Y_{1,t} = \frac{Kb_1(B)\omega_1(B)a_2(B)}{a_1(B)\delta_1(B)b_2(B)} Y_{2,t} + \frac{b_1(B)\theta_1(B)}{a_1(B)\delta_1(B)} \eta_{1,t}^* \qquad (7.7.9)$$

where $K = [\mathrm{Var}(\epsilon_{1,t})/\mathrm{Var}(\epsilon_{2,t})]^{1/2}$ since (7.7.6) is fitted to the standardized univariate residuals. Now, the great difficulty in practice with such a procedure, which yields the desired form (7.7.1), is that, unless the single series models contain very few coefficients, it is likely to yield a model with a great abundance of parameters. Two things can be done about this. First, one should be on the lookout for cancellations in the ratios of polynomials in B in (7.7.9), and second, on estimating the model, one should be prepared to drop coefficients that are very small compared to their standard errors.

Combining (7.7.4), (7.7.5) and (7.7.7), the amalgamated model for the unemployment–industrial production data is

$$Y_{1,t} = \frac{\omega_{1,0}^* + \omega_{1,1}^*B + \omega_{1,2}^*B^2 + \omega_{1,3}^*B^3}{1 - \delta_{1,1}^*B - \delta_{1,2}^*B^2 - \delta_{1,3}^*B^3 - \delta_{1,4}^*B^4} Y_{2,t}$$

$$+ \frac{1 - \theta_{1,1}^*B - \theta_{1,2}^*B^2}{1 - \phi_{1,1}^*B - \phi_{1,2}^*B^2 - \phi_{1,3}^*B^3 - \phi_{1,4}^*B^4} \eta_{1,t}^*$$

However, the term $\omega_{1,3}^*$ suggested by the merging of these equations is $(0.26)(0.25)K$, which is relatively small, and so this term was dropped before the model was estimated.

Multiplying through (7.7.1) by $\delta_1^*(B)\phi_1^*(B)$ produces

$$\phi_1^*(B)\delta_1^*(B)Y_{1,t} = \phi_1^*(B)\omega_1^*(B)Y_{2,t} + \delta_1^*(B)\theta_1^*(B)\eta_{1,t}^*$$

This can be written in the form

$$\eta_{1,t}^* = Y_{1,t} - \Phi_1 Y_{1,t-1} - \cdots - \Phi_H Y_{1,t-H} - \Omega_0 Y_{2,t} - \cdots - \Omega_J Y_{2,t-J}$$

$$+ \Theta_1 \eta_{1,t-1}^* + \cdots + \Theta_I \eta_{1,t-I}^* \qquad (7.7.10)$$

where H, J, and I are known integers and the Φ's, Ω's and Θ's are known functions of the ϕ_1^*'s, θ_1^*'s, ω_1^*'s, and δ_1^*'s. The coefficients of (7.7.1) are estimated by calculating, using (7.7.10), $\eta_{1,t}^*$, $t = \max(H + 1, J + 1), \ldots, n$, as a function of the data and the coefficients, setting the "starting up" values $\eta_{1,t}^*$, $t = \max(H, J), \ldots, \max(H, J) - I + 1$ equal to zero. The transient introduced by this approximation will typically not be of any great importance for moderately long series. The coefficients in (7.7.1) are then estimated by minimizing the sum of squares of the calculated $\eta_{1,t}^*$, using the nonlinear regression algorithm described in Section 3.5.

Unfortunately, there is no simple computational procedure available for optimal treatment of the starting value problem, analogous to that described in Section 3.5 for univariate time series model estimation. In fact, it has been shown by Newbold [1973b] that the optimal starting values depend on the unknown ratio of the variances of $\eta^*_{1,t}$ and $\epsilon_{2,t}$ of (7.7.1) and (7.7.2).

When the model relating unemployment and industrial production, with $Y_{1,t}$ and $Y_{2,t}$ denoting first differences of these series (the latter with mean subtracted), was estimated, the estimates of $\theta^*_{1,1}$ and $\theta^*_{1,2}$ were very small compared to their standard errors. These were dropped and the model reestimated, yielding estimates of $\phi^*_{1,2}$, $\phi^*_{1,3}$, $\phi^*_{1,4}$, and $\delta^*_{1,4}$ which were small compared to estimated standard errors. When the first three of these terms were dropped, the fourth remained very small, so that the model finally achieved was

$$Y_{1,t} = \frac{\overset{[13.2]}{(}\overset{[14.2]}{-28.9} \overset{}{-47.2B} \overset{[17.0]}{-36.8B^2})}{\underset{[0.18]}{(1+0.39B} \underset{[0.12]}{-0.53B^2} \underset{[0.12]}{-0.16B^3)}} Y_{2,t}$$

$$+ \frac{1}{\underset{[0.08]}{(1+0.24B)}} \eta^*_{1,t}; \qquad \hat{\sigma}^2_{\eta^*_1} = 10591 \qquad (7.7.11)$$

As a final step in the model building procedure, one should employ diagnostic checks on adequacy of representation. One possibility is the fitting of extra coefficients and application of tests of their statistical significance. However, in the present example, a good deal of this has already been done in arriving at (7.7.11). A second check is to examine the autocorrelations of the residuals from the fitted model. The values for (7.7.11) are shown in Table 7.2. On the hypothesis that $\eta^*_{1,t}$ is white noise, these quantities should be approximately normally distributed with zero means and standard deviations $n^{-1/2}$, where n is the number of residuals. In this case, $n = 139$ and so $n^{-1/2} = 0.085$. By this criterion, only the autocorrelations at lags 12 and 19 are more than two standard errors, which is not very suggestive, except that the former might be associated with some defect in the seasonal adjustment procedure. An overall test, due to Box and Jenkins [1970], on the size of these quantities is to compare the statistic $Q = n\sum_{\tau=1}^{M} r^2_\tau(\hat{\eta}^*_1)$ with tabulated values of χ^2 for $M - R$ degrees of freedom, where R is the sum of the orders of $\theta^*_1(B)$ and $\phi^*_1(B)$ in (7.7.1). For the data in Table 7.2, $Q = 139\sum_{\tau=1}^{20} r^2_\tau(\hat{\eta}^*_1) = 22.25$. This can be compared with tabulated values of χ^2 for 19 degrees of

Table 7.2   *Autocorrelations of residuals from (7.7.11)*

τ:	1	2	3	4	5	6	7	8	9	10
$r_\tau(\hat{\eta}^*_1)$:	−0.03	−0.05	0.00	0.08	0.05	0.03	−0.13	0.07	−0.04	−0.03

τ:	11	12	13	14	15	16	17	18	19	20
$r_\tau(\hat{\eta}^*_1)$:	−0.02	−0.23	0.01	0.01	0.09	−0.04	−0.08	−0.09	0.20	−0.07

freedom (since, from (7.7.11), $R = 1$), which are 30.14 at the 5% point and 27.20 at the 10% point. Thus, the test statistic is not unduly large, and the evidence does not contradict the hypothesis of white noise behavior in the residuals of (7.7.11).

Finally, one should check that the residuals from fitted models of the form (7.7.1) and (7.7.2) are mutually stochastically uncorrelated. For the unemployment–industrial production data, the cross correlations between the residuals from (7.7.11) and (7.7.5) are shown in Table 7.3. Under the null hypothesis of interest, Haugh [1972] has shown that, if the model is correct, the quantities should be approximately normally distributed with zero means and standard deviations $n^{-1/2}$ (≈ 0.085 in this case). By this criterion, none of the cross correlations in Table 7.3 seem unusually large. Haugh also derives overall tests on the size of these quantities, showing that the statistics

$$Q_1 = n \sum_{\tau=0}^{M_1} r^2(\hat{\eta}^*_{1,\,t},\, \hat{\epsilon}_{2,\,t-\tau}) \quad \text{and} \quad Q_2 = n \sum_{\tau=-M_2}^{M_2} r^2(\hat{\eta}^*_{1,\,t},\, \hat{\epsilon}_{2,\,t-\tau})$$

are, under the null hypothesis, approximately distributed as χ^2 with $M_1 - U$ and $2M_2 - U$ degrees of freedom, respectively, where U is the sum of the orders of $\omega^*_1(B)$ and $\delta^*_1(B)$ in (7.7.1). For the data in Table 7.3,

$$Q_1 = 139 \sum_{\tau=0}^{19} r^2(\hat{\eta}^*_{1,\,t},\, \hat{\epsilon}_{2,\,t-\tau}) = 12.15,$$

$$Q_2 = 139 \sum_{\tau=-12}^{12} r^2(\hat{\eta}^*_{1,\,t},\, \hat{\epsilon}_{2,\,t-\tau}) = 11.16$$

Q_1 can be compared with tabulated values of χ^2 for 14 degrees of freedom (since, from (7.7.11), $U = 5$), which are 23.68 at the 5% level and 21.06 at the 10% level. Q_2 is compared with values of χ^2 for 19 degrees of freedom, which were given in the previous paragraph. Neither test gives any cause to question the adequacy of the model.

Finally, the validity of the decision, based on Table 7.1, to employ a

Table 7.3   *Cross correlations between residuals from (7.7.11) and (7.7.5)*

τ:	0	1	2	3	4	5	6	7	8	9
$r(\hat{\eta}^*_{1,\,t},\, \hat{\epsilon}_{2,\,t-\tau})$:	-0.01	0.00	0.04	0.00	-0.05	-0.04	-0.03	0.04	0.01	-0.07
τ:	10	11	12	13	14	15	16	17	18	19
$r(\hat{\eta}^*_{1,\,t},\, \hat{\epsilon}_{2,\,t-\tau})$:	-0.01	0.06	0.14	-0.00	0.12	0.10	0.04	0.14	-0.04	0.06
τ:		-1	-2	-3	-4	-5	-6	-7	-8	-9
$r(\hat{\eta}^*_{1,\,t},\, \hat{\epsilon}_{2,\,t-\tau})$:		-0.06	-0.03	-0.04	-0.10	0.04	-0.09	-0.10	-0.05	-0.05
τ:	-10	-11	-12	-13	-14	-15	-16	-17	-18	-19
$r(\hat{\eta}^*_{1,\,t},\, \hat{\epsilon}_{2,\,t-\tau})$:	-0.04	0.02	0.03	-0.10	-0.01	-0.01	0.03	-0.06	-0.10	0.01

unidirectional causality model rather than a feedback model, can be checked by reference to the data in the bottom half of Table 7.3. The sample cross correlations here do not seem overly large, although they are persistently negative, which is what one would expect if there were feedback. However, the effect, if it exists, is certainly very weak, and these values give little justification for any elaboration of the fitted model. Haugh has shown that an overall test can be obtained by comparing

$$Q_3 = n \sum_{\tau = -M_3}^{-1} r^2(\hat{\eta}_{1,t}^*, \hat{\epsilon}_{2,t-\tau})$$

with tabulated χ^2 for M_3 degrees of freedom. For the data in Table 7.3,

$$Q_3 = 139 \sum_{\tau = -19}^{-1} r^2(\hat{\eta}_{1,t}^*, \hat{\epsilon}_{2,t-\tau}) = 9.52$$

which is small compared to tabulated values of χ^2 for 19 degrees of freedom.

The model composed of Eqs. (7.7.11) and (7.7.5)—it should be noted that *both* equations are required for a complete specification—seems, on the evidence of the data, to provide a satisfactory explanation of the relationship between unemployment and industrial production, and so will be retained.

7.8 Alternative Identification Procedure for Unidirectional Causality Models

The identification procedure described in the previous section is somewhat indirect. The construction of a model linking univariate residuals and the combination of this model with the two univariate models (7.7.2) and (7.7.3) is likely, unless the univariate models contain only a few coefficients, to lead to an amalgamated model that is overparametrized. Consequently, time may be wasted in fitting this model, dropping insignificant coefficients, and refitting the resulting equation. A more direct identification procedure, which does not suffer from this difficulty, was given by Box and Jenkins [1970], who first introduced models of the type (7.7.1), (7.7.2). It should be noted, however, that this procedure is not appropriate when there is feedback in the system.

Now, write (7.7.1) as

$$Y_{1,t} = V_1^*(B)Y_{2,t} + \psi_1^*(B)\eta_{1,t}^* \tag{7.8.1}$$

where $V_1^*(B) = (V_{1,0}^* + V_{1,1}^*B + V_{1,2}^*B^2 + \cdots)$, and define

$$Z_t = b_2^{-1}(B)a_2(B)Y_{1,t} \tag{7.8.2}$$

where $a_2(B)$ and $b_2(B)$ are given by (7.7.2). Multiplying through (7.8.1) by $b_2^{-1}(B)a_2(B)$ and substituting from (7.7.2) then yields

$$Z_t = V_1^*(B)\epsilon_{2,t} + b_2^{-1}(B)a_2(B)\psi_1^*(B)\eta_{1,t}^*$$

Multiplying through this equation by $\epsilon_{2,\,t-k}$, $k \geqslant 0$, and taking expectations, one has, since $\epsilon_{2,\,t}$ and $\eta^*_{1,\,t}$ are mutually stochastically uncorrelated white noise processes, $E[Z_t, \epsilon_{2,\,t-k}] = V^*_{1,\,k}\,\mathrm{var}(\epsilon_{2,\,t})$, $k = 0, 1, 2, \ldots$, and hence

$$V^*_{1,\,k} = \mathrm{corr}(Z_t, \epsilon_{2,\,t-k})\left[\frac{\mathrm{var}(Z_t)}{\mathrm{var}(\epsilon_{2,\,t})}\right]^{1/2} \tag{7.8.3}$$

Thus the $V^*_{1,\,k}$ are simply a constant multiple of the cross correlations between Z_t and $\epsilon_{2,\,t-k}$.

The suggested identification procedure for the transfer function is then as follows:

(i)   Difference the given series $X_{1,\,t}$ and $X_{2,\,t}$ a sufficient number of times to produce stationary series $Y_{1,\,t}$ and $Y_{2,\,t}$.

(ii)   Build a univariate model of the form (7.7.2) for the input series $Y_{2,\,t}$.

(iii)   Transform the output series $Y_{1,\,t}$, according to (7.8.2), using the coefficient estimates of (ii), to produce the series Z_t.

(iv)   Calculate the cross correlations between Z_t and $\hat{\epsilon}_{2,\,t-k}$, $k = 0, 1, 2, \ldots$, where the latter are the estimated residuals obtained from (ii), and hence obtain estimates of $V^*_{1,\,k}$, $k = 0, 1, 2, \ldots$.

(v)   Use these estimated $V^*_{1,\,k}$ to suggest an appropriate form for $V^*_1(B) = \omega^*_1(B)/\delta^*_1(B)$.

To identify the error structure in (7.8.1), Box and Jenkins recommend estimating the model $Y_{1,\,t} = [\omega^*_1(B)/\delta^*_1(B)]Y_{2,\,t} + e_t$, where the forms of $\omega^*_1(B)$ and $\delta^*_1(B)$ are those just identified. and where it is assumed for the present that e_t is white noise. The autocorrelation function of the residuals $\hat{e}_t$ from the fitted model is then employed to suggest an error structure $\phi^*_1(B)e_t = \theta^*_1(B)\eta^*_{1,\,t}$, using the methods of Chapter 3. The error structure so identified is then inserted into the model giving

$$Y_{1,\,t} = \frac{\omega^*_1(B)}{\delta^*_1(B)}\,Y_{2,\,t} + \frac{\theta^*_1(B)}{\phi^*_1(B)}\,\eta^*_{1,\,t}$$

the coefficients of which are then estimated in the usual way.

7.9 Outline of Bivariate Model Building Strategy: Feedback

The problem considered in the previous two sections is considerably less difficult than that of building a model of the form

$$Y_{1,\,t} = \frac{\omega^*_1(B)}{\delta^*_1(B)}\,Y_{2,\,t} + \frac{\theta^*_1(B)}{\phi^*_1(B)}\,\eta^*_{1,\,t}$$

$$Y_{2,\,t} = \frac{\omega^*_2(B)}{\delta^*_2(B)}\,Y_{1,\,t} + \frac{\theta^*_2(B)}{\phi^*_2(B)}\,\eta^*_{2,\,t} \tag{7.9.1}$$

where $\eta_{1,t}^*$ and $\eta_{2,t}^*$ are mutually stochastically uncorrelated white noise processes and $\omega_2^*(0) \equiv 0$. Indeed, we would not assert dogmatically that this model provides the best framework for practical analysis of the bivariate feedback problem. However, in our experience it has proved more satisfactory than any alternative.

We would strongly contend that if feedback exists, or is suspected, in the system, an important first step is to construct the single series models

$$a_j(B)Y_{j,t} = b_j(B)\epsilon_{j,t}, \qquad j = 1, 2 \qquad (7.9.2)$$

and to examine the relationship between the residuals from these equations. In addition, we have found that a well-worthwhile intermediate step, on the road to constructing (7.9.1), is to build a model linking the residuals, of the form

$$\epsilon_{1,t} = \frac{\omega_1(B)}{\delta_1(B)}\epsilon_{2,t} + \frac{\theta_1(B)}{\phi_1(B)}\eta_{1,t},$$

$$\epsilon_{2,t} = \frac{\omega_2(B)}{\delta_2(B)}\epsilon_{1,t} + \frac{\theta_2(B)}{\phi_2(B)}\eta_{2,t} \qquad (7.9.3)$$

where, in (7.9.3), the $\epsilon_{j,t}$ are standardized to have unit variances, $\eta_{1,t}$ and $\eta_{2,t}$ are mutually stochastically uncorrelated white noise series and $\omega_2(0) \equiv 0$.

The model building strategy, which will be described and illustrated in detail in the next three sections, proceeds in the following stages:

(i) Fit the single series models (7.9.2), where $Y_{j,t}$ may be an appropriately differenced form of a given series $X_{j,t}$. Calculate the residuals from each fitted model, and standardize so that each residual series has unit variance.

(ii) Calculate the cross correlogram between the univariate model residuals and use this to identify appropriate forms for the transfer functions $\omega_j(B)/\delta_j(B)$ of (7.9.3).

(iii) Identify the error structures—that is, the forms of $\theta_j(B)/\phi_j(B)$—in (7.9.3).

(iv) Estimate the model (7.9.3) relating the individual innovation series.

(v) Check the adequacy of representation of this fitted model, and, if necessary, modify and reestimate it.

(vi) Amalgamate the bivariate model fitted to the residuals (7.9.3) with the two univariate models (7.9.2) to suggest an appropriate structure, of the form (7.9.1), to describe the interrelationship between the original series.

(vii) Estimate the model (7.9.1) relating the original series.

(viii) Check the adequacy of representation of this fitted model, and, if necessary, modify and reestimate it.

7.10 Fitting Bivariate Models to Innovation Series: Feedback

In this section and the next, the model building strategy just described will be discussed in detail and illustrated by reference to U.S. data on total personal consumption expenditure $(X_{1,t})$ and gross private domestic investment $(X_{2,t})$, both seasonally adjusted and in current dollars, covering 68 quarterly observations. The univariate models fitted to this data were

$$(1 - 0.32B - 0.49B^2)(1 - B)X_{1,t} = 1.73 + \epsilon_{1,t}, \quad \hat{\sigma}_{\epsilon_1}^2 = 12.1 \quad (7.10.1)$$
$$\quad [0.11] \quad\quad [0.12] \quad\quad\quad\quad\quad [0.91]$$

and

$$(1 - B)X_{2,t} = 1.78 + \epsilon_{2,t}, \quad \hat{\sigma}_{\epsilon_2}^2 = 21.9 \quad\quad (7.10.2)$$
$$\quad\quad\quad [0.57]$$

Now, define $\lambda_k = E[\epsilon_{1,t}, \epsilon_{2,t-k}]$, $k = \ldots, -1, 0, 1, \ldots$, and, in (7.9.3), set

$$V_j(B) = \frac{\omega_j(B)}{\delta_j(B)}, \quad j = 1, 2$$

Then, since $\eta_{1,t}$ and $\eta_{2,t}$ are mutually stochastically uncorrelated, it follows from (7.9.3) that

$$E[(\epsilon_{1,t} - V_1(B)\epsilon_{2,t})(\epsilon_{2,t-k} - V_2(B)\epsilon_{1,t-k})] = 0, \quad k = \ldots, -1, 0, 1, \ldots$$

and hence

$$\lambda_k = V_{1,k} - \sum_{i=0}^{\infty} \sum_{j=1}^{\infty} V_{1,i} V_{2,j} \lambda_{i-j-k}, \quad k = 0, 1, 2, \ldots$$

$$\lambda_{-k} = V_{2,k} - \sum_{i=0}^{\infty} \sum_{j=1}^{\infty} V_{1,i} V_{2,j} \lambda_{k+i-j}, \quad k = 1, 2, \ldots \quad (7.10.3)$$

Given estimates of the λ_k, based on data, one might try to estimate $V_{1,k}$, $k = 0, 1, \ldots, M$ and $V_{2,k}$, $k = 1, 2, \ldots, M$, where M is some integer (say 10 or 15), by truncating the infinite sums in (7.10.3) at M, and using some iterative procedure to solve these equations. One possibility is to set $\hat{V}_{2,k}^{(1)} = \hat{\lambda}_{-k}$, $k = 1, 2, \ldots, M$, substitute these for $V_{2,k}$, in the first set of equations in (7.10.3) and solve the resulting linear system for $V_{1,k}$, giving estimates $\hat{V}_{1,k}^{(1)}$, $k = 0, 1, 2, \ldots, M$. These can be substituted for $V_{1,k}$ in the second set of equations in (7.10.3), and the resulting linear system solved for $V_{2,k}$, yielding new estimates $\hat{V}_{2,k}^{(2)}$, $k = 1, 2, \ldots, M$. The process is then continued iteratively until convergence is achieved.

Experience has indicated two difficulties with this approach. First, the iteration algorithm is not guaranteed to converge, and second, we have noted on occasion a tendency for sampling errors in the $\hat{\lambda}_k$ to sometimes cause an uncomfortable amount of distortion in the estimates of the V's. The following modifications considerably alleviate these difficulties. The equations (7.10.3) are solved, as indicated above, but taking in turn $M = 2, 3, 4, \ldots$. Initial estimates $\hat{V}_{2,k}^{(1)}$, $k = 1, 2, \ldots, M - 1$, are taken as the final estimates obtained at the previous stage, with $\hat{V}_{2,M}^{(1)} = 0$. Further, to increase the chances of convergence, the iterations are tried for a number of different weights W, setting

$$\hat{V}_{j,k}^{(i)} = W\hat{V}_{j,k}^{(i-1)} + (1 - W)(\text{value obtained by solving the equations})$$

We generally try $W = 0(0.2)0.8$. Thus one has a sequence of estimates of the V's to examine, for $M = 2, 3, 4, \ldots,$ and can come to some judgment as to whether abrupt changes in the sequence, for increasing M, are caused by sampling error or reflect some real underlying relationship. Of necessity, there is an element of subjectivity here as there is in any time series identification procedure.

In order to get some feeling for this proposed procedure for identifying the V's, a small simulation experiment was conducted. Time series of length 100 were generated from the model

$$\epsilon_{1,t} = \frac{0.7 - 0.7B}{1 - 0.6B} \epsilon_{2,t} + \psi_1(B)\eta_{1,t}, \qquad \epsilon_{2,t} = 0.8B\epsilon_{1,t} + \psi_2(B)\eta_{2,t} \qquad (7.10.4)$$

where the error structures $\psi_1(B)$ and $\psi_2(B)$ were determined, for data generation, from (7.6.18) and (7.6.21)–(7.6.23). For small values of M, estimates that were not too unreasonable resulted. Those for $M = 5$ are shown in Table 7.4, together with the corresponding true values. On the whole, these are reasonably accurate, though it is disappointing that $\hat{V}_{1,2}$ is so far off, making correct identification of the transfer function in the first equation of (7.10.4) unlikely at the first attempt. Nevertheless, we feel that results of this kind would provide sufficient information to enable a reasonable start to model building to be made. For values of M greater than 5, the estimates behaved rather erratically, reflecting some large sampling errors in the $\hat{\lambda}_k$. It is generally advisable to ignore such behavior in identifying a model. Thus, this simulation experiment underlined the desirability of estimating the V's for a sequence of values of M. Moreover, it indicates the necessity for subjecting the originally chosen model to checks on adequacy of representation, and one

Table 7.4 *Estimated and true V weights for generated data from (7.10.4)*

j	$\hat{V}_{1,j}$	$V_{1,j}$	j	$\hat{V}_{2,j}$	$V_{2,j}$
0	0.64	0.70			
1	−0.25	−0.28	1	0.62	0.80
2	0.12	−0.17	2	0.07	0.00
3	0.03	−0.10	3	0.20	0.00
4	0.07	−0.06	4	0.00	0.00
5	0.14	−0.04	5	0.15	0.00

should certainly be prepared to abandon this model if the checks should so indicate.

The sample cross correlations of the residuals from (7.10.1) and (7.10.2) are given in Table 7.5. These are based on $n = 65$ observations, so that the standard error appropriate if the series were independent would be $n^{-1/2} \approx 0.12$. Judging the size of the estimated cross correlations by this criterion, the value of $\hat{\lambda}_{-1}$ certainly looks large. Moreover, the moderately large values for $\hat{\lambda}_0$ and $\hat{\lambda}_1$, though each is less than $2n^{-1/2}$, suggest at least the possibility of feedback, particularly when it is noted that the signs are what economic theory would lead one to expect.

The solutions of Eqs. (7.10.3), taking in turn $M = 2, 3, 4, \ldots, 10$, are shown in Table 7.6. Perhaps the most striking feature of this table is the extremely erratic behavior of the estimates when M exceeds 7. The explanation for this phenomenon can be found in Table 7.5, which shows an unusually large value for $\hat{\lambda}_{-8}$. Keeping in mind that these are sample correlations between individual *innovation* series, it is to be doubted that this effect represents anything more than sampling error, or possibly an artifact created by the seasonal adjustment procedure.

Examination of Table 7.6 for $M = 7$ suggests that the $V_{1,j}$ might decay sinusoidally, perhaps immediately or after the first value, suggesting that, in (7.9.3), an appropriate form might be

$$V_1(B) = \frac{\omega_{1,0} + \omega_{1,1}B}{1 - \delta_{1,1}B - \delta_{1,2}B^2} \qquad (7.10.5)$$

Now, looking at Table 7.6 for $M = 7$, $\hat{V}_{2,1}$ is quite large, and the first few $\hat{V}_{2,j}$ are persistently positive, suggesting the possibility of first- or second-order decay. Accordingly, the form

$$V_2(B) = \frac{\omega_{2,1}B}{1 - \delta_{2,1}B - \delta_{2,2}B^2} \qquad (7.10.6)$$

seems appropriate.

Table 7.5 *Cross correlations of individual model residuals for consumption–investment data*

k	$\hat{\lambda}_k$	k	$\hat{\lambda}_k$	k	$\hat{\lambda}_k$	k	$\hat{\lambda}_k$
0	0.17	8	−0.10			−8	−0.25
1	0.20	9	0.03	−1	0.33	−9	0.14
2	0.03	10	0.03	−2	0.02	−10	0.05
3	−0.02	11	0.18	−3	0.16	−11	−0.08
4	−0.09	12	0.02	−4	0.15	−12	0.07
5	−0.19	13	−0.01	−5	−0.11	−13	0.14
6	0.07	14	0.13	−6	0.00	−14	0.02
7	0.19	15	0.04	−7	0.03	−15	0.13

Table 7.6 *Estimation of the V weights for use in identification of model linking innovation series for consumption–investment data for M = 2, 3, . . . , 10, respectively*

j	$\hat{V}_{1,j}$	$\hat{V}_{2,j}$	j	$\hat{V}_{1,j}$	$\hat{V}_{2,j}$	j	$\hat{V}_{1,j}$	$\hat{V}_{2,j}$
0	0.21		0	0.24		0	0.24	
1	0.23	0.36	1	0.26	0.39	1	0.24	0.40
2	0.05	0.04	2	0.06	0.07	2	0.03	0.10
			3	0.00	0.18	3	−0.01	0.21
						4	−0.11	0.16

j	$\hat{V}_{1,j}$	$\hat{V}_{2,j}$	j	$\hat{V}_{1,j}$	$\hat{V}_{2,j}$	j	$\hat{V}_{1,j}$	$\hat{V}_{2,j}$
0	0.25		0	0.26		0	0.23	
1	0.23	0.43	1	0.24	0.43	1	0.21	0.42
2	−0.01	0.10	2	0.00	0.10	2	−0.00	0.08
3	−0.05	0.18	3	−0.04	0.18	3	−0.02	0.19
4	−0.15	0.15	4	−0.15	0.14	4	−0.10	0.16
5	−0.29	−0.17	5	−0.28	−0.18	5	−0.23	−0.15
			6	0.06	−0.06	6	0.10	−0.03
						7	0.19	0.04

j	$\hat{V}_{1,j}$	$\hat{V}_{2,j}$	j	$\hat{V}_{1,j}$	$\hat{V}_{2,j}$	j	$\hat{V}_{1,j}$	$\hat{V}_{2,j}$
0	0.34		0	0.12		0	0.15	
1	0.36	0.47	1	0.40	0.52	1	0.21	0.43
2	0.08	0.07	2	0.00	0.04	2	−0.08	0.11
3	0.08	0.11	3	−0.14	0.14	3	−0.14	0.26
4	−0.07	0.04	4	0.10	0.16	4	−0.05	0.22
5	−0.34	−0.29	5	−0.40	−0.26	5	−0.29	−0.14
6	0.04	−0.11	6	−0.01	0.03	6	0.11	0.02
7	0.09	−0.02	7	0.38	−0.02	7	0.37	−0.02
8	−0.27	−0.29	8	−0.38	−0.32	8	−0.14	−0.31
			9	0.13	0.31	9	0.18	0.18
						10	0.19	−0.02

It remains now to identify appropriate error structures $\theta_j(B)/\phi_j(B)$ in (7.9.3). There are three possible approaches:

(i) For fairly simple models, (7.6.9) can be used directly, as illustrated in Section 7.6, given identified forms for $V_1(B)$ and $V_2(B)$.

(ii) Although the model actually identified for the transfer functions $V_1(B)$ and $V_2(B)$ may be too complicated for the solution of (7.6.9) to be analytically tractable, it may be possible to find some simple alternative model that could provide a reasonable approximation to the behavior of the $\hat{V}$'s, and hence allow method (i) to be applied to suggest appropriate error structures for the more complicated model.

(iii)   Consider the first equation of (7.9.3). The autocovariances for the error term can be written as

$$C_k = E[(\epsilon_{1,t} - V_1(B)\epsilon_{2,t})(\epsilon_{1,t-k} - V_1(B)\epsilon_{2,t-k})], \qquad k = 0, 1, 2, \ldots$$

Hence, since $\epsilon_{1,t}$ and $\epsilon_{2,t}$ have unit variances,

$$C_0 = 1 + \sum_{j=0}^{\infty} V_{1,j}^2 - 2 \sum_{j=0}^{\infty} V_{1,j}\lambda_j$$

and

$$C_k = \sum_{j=k}^{\infty} V_{1,j}V_{1,j-k} - \sum_{j=0}^{\infty} V_{1,j}\lambda_{j-k} - \sum_{j=0}^{\infty} V_{1,j}\lambda_{j+k}, \qquad k = 1, 2, \ldots$$

For practical use, the infinite sums in these expressions are truncated at some number M. The estimated λ_j can be substituted directly into the equations. For the $V_{1,j}$ it is best to use the values suggested by the identified transfer function, rather than those calculated from (7.10.3). That is to say, one solves the equation

$$\delta_1(B) = \omega_1(B)V_1(B) \tag{7.10.7}$$

by equating coefficients in B^0, B, $B^2, \ldots$. The first few equations will yield estimates of the ω's and δ's in the identified model, given the first few V's calculated from (7.10.3). Estimates of further V's are then obtained from the estimates of the ω's and δ's by equating coefficients in (7.10.7) for higher powers of B. Given the estimated autocovariances, one can estimate the corresponding autocorrelations and hence, using the methodology of Chapter 3, attempt to identify an appropriate error structure. This procedure will be illustrated in Section 7.12.

For the consumption–investment data, procedure (ii) was used. As a crude approximation, it appears from Table 7.6 that the forms

$$V_1(B) = 0.23 + 0.21B, \qquad V_2(B) = 0.42B$$

might be adequate. Using (7.6.18) and (7.6.21)–(7.6.23), the implied error structure is

$$\psi_1(B) = \frac{1 - 0.06B - 0.11B^2 - 0.004B^3}{1 + 0.01B}, \qquad \psi_2(B) = \frac{1 - 0.10B - 0.09B^2}{1 + 0.01B}$$

Dropping the very small terms in these equations, the forms finally employed were

$$\psi_1(B) = 1 - \theta_{1,1}B - \theta_{1,2}B^2 \tag{7.10.8}$$

$$\psi_2(B) = 1 - \theta_{2,1}B - \theta_{2,2}B^2 \tag{7.10.9}$$

Combining (7.10.8) and (7.10.9) with (7.10.5) and (7.10.6) gives the model structure to be estimated. The estimation procedure employed here is identical to that used for the model relating the actual series, which will be described in the following section.

The estimated equations were

$$\epsilon_{1,t} = \frac{\overset{[0.10]}{0.24} - \overset{[0.13]}{0.01}B}{1 - \underset{[0.20]}{1.03}B + \underset{[0.15]}{0.74}B^2} \epsilon_{2,t} + (1 - \underset{[0.10]}{0.22}B - \underset{[0.14]}{0.32}B^2)\eta_{1,t} \quad (7.10.10)$$

and

$$\epsilon_{2,t} = \frac{\overset{[0.12]}{0.51}B}{1 - \underset{[0.18]}{0.49}B + \underset{[0.13]}{0.00}B^2} \epsilon_{1,t} + (1 - \underset{[0.12]}{0.09}B - \underset{[0.12]}{0.39}B^2)\eta_{2,t} \quad (7.10.11)$$

One check on the adequacy of (7.10.10) and (7.10.11) is to add further coefficients, reestimate the equations, and test the statistical significance of the extra coefficients. To some extent this has already been done since the size of $\hat{\omega}_{1,1}$ in (7.10.10) and $\hat{\delta}_{2,2}$ in (7.10.11) strongly suggest their probable redundancy. Accordingly, in the further analysis of the following section, these terms will be dropped. It should be noted that, in adding and dropping coefficients from these equations, the restrictions imposed by (7.6.9) may be violated. However, in practice, the residual series achieved should be indistinguishable from white noise.

A further check on model adequacy is to calculate the autocorrelations of and cross correlations between the residuals $\eta_{1,t}$ and $\eta_{2,t}$ from the fitted equations. For the model (7.10.10), (7.10.11), these quantities are given in Tables 7.7 and 7.8. When compared with standard errors of 0.12, appropriate under the hypothesis that $\eta_{1,t}$ and $\eta_{2,t}$ are mutually stochastically uncorrelated white noise processes, none of these values seem alarmingly large, particularly since the present is only an intermediate stage in the whole model building procedure.

Table 7.7 *Autocorrelations of estimated residuals from* (7.10.10) *and* (7.10.11)

τ	$r_\tau(\hat{\eta}_1)$	τ	$r_\tau(\hat{\eta}_1)$	τ	$r_\tau(\hat{\eta}_2)$	τ	$r_\tau(\hat{\eta}_2)$
1	−0.14	11	0.01	1	−0.01	11	0.16
2	0.04	12	0.24	2	0.00	12	−0.01
3	0.08	13	−0.00	3	−0.05	13	−0.02
4	0.08	14	−0.08	4	−0.07	14	−0.08
5	0.08	15	0.01	5	0.04	15	−0.02
6	−0.04	16	0.18	6	0.08	16	−0.13
7	0.03	17	−0.05	7	0.02	17	−0.01
8	0.19	18	0.10	8	−0.26	18	−0.03
9	−0.08	19	−0.03	9	−0.07	19	−0.20
10	0.15	20	0.08	10	0.04	20	0.01

Table 7.8 *Cross correlations between estimated residuals from (7.10.10) and (7.10.11)*

τ	$r(\hat{\eta}_{1,t}, \hat{\eta}_{2,t-\tau})$	τ	$r(\hat{\eta}_{1,t}, \hat{\eta}_{2,t-\tau})$	τ	$r(\hat{\eta}_{1,t}, \hat{\eta}_{2,t-\tau})$	τ	$r(\hat{\eta}_{1,t}, \hat{\eta}_{2,t-\tau})$
0	0.04	8	−0.15			−8	−0.16
1	−0.00	9	−0.01	−1	−0.03	−9	0.08
2	−0.15	10	−0.04	−2	−0.07	−10	−0.05
3	−0.07	11	0.05	−3	0.16	−11	−0.16
4	−0.02	12	−0.05	−4	0.10	−12	0.00
5	−0.05	13	−0.00	−5	−0.14	−13	0.02
6	−0.00	14	−0.04	−6	−0.02	−14	−0.00
7	0.06	15	0.00	−7	0.14	−15	0.23

Accordingly, the model (7.10.10) and (7.10.11), with insignificant coefficients dropped, will be carried forward to the next section, where a model for the actual consumption and investment series will be constructed.

It is, of course, true that in principle one need not go to the trouble of estimating and checking a bivariate model for the residuals. One could simply amalgamate the identified model for the residuals with the two univariate models, and immediately proceed to estimate and check the resulting equations. Nevertheless, we feel that this intermediate step is frequently likely to be well worthwhile. In the first place, any simplifications that result from the dropping of redundant coefficients here, will lead to a simpler form for the amalgamated model. Further, any inadequacies revealed in the model structure are more easily rectified in terms of the model linking the innovations, as one should get a much clearer idea here of just where these defects lie.

7.11 Fitting Bivariate Models to Observed Series: Feedback

The final stages of the feedback model building procedure involve the identification, based on results from earlier stages, of a model of the form (7.9.1), linking the given series, together with estimation and checking of that model. The first step, then, is to merge the univariate models (7.9.2) with the model (7.9.3) relating the innovation series. In general, this yields

$$Y_{1,t} = K_1 \frac{b_1(B)\omega_1(B)a_2(B)}{a_1(B)\delta_1(B)b_2(B)} Y_{2,t} + \frac{b_1(B)\theta_1(B)}{a_1(B)\phi_1(B)} \eta_{1,t}^*$$

$$Y_{2,t} = K_2 \frac{b_2(B)\omega_2(B)a_1(B)}{a_2(B)\delta_2(B)b_1(B)} Y_{1,t} + \frac{b_2(B)\theta_2(B)}{a_2(B)\phi_2(B)} \eta_{2,t}^*$$

where

$$K_1 = \left[\frac{\text{var}(\epsilon_{1,t})}{\text{var}(\epsilon_{2,t})} \right]^{1/2}, \qquad K_2 = K_1^{-1}$$

The danger here is of obtaining an overparametrized structure, and one should be on the lookout for possible cancellations in the various ratios of polynomials in B in the amalgamated structure, and also be prepared to drop redundant coefficients when the model is estimated.

Returning to the example of the previous section, let $Y_{1,t}$ and $Y_{2,t}$ denote respectively first differences, with means subtracted, of consumption and investment. Combining (7.10.1) and (7.10.2) with (7.10.10) and (7.10.11) (with insignificant coefficients dropped) then suggests equations of the form

$$Y_{1,t} = \frac{\omega^*_{1,0}}{1 - \delta^*_{1,1}B - \delta^*_{1,2}B^2 - \delta^*_{1,3}B^3 - \delta^*_{1,4}B^4} Y_{2,t}$$

$$+ \frac{1 - \theta^*_{1,1}B - \theta^*_{1,2}B^2}{1 - \phi^*_{1,1}B - \phi^*_{1,2}B^2} \eta^*_{1,t} \tag{7.11.1}$$

and

$$Y_{2,t} = \frac{\omega^*_{2,1}B + \omega^*_{2,2}B^2 + \omega^*_{2,3}B^3}{1 - \delta^*_{2,1}B} Y_{1,t} + (1 - \theta^*_{2,1}B - \theta^*_{2,2}B^2)\eta^*_{2,t} \tag{7.11.2}$$

Now, since $\eta^*_{1,t}$ and $\eta^*_{2,t}$ are mutually stochastically uncorrelated white noise processes, and since $\omega^*_2(0) = 0$, one can treat the equations in (7.9.1) separately for estimation purposes. The estimation procedure then involves application of nonlinear least squares, exactly as in the unidirectional causality case described in Section 7.7, the errors being calculated in terms of the data and coefficients using equations of the form (7.7.10), with unknown "starting up" values again set equal to zero.

When Eq. (7.11.1) was estimated, the estimates of $\delta^*_{1,3}$, $\delta^*_{1,4}$, and $\phi^*_{1,2}$ did not differ significantly from zero. Moreover, when the last two of these terms were dropped, the estimate of the first remained insignificant. The equation finally fitted was

$$Y_{1,t} = \frac{\overset{[0.06]}{0.20}}{\underset{[0.17]\quad[0.16]}{1 - 0.94B + 0.41B^2}} Y_{2,t} + \frac{\overset{[0.15]\quad\ \ [0.14]}{1 - 1.17B + 0.39B^2}}{\underset{[0.09]}{1 - 0.95B}} \eta^*_{1,t}, \quad \hat{\sigma}^2_{\eta^*_1} = 8.54$$

$$\tag{7.11.3}$$

When Eq. (7.11.2) was estimated, the coefficient estimate $\hat{\omega}^*_{2,3}$ was small compared to its estimated standard error. Dropping this term from the equation, and reestimating, yielded

$$Y_{2,t} = \frac{\overset{[0.17]\quad\ \ [0.16]}{0.62B - 0.57B^2}}{\underset{[0.18]}{1 - 0.83B}} Y_{1,t} + (1 - \underset{[0.13]}{0.01}B - \underset{[0.13]}{0.28}B^2)\eta^*_{2,t}, \quad \hat{\sigma}^2_{\eta^*_2} = 17.52$$

$$\tag{7.11.4}$$

It remains only to check the adequacy of the fitted model. As is very

Table 7.9   *Autocorrelations of estimated residuals from (7.11.3) and (7.11.4)*

τ	$r_\tau(\hat{\eta}_1^*)$	τ	$r_\tau(\hat{\eta}_1^*)$	τ	$r_\tau(\hat{\eta}_2^*)$	τ	$r_\tau(\hat{\eta}_2^*)$
1	−0.01	11	0.07	1	−0.01	11	0.19
2	−0.06	12	0.20	2	0.04	12	−0.01
3	−0.09	13	−0.14	3	0.00	13	0.07
4	0.05	14	−0.20	4	−0.12	14	−0.03
5	0.07	15	−0.02	5	0.02	15	0.01
6	−0.06	16	0.23	6	−0.02	16	−0.12
7	−0.08	17	0.04	7	−0.03	17	−0.03
8	0.05	18	0.07	8	−0.22	18	−0.04
9	−0.11	19	−0.04	9	−0.09	19	−0.17
10	0.13	20	0.08	10	0.07	20	0.00

common in this kind of approach, quite a lot of overfitting was already done before achieving (7.11.3) and (7.11.4), and it was not felt that the addition of further coefficients was likely to prove useful. The autocorrelations and cross correlations of the residuals from these equations are given in Tables 7.9 and 7.10. On the hypothesis that the error series in (7.11.3) and (7.11.4) are mutually stochastically uncorrelated white noise processes, the quantities in these tables should be asymptotically distributed as independent normal, with zero means and standard deviations approximately 0.125. The results of Table 7.9 are pretty reassuring, and the same can be said for the left-hand half of Table 7.10. The right-hand half of Table 7.10 does, however, call for further comment, although it should be noted that none of these sample cross correlations is as big as two standard errors. What is slightly worrying is that moderately large positive values are obtained for $\tau = -2$, -3, and -4, and an even larger negative value for $\tau = -5$. One would have hoped to clear out cross correlations at small lags rather more successfully than this. The explanation for these results can be seen in Table 7.5 and, even more clearly,

Table 7.10   *Cross correlations between estimated residuals from (7.11.3) and (7.11.4)*

τ	$r(\hat{\eta}_{1,t}^*, \hat{\eta}_{2,t-\tau}^*)$	τ	$r(\hat{\eta}_{1,t}^*, \hat{\eta}_{2,t-\tau}^*)$	τ	$r(\hat{\eta}_{1,t}^*, \hat{\eta}_{2,t-\tau}^*)$	τ	$r(\hat{\eta}_{1,t}^*, \hat{\eta}_{2,t-\tau}^*)$
0	−0.05	8	−0.01			− 8	−0.14
1	0.01	9	0.01	−1	−0.01	− 9	0.07
2	0.03	10	−0.04	−2	0.15	−10	−0.10
3	0.06	11	−0.01	−3	0.15	−11	−0.21
4	0.02	12	−0.04	−4	0.13	−12	0.04
5	−0.11	13	0.03	−5	−0.21	−13	0.04
6	0.01	14	0.10	−6	−0.04	−14	0.10
7	0.11	15	0.08	−7	0.05	−15	0.19

in Table 7.6, where persistently positive estimates $\hat{V}_{2,j}$, $j = 1, 2, 3, 4$, are abruptly followed by the moderately large and negative $\hat{V}_{2,5}$. This unusual value appears to be exerting an influence on the model estimates, and this effect is reflected in Table 7.10. This leads to the question of what, if anything, should be done about it. Our inclination is to leave the estimated model as it stands, for after all the sample cross correlations in question are not statistically significant at the usual levels. Moreover no simple modification of the model would seem likely to make things much better. Doubtless, in the previous section, one could have identified

$$V_2(B) = \omega_{2,1}B + \omega_{2,2}B^2 + \omega_{2,3}B^3 + \omega_{2,4}B^4 + \omega_{2,5}B^5$$

but this hardly seems warranted by the evidence, and positive values for the first four ω's, followed by a negative value for the fifth would seem intuitively implausible, and extremely difficult to justify. Accordingly (7.11.3) and (7.11.4) are accepted as providing a reasonable model to describe the interrelationship of the two series.

7.12 A Further Illustration of Bivariate Model Building

In order to illustrate the points made in the previous sections, a further example will now be considered. It seems possible that a feedback relationship exists between sales and interest rates. Interest rates might be expected to increase when sales are booming, and a rise in interest rates to lead to a drop in sales. United States data on manufacturer's sales and discount rate on new issues of treasury bills (as an interest rate series), each covering 68 quarterly observations, were employed. Denote discount rate by $X_{1,t}$ and sales by $X_{2,t}$. The univariate models fitted to these series were

$$(1 - B)X_{1,t} = (1 + 0.48B - 0.09B^2)\epsilon_{1,t}, \qquad \hat{\sigma}_{\epsilon_1}^2 = 0.20 \qquad (7.12.1)$$
$$\phantom{(1 - B)X_{1,t} = (1 + }{\scriptstyle[0.13]}{\scriptstyle[0.13]}$$

and

$$(1 - B)X_{2,t} = 1.44 + (1 + 0.13B)\epsilon_{2,t}, \qquad \hat{\sigma}_{\epsilon_2}^2 = 5.97 \qquad (7.12.2)$$
$$\phantom{(1 - B)X_{2,t} = 1.44 + (1 + }{\scriptstyle[0.12]}$$

The cross correlations of the residuals from these two fitted equations are given in Table 7.11, and the corresponding estimated V weights, obtained from the solution of (7.10.3), in Table 7.12. On the hypothesis that the innovation series are mutually stochastically uncorrelated, many of the values in Table 7.11 certainly seem unusually large, since $n^{-1/2} \approx 0.12$. The values $\hat{\lambda}_0$ and $\hat{\lambda}_1$ are bigger than two standard errors, and of the expected sign. Rather perplexingly, the largest value on the left-hand side of this table is $\hat{\lambda}_5 = -0.31$, which certainly seems odd, and is likely to cause trouble. Strangely enough, a similar phenomenon can be seen on the other side of the table, for moderately large negative values around $\hat{\lambda}_{-4}$ are abruptly followed by $\hat{\lambda}_{-6} = 0.17$. If nothing else this example seems destined to show that all time series model building problems that occur in practice are not straightforward!

Table 7.11   Cross correlations of individual model residuals for discount rate–sales data

k	$\hat{\lambda}_k$	k	$\hat{\lambda}_k$	k	$\hat{\lambda}_k$	k	$\hat{\lambda}_k$
0	0.26	8	−0.00			−8	−0.11
1	0.26	9	−0.10	−1	0.01	−9	0.14
2	0.06	10	0.10	−2	0.03	−10	−0.06
3	0.08	11	−0.03	−3	−0.18	−11	0.02
4	0.12	12	−0.03	−4	−0.27	−12	0.24
5	−0.31	13	0.01	−5	−0.16	−13	−0.06
6	−0.04	14	0.06	−6	0.17	−14	0.04
7	0.08	15	−0.01	−7	−0.06	−15	0.01

Table 7.12   Estimation of the V weights for use in identification of model linking innovation series for discount rate–sales data

j	$\hat{V}_{1,j}$	$\hat{V}_{2,j}$	j	$\hat{V}_{1,j}$	$\hat{V}_{2,j}$	j	$\hat{V}_{1,j}$	$\hat{V}_{2,j}$
0	0.26		0	0.26		0	0.26	
1	0.26	0.01	1	0.28	−0.00	1	0.33	−0.05
2	0.06	0.04	2	0.09	0.01	2	0.13	−0.05
			3	0.08	−0.21	3	0.12	−0.29
						4	0.18	−0.36

j	$\hat{V}_{1,j}$	$\hat{V}_{2,j}$	j	$\hat{V}_{1,j}$	$\hat{V}_{2,j}$	j	$\hat{V}_{1,j}$	$\hat{V}_{2,j}$
0	0.23		0	0.23		0	0.27	
1	0.39	−0.21	1	0.33	−0.21	1	0.39	−0.24
2	0.27	−0.24	2	0.18	−0.18	2	0.18	−0.20
3	0.31	−0.45	3	0.26	−0.32	3	0.27	−0.34
4	0.31	−0.48	4	0.28	−0.39	4	0.28	−0.48
5	−0.24	−0.23	5	−0.31	−0.17	5	−0.36	−0.18
			6	−0.03	0.31	6	−0.04	0.35
						7	0.18	−0.02

j	$\hat{V}_{1,j}$	$\hat{V}_{2,j}$	j	$\hat{V}_{1,j}$	$\hat{V}_{2,j}$	j	$\hat{V}_{1,j}$	$\hat{V}_{2,j}$
0	0.32		0	0.32		0	0.30	
1	0.34	−0.15	1	0.24	−0.16	1	0.23	−0.16
2	0.19	−0.21	2	0.11	−0.26	2	0.11	−0.25
3	0.31	−0.29	3	0.35	−0.25	3	0.34	−0.27
4	0.23	−0.42	4	0.23	−0.39	4	0.23	−0.39
5	−0.35	−0.27	5	−0.32	−0.33	5	−0.30	−0.32
6	0.01	0.28	6	0.14	0.31	6	0.14	0.30
7	0.10	−0.00	7	0.10	0.08	7	0.10	0.08
8	−0.11	−0.15	8	−0.15	−0.17	8	−0.13	−0.15
			9	−0.08	0.24	9	−0.09	0.24
						10	−0.01	0.07

Not surprisingly, the odd features of Table 7.11 are reflected in Table 7.12. It is seen that $\hat{V}_{1,j}$, $j = 0, 1, \ldots, 4$, are large and positive, while $\hat{V}_{1,5}$ is large and negative. The best that can be done here, it seems, is to tentatively identify the form

$$V_1(B) = \omega_{1,0} + \omega_{1,1}B + \omega_{1,2}B^2 + \omega_{1,3}B^3 + \omega_{1,4}B^4 \qquad (7.12.3)$$

Doubtless, the addition of a further term $\omega_{1,5}B^5$ would lead to a significant coefficient estimate, of the opposite sign to all the others, but we cannot justify doing this, preferring rather to think of $\hat{\lambda}_5$ as an unfortunate consequence of sampling error.

In the same way, a term $\omega_{2,6}B^6$ could be added to

$$V_2(B) = \omega_{2,1}B + \omega_{2,2}B^2 + \omega_{2,3}B^3 + \omega_{2,4}B^4 + \omega_{2,5}B^5 \qquad (7.12.4)$$

which seems the only logical choice for the second transfer function, but again we prefer to think of $\hat{\lambda}_{-6}$ as reflecting sampling error. It should be added that such sampling error in the $\hat{\lambda}$'s is sure to be reflected in all the $\hat{V}$'s of Table 7.12. The identified forms (7.12.3) and (7.12.4) are rather crude guesses, though it is difficult to see what else could be done in the circumstances. It is to be hoped that some additional insight may come from the estimation stage.

It remains to tentatively identify an error structure for the model linking the residuals. Write

$$\epsilon_{1,t} = V_1(B)\epsilon_{2,t} + e_{1,t}, \qquad \epsilon_{2,t} = V_2(B)\epsilon_{1,t} + e_{2,t}$$

where $e_{j,t} = \psi_j(B)\eta_{j,t}$, $j = 1, 2$. The procedure employed here was that designated method (iii) in Section 7.10. The quantities required for $\hat{V}_{1,j}$, $j = 0, 1, \ldots, 4$, and $\hat{V}_{2,j}$, $j = 1, 2, \ldots, 5$, are given in Table 7.12 for $M = 10$, with $\hat{V}_{1,j}$, $j = 5, 6, \ldots$, and $\hat{V}_{2,j}$, $j = 6, 7, \ldots$, set equal to zero. The approximation involved here is rather crude, but should at least give a starting point. The calculated sample autocorrelations for $e_{1,t}$ and $e_{2,t}$ suggested the forms

$$\psi_1(B) = 1 - \theta_{1,1}B - \theta_{1,2}B^2 - \theta_{1,3}B^3 - \theta_{1,4}B^4 - \theta_{1,5}B^5$$

$$\psi_2(B) = 1 - \theta_{2,1}B - \theta_{2,2}B^2 - \theta_{2,3}B^3 - \theta_{2,4}B^4 - \theta_{2,5}B^5$$

Combining these with the identified transfer functions produces the models

$$\epsilon_{1,t} = (\omega_{1,0} + \omega_{1,1}B + \omega_{1,2}B^2 + \omega_{1,3}B^3 + \omega_{1,4}B^4)\epsilon_{2,t}$$
$$+ (1 - \theta_{1,1}B - \theta_{1,2}B^2 - \theta_{1,3}B^3 - \theta_{1,4}B^4 - \theta_{1,5}B^5)\eta_{1,t} \qquad (7.12.5)$$

and

$$\epsilon_{2,t} = (\omega_{2,1}B + \omega_{2,2}B^2 + \omega_{2,3}B^3 + \omega_{2,4}B^4 + \omega_{2,5}B^5)\epsilon_{1,t}$$
$$+ (1 - \theta_{2,1}B - \theta_{2,2}B^2 - \theta_{2,3}B^3 - \theta_{2,4}B^4 - \theta_{2,5}B^5)\eta_{2,t} \qquad (7.12.6)$$

When (7.12.5) was estimated, the terms $\hat{\omega}_{1,2}$, $\hat{\omega}_{1,3}$, $\hat{\omega}_{1,4}$, $\hat{\omega}_{1,5}$, and $\hat{\theta}_{1,5}$ were all small compared to their estimated standard errors. Dropping these terms and reestimating led to the fitted equation

$$\epsilon_{1,t} = \underset{[0.11]}{(\,0.29} + \underset{[0.11]}{0.37\,B\,)}\epsilon_{2,t} + (1 - \underset{[0.13]}{0.14\,B} - \underset{[0.13]}{0.15\,B^2} + \underset{[0.13]}{0.03\,B^3} + \underset{[0.15]}{0.30\,B^4}\,)\eta_{1,t}$$

(7.12.7)

When (7.12.6) was estimated, the equation originally achieved was

$$\epsilon_{2,t} = (\underset{[0.14]}{0.08\,B} + \underset{[0.12]}{0.09\,B^2} - \underset{[0.10]}{0.29\,B^3} - \underset{[0.10]}{0.20\,B^4} - \underset{[0.12]}{0.09\,B^5}\,)\epsilon_{2,t}$$

$$+ (1 - \underset{[0.12]}{0.00\,B} + \underset{[0.12]}{0.20\,B^2} + \underset{[0.13]}{0.33\,B^3} - \underset{[0.13]}{0.60\,B^4} + \underset{[0.17]}{0.05\,B^5}\,)\eta_{2,t}$$

Notice that $\hat{\omega}_{2,1}$ and $\hat{\omega}_{2,2}$ are positive, contradicting the negative values for $\hat{V}_{2,1}$ and $\hat{V}_{2,2}$ in Table 7.12, which must be put down to sampling error. Dropping these terms, together with $\omega_{2,5}$ and $\theta_{2,5}$ and reestimating, yielded

$$\epsilon_{2,t} = (- \underset{[0.07]}{0.27\,B^3} - \underset{[0.07]}{0.27\,B^4})\,\epsilon_{1,t}$$

$$+ (1 + \underset{[0.12]}{0.01\,B} + \underset{[0.12]}{0.27\,B^2} + \underset{[0.11]}{0.29\,B^3} - \underset{[0.12]}{0.54\,B^4})\eta_{2,t}$$ (7.12.8)

Although several redundant coefficients were dropped in arriving at (7.12.7) and (7.12.8), it was felt that the original model identification was so tenuous that some more overfitting ought to be tried. The following equations were then estimated:

$$\epsilon_{1,t} = \frac{\overset{[0.11]\quad[0.16]}{0.28 + 0.26\,B}}{\underset{[0.24]}{1 - 0.32\,B}}\,\epsilon_{2,t} + (1 - \underset{[0.13]}{0.14\,B} - \underset{[0.13]}{0.15\,B^2} + \underset{[0.13]}{0.03\,B^3} + \underset{[0.14]}{0.29\,B^4})\eta_{1,t}$$

$$\epsilon_{2,t} = \frac{\overset{[0.08]\quad\quad[0.11]}{- 0.34\,B^3 - 0.22\,B^4}}{\underset{[0.28]}{1 - 0.24\,B}}\,\epsilon_{1,t} + (1 + \underset{[0.12]}{0.07\,B} + \underset{[0.12]}{0.28\,B^2} + \underset{[0.12]}{0.30\,B^3} - \underset{[0.12]}{0.58\,B^4}\,)\eta_{2,t}$$

Neither added coefficient was clearly significant, and so they were not retained.

The autocorrelations of the residuals from (7.12.7) and (7.12.8) are shown in Table 7.13 and the cross correlations between these quantities in Table 7.14. Compared to standard errors of $n^{-1/2} \approx 0.125$, the quantities in Table 7.13 look very small and certainly do not suggest any promising modifications to the error structure. Moreover, none of the estimates in Table 7.14 are very close to $2n^{-1/2}$. However, one can hardly be complacent about the value for $\tau = 1$, which one certainly would have liked to be a good deal smaller for comfort. It is worth inquiring further about what, at first sight, seems a

Table 7.13   *Autocorrelations of estimated residuals from (7.12.7) and (7.12.8)*

τ	$r_\tau(\hat{\eta}_1)$	τ	$r_\tau(\hat{\eta}_1)$	τ	$r_\tau(\hat{\eta}_2)$	τ	$r_\tau(\hat{\eta}_2)$
1	0.00	11	-0.06	1	-0.02	11	-0.01
2	-0.02	12	0.01	2	-0.11	12	0.09
3	-0.05	13	-0.03	3	0.11	13	0.11
4	0.05	14	0.02	4	0.08	14	0.02
5	0.03	15	0.04	5	-0.04	15	0.01
6	-0.07	16	0.13	6	0.19	16	-0.12
7	-0.17	17	-0.11	7	0.06	17	-0.03
8	0.17	18	-0.06	8	-0.07	18	0.05
9	-0.08	19	0.05	9	0.03	19	-0.06
10	0.06	20	0.05	10	0.11	20	-0.00

Table 7.14   *Cross correlations between estimated residuals from (7.12.7) and (7.12.8)*

τ	$r(\hat{\eta}_{1,t}, \hat{\eta}_{2,t-\tau})$	τ	$r(\hat{\eta}_{1,t}, \hat{\eta}_{2,t-\tau})$	τ	$r(\hat{\eta}_{1,t}, \hat{\eta}_{2,t-\tau})$	τ	$r(\hat{\eta}_{1,t}, \hat{\eta}_{2,t-\tau})$
0	0.03	8	-0.10			-8	-0.15
1	-0.19	9	-0.19	-1	0.15	-9	-0.02
2	0.12	10	0.05	-2	0.03	-10	-0.01
3	0.05	11	0.07	-3	-0.07	-11	-0.06
4	0.08	12	-0.02	-4	0.03	-12	0.17
5	-0.20	13	-0.09	-5	-0.19	-13	-0.07
6	-0.01	14	-0.05	-6	0.12	-14	-0.08
7	0.04	15	-0.00	-7	-0.07	-15	0.01

surprisingly high value. It appears to suggest that, for the error series of (7.12.7) and (7.12.8), a plausible model might be $\eta_{1,t} = \beta\eta_{2,t-1} + \text{error}$, where β is negative. Now, the leading term in the expression for $\eta_{2,t-1}$, which can be obtained from (7.12.8), is of course, $\epsilon_{2,t-1}$. Hence, it seems that substitution of $\beta\epsilon_{2,t-1}$ for $\eta_{1,t}$ in (7.12.7) should give a crude modification of the transfer function, which should lead to a fitted model such that the residual cross correlation at $\tau = 1$ is considerably dampened. Thus, from (7.12.7), one has a term

$$(1 - 0.14B - 0.15B^2 + 0.03B^3 + 0.30B^4)\beta\epsilon_{2,t-1}$$

the dominant term of which is $0.30\beta\epsilon_{2,t-5}$, so that adding a term $\omega_{1,5}B^5\epsilon_{2,t}$ to the identified model, with $\omega_{1,5}$ negative, should give better results. But this is precisely the path we rejected earlier, on grounds of common sense and plausibility, in identifying the model in the first place. Thus, if we stick to our guns on this point, and we are inclined to do so, the consequence appears to be that we must live with a residual cross correlation that is higher than we would wish. The value of 0.15 for $\tau = -1$ in Table 7.14 is less worrying since no term $\omega_{2,1}$ is included in (7.12.8)—one was tried, and the resulting estimate found to be not statistically significant.

Now, let $Y_{1,t}$ and $Y_{2,t}$ denote first differences of discount rate and sales respectively, the latter with mean subtracted. Merging Eqs. (7.12.1) and (7.12.2) with (7.12.7) and (7.12.8) then suggests a model

$$Y_{1,t} = \frac{\omega_{1,0}^* + \omega_{1,1}^* B + \omega_{1,2}^* B^2 + \omega_{1,3}^* B^3}{1 - \delta_{1,1}^* B} Y_{2,t}$$

$$+ \left(1 - \theta_{1,1}^* B - \theta_{1,2}^* B^2 - \theta_{1,3}^* B^3 - \theta_{1,4}^* B^4 - \theta_{1,5}^* B^5 - \theta_{1,6}^* B^6\right)\eta_{1,t}^*$$

$$Y_{2,t} = \frac{\omega_{2,3}^* B^3 + \omega_{2,4}^* B^4 + \omega_{2,5}^* B^5}{1 - \delta_{2,1}^* B - \delta_{2,2}^* B^2} Y_{1,t}$$

$$+ \left(1 - \theta_{2,1}^* B - \theta_{2,2}^* B^2 - \theta_{2,3}^* B^3 - \theta_{2,4}^* B^4 - \theta_{2,5}^* B^5\right)\eta_{2,t}^*$$

However, the values suggested by this amalgamation for $\omega_{1,3}^*$, $\theta_{1,6}^*$, $\omega_{2,5}^*$, and $\theta_{2,5}^*$ are all very small. Accordingly, these were dropped before the equations were estimated. The first fitted equation was then

$$Y_{1,t} = \frac{\overset{[0.021]}{0.053} + \overset{[0.028]}{0.081} B + \overset{[0.028]}{0.044} B^2}{1 + \underset{[0.14]}{0.003} B} Y_{2,t}$$

$$+ \left(1 + \underset{[0.15]}{0.34} B - \underset{[0.15]}{0.34} B^2 - \underset{[0.15]}{0.01} B^3 + \underset{[0.16]}{0.32} B^4 + \underset{[0.17]}{0.12} B^5\right)\eta_{1,t}^*$$

Dropping the terms $\delta_{1,1}^*$ and $\theta_{1,5}^*$, whose estimates were not close to significance, then yielded

$$Y_{1,t} = \left(\underset{[0.020]}{0.046} + \underset{[0.020]}{0.080} B + \underset{[0.020]}{0.039} B^2\right)Y_{2,t}$$

$$+ \left(1 + \underset{[0.13]}{0.30} B - \underset{[0.14]}{0.36} B^2 + \underset{[0.10]}{0.00} B^3 + \underset{[0.13]}{0.28} B^4\right)\eta_{1,t}^*, \qquad \hat{\sigma}_{\eta_1^*}^2 = 0.157 \quad (7.12.9)$$

The second estimated equation was

$$Y_{2,t} = \frac{\overset{[0.39]}{-1.73} B^3 - \overset{[0.43]}{1.70} B^4}{1 + \underset{[0.19]}{0.43} B - \underset{[0.17]}{0.42} B^2} Y_{1,t}$$

$$+ \left(1 + \underset{[0.12]}{0.24} B + \underset{[0.12]}{0.18} B^2 + \underset{[0.12]}{0.44} B^3 - \underset{[0.13]}{0.58} B^4\right)\eta_{2,t}^*, \qquad \hat{\sigma}_{\eta_2^*}^2 = 3.67 \quad (7.12.10)$$

In addition to the overfitting done in arriving at these equations, the coefficient $\omega_{1,3}^*$ was added to (7.12.9) and $\omega_{2,5}^*$ to (7.12.10), but when the augmented equations were reestimated neither extra estimated coefficient turned out to be statistically significant.

The autocorrelations of the residuals from (7.12.9) and (7.12.10) are given in Table 7.15, and the cross correlations between these quantities in Table 7.16. The quantities in Table 7.15 are generally very small compared to standard errors of $n^{-1/2} \approx 0.125$, and provide little basis for doubting the error structures in the fitted model. Only one of the cross correlations in Table 7.16 is bigger than two standard errors, and that, not surprisingly, is at $\tau = 5$. The moderately high negative value at $\tau = 1$ is not unexpected, and merely parallels the result in Table 7.14. Faced, as we are, with the choice of damping this quantity at the expense of fitting an intuitively implausible model or leaving things as they stand, our preference is to maintain Eqs. (7.12.9) and (7.12.10) as a working hypothesis. It must be added that, in such circumstances, the forecasting performance of the fitted model ought to be

Table 7.15   *Autocorrelations of estimated residuals from (7.12.9) and (7.12.10)*

τ	$r_\tau(\hat{\eta}_1^*)$	τ	$r_\tau(\hat{\eta}_1^*)$	τ	$r_\tau(\hat{\eta}_2^*)$	τ	$r_\tau(\hat{\eta}_2^*)$
1	0.03	11	-0.06	1	-0.00	11	-0.01
2	-0.04	12	0.02	2	-0.15	12	0.06
3	-0.05	13	-0.05	3	0.12	13	0.12
4	0.08	14	-0.01	4	0.09	14	0.04
5	0.08	15	0.04	5	-0.05	15	-0.01
6	-0.10	16	0.12	6	0.19	16	-0.12
7	-0.10	17	-0.11	7	0.13	17	-0.03
8	0.10	18	-0.01	8	-0.09	18	0.06
9	-0.04	19	0.09	9	0.03	19	-0.06
10	0.11	20	0.05	10	0.09	20	0.00

Table 7.16   *Cross correlations between estimated residuals from (7.12.9) and (7.12.10)*

τ	$r(\hat{\eta}_{1,t}^*, \hat{\eta}_{2,t-\tau}^*)$	τ	$r(\hat{\eta}_{1,t}^*, \hat{\eta}_{2,t-\tau}^*)$	τ	$r(\hat{\eta}_{1,t}^*, \hat{\eta}_{2,t-\tau}^*)$	τ	$r(\hat{\eta}_{1,t}^*, \hat{\eta}_{2,t-\tau}^*)$
0	0.10	8	-0.10			-8	-0.22
1	-0.19	9	-0.19	-1	0.15	-9	0.01
2	-0.05	10	0.04	-2	-0.05	-10	-0.08
3	0.08	11	0.01	-3	-0.05	-11	-0.03
4	0.08	12	0.02	-4	0.01	-12	0.11
5	-0.27	13	-0.10	-5	-0.12	-13	-0.02
6	0.01	14	-0.04	-6	0.10	-14	-0.07
7	0.05	15	-0.07	-7	-0.07	-15	-0.01

closely watched, and, if unsatisfactory forecasts were to be produced, the model could be modified. The question of forecast evaluation in general is considered in some detail in Chapter 8.

Just as in univariate model building the objective is to find a filter that transforms a given series to white noise, the objective in bivariate model building is to transform a pair of given series to a pair of mutually stochastically uncorrelated white noise processes. In practice the analyst sets about this task by fitting to actual data a model whose residuals appear, as closely as possible, to have the properties of mutually stochastically uncorrelated white noise series. However the analyst has other tools than the blind ability to derive a model from a given set of estimated second-order moments. He presumably has some knowledge of economic theory to suggest to him which sample phenomena are likely to reflect real relationships and which are likely to have arisen by chance. He also has his own common sense to distinguish which apparent relationships are plausible and which are not. Thus, on occasion, the desire to produce residual series with very small sample cross correlations, and the desire to produce sensible models may, to some extent, come into conflict, leading to a delicate choice. We feel that, if in addition to Box and Jenkins' delightfully named and extremely important principle of "parsimonious parametrization," the analyst embraces the principle of "plausible parametrization," he will not go far wrong. Of course, any principle blindly followed is worthless, and any theory ought to be tested, so that a plausible model is a poor substitute for an "implausible" one that constantly outpredicts it. In circumstances such as this, model building should be regarded as a learning process, with postsample data providing excellent additional information. If the plausible model can predict these data satisfactorily, then all well and good. If not, the analyst might be forced to question the economic theory, or his own common sense, on which his notion of plausibility was founded. However much effort is put into building a model—econometric or time series—it is, in our opinion, not worth a great deal until its forecasting ability has been validated.

7.13 Forecasting from Bivariate Time Series Models

In this section the procedure for computing forecasts from a fitted bivariate feedback model is illustrated. The approach used here can also be applied, in an obvious fashion, to unidirectional causality models.

The first step is to transpose the fitted equations to convenient form for forecasting by multiplying through by the denominator polynomials. Thus, for example, for the consumption–investment equations, one has from (7.11.3) and (7.11.4)

$$(1 - 1.89B + 1.303B^2 - 0.3895B^3)Y_{1,t} = (0.20 - 0.19B)Y_{2,t}$$

$$+ (1 - 2.11B + 1.8998B^2 - 0.8463B^3 + 0.1599B^4)\eta^*_{1,t}, \qquad (7.13.1)$$

and

$$(1 - 0.83B)Y_{2,t} = (0.62B - 0.57B^2)Y_{1,t}$$

$$+ (1 - 0.84B - 0.2717B^2 + 0.2324B^3)\eta_{2,t}^* \qquad (7.13.2)$$

Then, setting

$$Y_{1,t} = (1 - B)X_{1,t} - 9.10, \qquad Y_{2,t} = (1 - B)X_{2,t} - 1.78$$

in (7.13.1) and (7.13.2), one has

$$(1 - 2.89B + 3.193B^2 - 1.6925B^3 + 0.3895B^4)X_{1,t}$$

$$= 0.0036 + (0.20 - 0.39B + 0.19B^2)X_{2,t}$$

$$+ (1 - 2.11B + 1.8998B^2 - 0.8463B^3 + 0.1599B^4)\eta_{1,t}^* \qquad (7.13.3)$$

and

$$(1 - 1.83B + 0.83B^2)X_{2,t} = -0.1524 + (0.62B - 1.19B^2 + 0.57B^3)X_{1,t}$$

$$+ (1 - 0.84B - 0.2717B^2 + 0.2324B^3)\eta_{2,t}^* \qquad (7.13.4)$$

These equations are now easily used for forecasting since writing $f_{n,h}^{(1)}$ and $f_{n,h}^{(2)}$ as the optimal h-step forecasts of $X_{1,n+h}$ and $X_{2,n+h}$, standing at time n, one has

$$f_{n,h}^{(1)} = 2.89f_{n,h-1}^{(1)} - 3.193f_{n,h-2}^{(1)} + 1.6925f_{n,h-3}^{(1)} - 0.3895f_{n,h-4}^{(1)} + 0.0036$$

$$+ 0.20f_{n,h}^{(2)} - 0.39f_{n,h-1}^{(2)} + 0.19f_{n,h-2}^{(2)}$$

$$+ \hat{\eta}_{1,n+h}^* - 2.11\hat{\eta}_{1,n+h-1}^* + 1.8998\hat{\eta}_{1,n+h-2}^* - 0.8463\hat{\eta}_{1,n+h-3}^*$$

$$+ 0.1599\hat{\eta}_{1,n+h-4}^* \qquad (7.13.5)$$

and

$$f_{n,h}^{(2)} = 1.83f_{n,h-1}^{(2)} - 0.83f_{n,h-2}^{(2)} - 0.1524$$

$$+ 0.62f_{n,h-1}^{(1)} - 1.19f_{n,h-2}^{(1)} + 0.57f_{n,h-3}^{(1)}$$

$$+ \hat{\eta}_{2,n+h}^* - 0.84\hat{\eta}_{2,n+h-1}^* - 0.2717\hat{\eta}_{2,n+h-2}^* + 0.2324\hat{\eta}_{2,n+h-3}^* \qquad (7.13.6)$$

where

$$f_{n,k}^{(j)} = x_{j,n+k}, \qquad k \leqslant 0, \quad j = 1, 2$$

$$\hat{\eta}_{j,n+k}^* = 0, \qquad k > 0$$

$$= \text{estimate of } \eta_{j,n+k}^*, \qquad k \leqslant 0, \quad j = 1, 2 \qquad (7.13.7)$$

Table 7.17   *Quantities required to compute forecasts of comsumption and investment*

t	$x_{1,t}$	$\hat{\eta}^*_{1,t}$	$x_{2,t}$	$\hat{\eta}^*_{2,t}$
$n-3$	700.2	3.1	167.5	5.2
$n-2$	719.2	6.9	174.7	2.6
$n-1$	734.1	2.8	181.5	1.5
n	752.6	4.1	189.4	4.9

Table 7.18   *Forecasts of consumption and investment made up to six quarters ahead*

h:	1	2	3	4	5	6
$f^{(1)}_{n,h}$:	766.5	780.2	793.7	807.6	821.6	835.3
$f^{(2)}_{n,h}$:	194.9	196.3	199.1	201.9	204.9	208.0

Forecasts are then obtained from (7.13.5) and (7.13.6). Because of the triangularity imposed on the system, a specific order must be used. First $f^{(2)}_{n,1}$ is calculated from (7.13.6). This is used to obtain $f^{(1)}_{n,1}$ from (7.13.5) and hence $f^{(2)}_{n,2}$ from (7.13.6), and so on as far ahead as forecasts are required.

The necessary quantities required for these calculations are shown in Table 7.17. The estimated residuals were obtained when Eqs. (7.11.3) and (7.11.4) were estimated. Forecasts made up to six quarters ahead are shown in Table 7.18.

In order to get an estimate of forecast error variance, the model should be transformed to

$$X_{1,t} = C_{11}(B)\eta^*_{1,t} + C_{12}(B)\eta^*_{2,t}, \qquad X_{2,t} = C_{21}(B)\eta^*_{1,t} + C_{22}(B)\eta^*_{2,t}$$
$$(7.13.8)$$

where here, and throughout the rest of this section, constant terms (which are irrelevant in this context) are ignored. It then follows from (7.5.11) that the respective error variances are given by

$$V^{(1)}(h) = \sigma^2_{\eta^*_1} \sum_{j=0}^{h-1} C^2_{11,j} + \sigma^2_{\eta^*_2} \sum_{j=0}^{h-1} C^2_{12,j}$$

$$V^{(2)}(h) = \sigma^2_{\eta^*_1} \sum_{j=0}^{h-1} C^2_{21,j} + \sigma^2_{\eta^*_2} \sum_{j=0}^{h-1} C^2_{22,j} \qquad (7.13.9)$$

To get the C weights of (7.13.8), for the consumption–investment model, one first substitutes for $Y_{2,t}$ from (7.11.4) in (7.11.3), and for $Y_{1,t}$ from

(7.11.3) in (7.11.4). This gives

$$\frac{1 - 1.894\,B + 1.3042\,B^2 - 0.3403\,B^3}{1 - 1.77\,B + 1.1902\,B^2 - 0.3403\,B^3}\,Y_{1,\,t} = \frac{1 - 1.17\,B + 0.39\,B^2}{1 - 0.95\,B}\,\eta_{1,\,t}^*$$

$$+ \frac{0.20 - 0.002\,B - 0.056\,B^2}{1 - 0.94\,B + 0.41\,B^2}\,\eta_{2,\,t}^* \qquad (7.13.10)$$

and

$$\frac{1 - 1.894\,B + 1.3042\,B^2 - 0.3403\,B^3}{1 - 1.77\,B + 1.1902\,B^2 - 0.3403\,B^3}\,Y_{2,\,t}$$

$$= \frac{0.62\,B - 1.2954\,B^2 + 0.9087\,B^3 - 0.2223\,B^4}{1 - 1.78\,B + 0.7885\,B^2}\,\eta_{1,\,t}^*$$

$$+ (1 - 0.01\,B - 0.28\,B^2)\eta_{2,\,t}^* \qquad (7.13.11)$$

To illustrate the calculations, only the consumption equation will be considered in any detail. From (7.13.10) it follows that

$$Y_{1,\,t} = \frac{1 - 2.94\,B + 3.6511\,B^2 - 2.4231\,B^3 + 0.8573\,B^4 - 0.1327\,B^5}{1 - 2.844\,B + 3.1035\,B^2 - 1.5793\,B^3 + 0.3233\,B^4}\,\eta_{1,\,t}^*$$

$$+ \frac{0.20 - 0.356\,B + 0.1856\,B^2 + 0.0287\,B^3 - 0.066\,B^4 + 0.0191\,B^5}{1 - 2.834\,B + 3.4946\,B^2 - 2.3428\,B^3 + 0.8546\,B^4 - 0.1395\,B^5}\,\eta_{2,\,t}^*$$

Multiplying through by $(1 - B)^{-1}$, and ignoring the constant term, then gives

$$X_{1,\,t} = \frac{1 - 2.94\,B + 3.6511\,B^2 - 2.4231\,B^3 + 0.8573\,B^4 - 0.1327\,B^5}{1 - 3.844\,B + 5.9475\,B^2 - 4.6828\,B^3 + 1.9026\,B^4 - 0.3233\,B^5}\,\eta_{1,\,t}^*$$

$$+ \frac{0.20 - 0.356\,B + 0.1856\,B^2 + 0.0287\,B^3 - 0.066\,B^4 + 0.0191\,B^5}{1 - 3.834\,B + 6.3286\,B^2 - 5.8374\,B^3 + 3.1974\,B^4 - 0.9941\,B^5 + 0.1395\,B^6}\,\eta_{2,\,t}^*$$

$$(7.13.12)$$

The C weights are then obtained by equating coefficients in B^j for each of the ratios of polynomials in (7.3.12). The first few values, calculated in this way, are shown in Table 7.19.

Table 7.19 C weights required for calculation of error variance of forecasts of consumption derived from bivariate model

j:	0	1	2	3	4	5
$C_{11,\,j}$:	1.00	0.90	1.18	1.42	1.62	1.79
$C_{12,\,j}$:	0.20	0.41	0.50	0.50	0.47	0.44

Table 7.20   *Error variances of forecasts of consumption derived from bivariate model*

h:	1	2	3	4	5	6
$V^{(1)}(h)$:	9.24	19.10	35.37	56.97	83.25	113.89

The forecast error variances are then calculated from the first equation of (7.13.9), using the estimates $\hat{\sigma}_{\eta_1^*}^2 = 8.54$ and $\hat{\sigma}_{\eta_2^*}^2 = 17.52$ obtained from fitting (7.11.3) and (7.11.4). The resulting error variances, particularly for longer lead times, are likely to be under estimates for two reasons. First, it has been assumed that the model achieved is correctly identified. This is much less likely here than in the univariate case, where identification is relatively much simpler. We feel that the model building procedure described in earlier sections is likely to produce a model that will forecast quite well. However, it would be extremely rash for one to assert with complete confidence that the model achieved was the correct form. Second, although the above calculations assume the coefficients of the model are given, in practice these must be estimated from data and, of course, the estimates are subject to error.

The first few forecast error variances for the consumption series are shown in Table 7.20.

The one-step variance 9.24 represents a 24% improvement over the one-step error variance of 12.1 for univariate forecasts of consumption based on (7.10.1). Further, it follows from (7.13.11) that the one-step forecast error variance for investment is $\sigma_{\eta_2^*}^2$, which is estimated by 17.52. This represents a 20% improvement over the value of 21.9 for univariate forecasts obtained from (7.10.2).

7.14   Multivariate Time Series Model Building

In principle, the problem of building a multivariate time series model to describe the interrelationships between series $Y_1, Y_2, \ldots, Y_m$ is similar to the bivariate case. The analyst's objective should be to construct equations in such a way that the m given time series are transformed to m series that have the properties of mutually stochastically uncorrelated white noise processes. However, it should be clear from earlier sections of this chapter that, for sample sizes likely to occur in practice, as m increases, the practical difficulties will quickly become unmanageable.

Given a set of time series, perhaps the best strategy is to look for simplifications to the problem. For example, the series may consist of two or more groups of series, each group unrelated to the other. Again, some of the interrelationships may be unidirectional rather than feedback, which would greatly simplify the analysis. To assess these possibilities, one should first build univariate models

$$a_j(B)Y_{j,t} = b_j(B)\epsilon_{j,t}, \qquad j = 1, 2, \ldots, m \qquad (7.14.1)$$

and examine the cross correlations between all pairs of estimated residuals.

As a simple example of what might be done, consider the case where the pair of series Y_1 and Y_2, themselves exhibiting feedback, jointly cause, but are not caused by a third series Y_3. Behavior of this kind may be detected by examining the cross correlations of the univariate residuals. As a first step in this problem, a bivariate model, linking ϵ_1 and ϵ_2, and producing mutually stochastically uncorrelated white noise series $\eta_{1,t}$ and $\eta_{2,t}$ can be built along the lines suggested in Section 7.10. One can then either relate $(\eta_{1,t}, \eta_{2,t})$ to $\epsilon_{3,t}$ or alternatively build, along the lines of Section 7.11, a model linking $Y_{1,t}$ and $Y_{2,t}$, giving $\eta_{1,t}^*$ and $\eta_{2,t}^*$. In this case the second alternative is probably better. Then a model of the form

$$\epsilon_{3,t} = V_1(B)\eta_{1,t}^* + V_2(B)\eta_{2,t}^* + \psi(B)\eta_{3,t}^* \qquad (7.14.2)$$

can easily be identified using the procedures of Section 7.7 since $\eta_{1,t}^*$ and $\eta_{2,t}^*$ are unrelated series. For example, multiplying through (7.14.2) by $\eta_{1,t-j}^*$ and taking expectations gives $E[\epsilon_{3,t}\eta_{1,t-j}^*] = V_{1,j}\sigma_{\eta_1^*}^2$.

More general multivariate models may be constructed by building on such simple examples as that just described. Whatever the problem, however, we feel that initial prewhitening of the form (7.14.1) is a vital starting point.

It should be obvious that complicated econometric models, of the type described in the previous chapter, cannot be built using the principles of time series analysis alone. Some sort of marriage of the two approaches is clearly desirable. For example, economic theory may be employed by the time series analyst to suggest a subclass of all possible models, which accords with the theory. This should greatly reduce the number of models the analyst needs to contemplate, and thus make his identification problem correspondingly simpler. One possibility might be to examine the time series properties (multivariate) of residuals from fitted econometric models, and construct a model with more appropriate time series error structure.

A simpler alternative to combining time series and econometric methodologies would be to combine the forecasts produced by the two approaches. This possibility is considered in the following chapter.

CHAPTER EIGHT

THE COMBINATION AND EVALUATION OF FORECASTS

A: "How is your wife?"
B: "Compared to what?"

8.1 Typical Suboptimality of Economic Forecasts

Given a fairly limited information set, for example all past values of a time series, one might reasonably hope in practice to achieve a forecast that at least closely approximates the optimum. Indeed, the univariate Box–Jenkins procedure described in Chapters 3 and 5 might well be expected to achieve this goal frequently, given a sufficiently large sample. Even in this situation, however, further improvement may still be achievable through consideration of the possibilities of nonlinearity and nonstationarity, as will be seen in Chapter 9. More importantly, though, the forecaster in the real world is not, and should not be, constrained to employ an information set of this kind. The information considered by many macroeconomic forecasters is indeed vast. Insofar as forecasts from quantitative models are modified judgmentally, the information taken into account is often of a qualitative nature, being a sublimation of the practitioner's experience of the variable under study and its principal determinants. Indeed, in forming economic forecasts, it is often the case that noneconomic information could profitably be taken into account. Thus, the supply of grains to the market in any given year, and hence prices of food, is greatly affected by the weather. Again, the supply and price of oil in the past two or three years have been determined as much by political as economic factors. Thus, given a world in which the amount of potential information is vast and the number of potential ways of employing it enormous (partly because in many areas economic theory is ill structured, or insufficient data are available to distinguish with any high degree of confidence between competing theories), it makes very little sense to view forecast optimality as a useful working concept. Put another way, we doubt whether any economic forecaster would view his product as the best that could possibly be achieved given all the information in the universe.

Once it is recognized that the typical economic forecast cannot generally be thought of as in any sense optimal, two consequences follow. First, a forecast may well be able to be improved. One way to achieve this would be to consider two or more forecasts of the same quantity since it is often the case in macroeconomics that competing forecasts are available. Suppose one has several forecasts, then it is quite possible that any one of them contains useful information absent in the others, and so rather than discard all but one forecast, it could well be profitable to incorporate them all into an overall combined forecast. A particularly simple way to achieve this is to let the combined forecast be a weighted average, with appropriately chosen weights, of the individual forecasts. Since the economic forecaster will not be in the position to assert reasonably that his forecasts are the best possible, it is important that economic forecasts be critically evaluated. An evaluation exercise, as well as providing information about the relative worth of a set of forecasts, may well suggest directions in which the forecast-generating mechanism can be improved, and hence act as a stimulus to future productive research. With this in mind, it follows that the evaluation criteria employed should be as demanding as possible since the object ought to be self-criticism rather than self-congratulation.

In the remainder of this chapter methods that have been employed in the combination and evaluation of forecasts are discussed, and details of some actual evaluation exercises, applied to macroeconomic forecasts in the past few years, are given.

8.2 The Combination of Forecasts

As a simple example of the potential of combining forecasts, an examination of data given in Barnard [1963] is instructive. One-step ahead forecasts of world airline passenger miles per month over the period 1951–1960 are given for both a Box–Jenkins model and an exponential smoothing model. The former yielded an error variance of 148.6 and the latter 177.7. However, an alternative forecast, which is simply the average of the two individual forecasts, can be shown to have an error variance of 130.2. Thus, in this particular instance, a combined forecast that outperforms both individual forecasts can readily be found. One is led to ask the question as to whether more sophisticated combination rules might lead to further improvement and produce a procedure of wider applicability.

Following Bates and Granger [1969], consider first the case of combining two *unbiased* one-step ahead forecasts. Let $f_n^{(1)}$ and $f_n^{(2)}$ be forecasts of X_n with errors

$$e_n^{(j)} = X_n - f_n^{(j)}, \qquad j = 1, 2$$

such that

$$E(e_n^{(j)}) = 0, \quad E(e_n^{(j)2}) = \sigma_j^2, \quad j = 1, 2$$

and

$$E(e_n^{(1)}e_n^{(2)}) = \rho\sigma_1\sigma_2$$

Consider now a combined forecast, taken to be a weighted average of the two individual forecasts (since both are unbiased),

$$C_n = kf_n^{(1)} + (1 - k)f_n^{(2)}$$

The forecast error is

$$e_n^{(c)} = X_n - C_n = ke_n^{(1)} + (1 - k)e_n^{(2)}$$

Hence the error variance is

$$\sigma_c^2 = k^2\sigma_1^2 + (1 - k)^2\sigma_2^2 + 2k(1 - k)\rho\sigma_1\sigma_2 \qquad (8.2.1)$$

This expression is minimized for the value of k given by

$$k_0 = \frac{\sigma_2^2 - \rho\sigma_1\sigma_2}{\sigma_1^2 + \sigma_2^2 - 2\rho\sigma_1\sigma_2} \qquad (8.2.2)$$

and substitution into (8.2.1) yields the minimum achievable error variance as

$$\sigma_{c,0}^2 = \frac{\sigma_1^2\sigma_2^2(1 - \rho^2)}{\sigma_1^2 + \sigma_2^2 - 2\rho\sigma_1\sigma_2} \qquad (8.2.3)$$

Note $\sigma_{c,0}^2 < \min(\sigma_1^2, \sigma_2^2)$ unless either ρ is exactly equal to σ_1/σ_2 or to σ_2/σ_1. If either equality holds, then the variance of the combined forecast is equal to the smaller of the two error variances. Thus, a priori, it is reasonable to expect in most practical situations that the best available combined forecast will outperform the better individual forecast—it cannot, in any case, do worse.

Two results of passing interest can be derived from expressions (8.2.2) and (8.2.3). First it can be seen that

$$k_0 \gtreqless 0 \qquad \text{if and only if} \qquad \sigma_2/\sigma_1 \gtreqless \rho$$

It thus follows that, if $f_n^{(2)}$ is the optimal forecast based on a particular information set, any other forecast $f_n^{(1)}$ based on the same information set must be such that $\rho = \sigma_2/\sigma_1$, exactly. The situation $k_0 < 0$ is interesting. In much of our empirical work on combining we have constrained the weights to be nonnegative, it being rather difficult to justify on general grounds the assignment of a negative weight to a particular forecast. However, in light of the above conditions, it appears that an inferior forecast may still be worth including with negative weight on the grounds that its relatively high error variance is outweighed by a large ρ value—that is to say, the part of the variable of interest left unexplained by it is sufficiently strongly related to the part left unexplained by the better forecast. A second point concerns the behavior of (8.2.3) as ρ approaches -1 or 1. In the former case $\sigma_{c,0}^2$ tends to zero, implying a perfect forecast is obtainable. As ρ approaches 1, $\sigma_{c,0}^2$ also tends to zero except when $\sigma_1 = \sigma_2$, in which case its limit is σ_1^2. In interpreting the result for ρ tending to 1, which at first sight seems counterintuitive, it should be borne in mind that ρ is the correlation between the two forecast errors, not between the forecasts themselves. It is simply explained, as follows. Consider two forecasts producing errors $e_n^{(1)}$ and $Ae_n^{(1)}$, where A is positive. Then

$$e_n^{(1)} = X_n - f_n^{(1)} \qquad \text{and} \qquad e_n^{(2)} = Ae_n^{(1)} = A(X_n - f_n^{(1)})$$

Hence, the second forecast is

$$f_n^{(2)} = X_n - e_n^{(2)} = (1 - A)X_n + Af_n^{(1)}$$

Thus, this forecast, for $A \neq 1$, involves the quantity to be forecast X_n and one has exactly

$$X_n = -\frac{A}{1 - A} f_n^{(1)} + \frac{1}{1 - A} f_n^{(2)}$$

Now, as it stands, formula (8.2.2) is of no operational value since in practice one would never know the values σ_1^2, σ_2^2, and ρ. However, suppose that the two forecasting procedures have been observed over the previous $n - 1$ time periods, yielding errors $e_t^{(j)}$, $t = 1, 2, \ldots, n - 1$, $j = 1, 2$. Denoting the combining weight to be employed at time n by $\hat{k}_n$, and replacing quantities in (8.2.2) by their sample estimates, a natural choice is

$$\hat{k}_n = \frac{\displaystyle\sum_{t=n-v}^{n-1} \left(e_t^{(2)2} - e_t^{(1)}e_t^{(2)} \right)}{\displaystyle\sum_{t=n-v}^{n-1} \left(e_t^{(1)2} + e_t^{(2)2} - 2e_t^{(1)}e_t^{(2)} \right)} \qquad (8.2.4)$$

Given a sample $e_t^{(j)}$, $t = n - v, \ldots, n - 1$, $j = 1, 2$, from a bivariate normal distribution, the quantity (8.2.4) is the maximum likelihood estimator of k_0 of (8.2.2). Alternatively, it can be viewed as a least-squares estimator based on the same sample since the combined forecast at time t can be written

$$C_t - f_t^{(2)} = k(f_t^{(1)} - f_t^{(2)})$$

or alternatively

$$e_t^{(2)} = k(e_t^{(2)} - e_t^{(1)}) + e_t^{(c)}$$

where $e_t^{(c)}$ is the error of the combined forecast. Estimation of k by least squares then yields (8.2.4). Although this choice of combining weights is a very natural one, it is subject to two difficulties. First, population correlation coefficients are generally not well estimated in small samples; and second it may be that the relative performance of the two forecasting methods is nonstationary, suggesting the use of a system of weights that can adapt fairly quickly through time. With these considerations in mind, Bates and Granger were led to consider a number of alternative choices of weights, including the following:

$$\hat{k}_n = \frac{\displaystyle\sum_{t=n-v}^{n-1} e_t^{(2)2}}{\displaystyle\sum_{t=n-v}^{n-1} \left(e_t^{(1)2} + e_t^{(2)2}\right)} \tag{8.2.5}$$

$$\hat{k}_n = \alpha \hat{k}_{n-1} + \frac{(1-\alpha)\displaystyle\sum_{t=n-v}^{n-1} e_t^{(2)2}}{\displaystyle\sum_{t=n-v}^{n-1} \left(e_t^{(1)2} + e_t^{(2)2}\right)}, \qquad 0 < \alpha < 1 \tag{8.2.6}$$

$$\hat{k}_n = \frac{\displaystyle\sum_{t=1}^{n-1} W^t\left(e_t^{(2)2} - e_t^{(1)}e_t^{(2)}\right)}{\displaystyle\sum_{t=1}^{n-1} W^t\left(e_t^{(1)2} + e_t^{(2)2} - 2e_t^{(1)}e_t^{(2)}\right)}, \qquad W \geqslant 1 \tag{8.2.7}$$

$$\hat{k}_n = \frac{\displaystyle\sum_{t=1}^{n-1} W^t e_t^{(2)2}}{\displaystyle\sum_{t=1}^{n-1} W^t\left(e_t^{(1)2} + e_t^{(2)2}\right)}, \qquad W \geqslant 1 \tag{8.2.8}$$

If one wishes to restrict the weights employed to lie between zero and one, then appropriate end points of this range can be substituted for any calculated values from (8.2.4) or (8.2.7) falling outside the range.

Bates and Granger applied the various combining formulas to the world airline passenger data, looking at several pairs of univariate one-step ahead forecasts, with generally successful results. This prompted a wider study by Newbold and Granger [1974], who considered the combination of one-step ahead Box–Jenkins, Holt–Winters, and stepwise autoregressive forecasts for the 80 monthly series in the collection described in Section 5.6. Forecasts were combined in pairs, using (8.2.4) and (8.2.5) with $v = 1, 3, 6, 9$, and 12; (8.2.6) for combinations of $\alpha = 0.5, 0.7$, and 0.9 and $v = 1, 3, 6, 9, 12$; and (8.2.7) and (8.2.8) for $W = 1, 1.5, 2$, and 2.5. The weights were constrained to lie between zero and unity. Table 8.1 shows the percentage of series for which

Table 8.1   *Combination of pairs of forecasts; percentage number of series for which the combined forecast outperforms both individual forecasts for: (A) Box–Jenkins combined with Holt–Winters, (B) Box–Jenkins combined with stepwise autoregressive, (C) Holt–Winters combined with stepwise autoregressive*

Formula (8.2.4)			ν	Formula (8.2.5)			ν	α	Formula (8.2.6)		
A	B	C		A	B	C			A	B	C
37.50	25.00	37.50	1	37.50	27.50	37.50	1	0.5	40.00	41.25	46.25
26.25	26.25	30.00	3	32.50	38.75	46.25	1	0.7	41.25	38.75	46.25
20.00	25.00	28.75	6	37.50	38.75	46.25	1	0.9	37.50	40.00	48.75
25.00	20.00	33.75	9	40.00	40.00	46.25	3	0.5	32.50	40.00	45.00
21.25	18.75	35.00	12	37.50	40.00	45.00	3	0.7	38.75	41.25	47.50
Formula (8.2.7)				Formula (8.2.8)			3	0.9	37.50	40.00	50.00
A	B	C	W	A	B	C	6	0.5	41.25	43.75	48.75
18.75	16.25	30.00	1.00	35.00	41.25	43.75	6	0.7	41.25	40.00	48.75
30.00	26.25	40.00	1.50	35.00	41.25	48.75	6	0.9	37.50	37.50	47.50
35.00	25.00	37.50	2.00	37.50	41.25	48.75	9	0.5	40.00	40.00	45.00
36.25	31.25	38.75	2.50	37.50	40.00	50.00	9	0.7	40.00	36.25	48.75
							9	0.9	36.25	37.50	48.75
							12	0.5	38.75	41.25	50.00
							12	0.7	37.50	38.75	48.75
							12	0.9	33.75	36.75	45.00

the combination of a pair of forecasts yielded an overall forecast superior to *both* individual forecasts, employing here and throughout this section the criterion of mean squared error. It appears from this table that those procedures that ignore correlation between the forecast errors (formulas (8.2.5), (8.2.6), and (8.2.8)) are considerably more successful than those that attempt to take account of it. It emerges from the table that when Box–Jenkins is combined with one of the fully automatic procedures, the resulting forecast outperforms both individual forecasts on about 40% of all occasions for the more successful combining methods. Table 8.2 compares these combined forecasts with the individual Box–Jenkins forecasts. It is seen that, for the combining methods that ignore correlation between the individual forecast errors, the combined forecasts outperform the individual Box–Jenkins forecasts on a slight majority of the series in the collection. The only exception is for formula (8.2.5) with $\nu = 1$. However, this approach would be employed only if one suspected extreme nonstationarity in the relative performance of the individual forecasting methods—a situation for which one could not make a strong a priori argument in the present context.

In Section 5.5, the notion of a fully automatic predictor that would be a weighted average of a forecast obtained by stepwise autoregression and one derived from an exponential smoothing procedure was briefly introduced. To examine the potential usefulness of such a combined forecast, Table 8.3 gives results for the combination of Holt–Winters and stepwise autoregressive

Table 8.2　*Percentage number of series for which Box–Jenkins is outperformed by (A) Box–Jenkins combined with Holt–Winters, (B) Box–Jenkins combined with stepwise autoregressive*

Formula (8.2.4)			Formula (8.2.5)			Formula (8.2.6)			
A	B	ν	A	B	ν	α	A	B	
53.75	41.25	1	55.00	48.75	1	0.5	57.50	58.75	
46.25	42.50	3	50.00	53.75	1	0.7	57.50	57.50	
41.25	38.75	6	55.00	52.50	1	0.9	55.00	56.25	
47.50	38.75	9	57.50	53.75	3	0.5	50.00	53.75	
45.00	37.50	12	56.25	55.00	3	0.7	56.25	56.25	
Formula (8.2.7)			Formula (8.2.8)		3	0.9	55.00	53.75	
A	B	W	A	B	6	0.5	58.75	56.25	
42.50	37.50	1.00	53.75	55.00	6	0.7	58.75	55.00	
46.25	41.25	1.50	52.50	56.25	6	0.9	55.00	52.50	
48.75	41.25	2.00	55.00	57.50	9	0.5	57.50	53.75	
50.00	46.25	2.50	55.00	56.25	9	0.7	57.50	52.50	
					9	0.9	53.75	53.75	
					12	0.5	57.50	55.00	
					12	0.7	56.25	53.75	
					12	0.9	51.25	53.75	

Table 8.3　*Percentage number of series for which the combined Holt–Winters and stepwise autoregressive forecast outperforms: (A) Holt–Winters, (B) stepwise autoregressive*

Formula (8.2.4)			Formula (8.2.5)			Formula (8.2.6)			
A	B	ν	A	B	ν	α	A	B	
63.75	65.25	1	63.75	71.25	1	0.5	68.75	77.50	
60.00	65.25	3	67.50	78.75	1	0.7	71.25	75.00	
57.50	65.25	6	67.50	78.75	1	0.9	72.50	76.25	
57.50	71.25	9	68.75	77.50	3	0.5	67.50	77.50	
60.00	71.25	12	68.75	76.25	3	0.7	70.00	77.50	
Formula (8.2.7)			Formula (8.2.8)		3	0.9	71.25	78.75	
A	B	W	A	B	6	0.5	70.00	78.75	
57.50	70.00	1.00	68.75	75.00	6	0.7	70.00	78.75	
62.50	73.75	1.50	68.75	80.00	6	0.9	68.75	78.75	
62.50	72.50	2.00	68.75	80.00	9	0.5	68.75	76.25	
63.75	70.00	2.50	68.75	81.25	9	0.7	70.00	78.75	
					9	0.9	70.00	78.75	
					12	0.5	70.00	80.00	
					12	0.7	70.00	78.75	
					12	0.9	67.50	77.50	

forecasts. The results of this table are much more conclusive than those of Table 8.2. In this situation, combining definitely appears to be worthwhile, and hence the resulting fully automatic forecast seems to be a potentially useful tool in those situations where for one reason or another a full Box–Jenkins analysis is not possible.

The evidence of Tables 8.1–8.3 points strongly in favor of the combining procedures that ignore correlation over those that attempt to take correlation into account. To confirm this indication, and to attempt to see if one particular procedure could be regarded as a "best bet," each method was ranked (giving rank 1 to the best and rank 33 to the worst) for each of the 240 pairwise combinations. Formula (8.2.5) with $v = 12$ had the lowest average rank, though there was remarkably little to choose between the best few methods. Average ranks for methods that attempted to account for correlation were a good deal higher than for those which did not. Full details of these results are given in Newbold and Granger [1974].

Using formula (8.2.5) with $v = 12$, the combined forecasts were compared in terms of ratios of average squared forecast errors with the Box–Jenkins forecasts. The combined Holt–Winters and stepwise autoregressive forecasts outperformed Box–Jenkins on 46.25% of series in the sample. The geometric means of the ratios of average squared forecast errors were

$$\frac{\text{average squared error Box–Jenkins and Holt–Winters combined}}{\text{average squared error Box–Jenkins}} = 0.94$$

$$\frac{\text{average squared error Box–Jenkins and stepwise autoregressive combined}}{\text{average squared error Box–Jenkins}}$$
$$= 0.98$$

$$\frac{\text{average squared error Box–Jenkins}}{\text{average squared error Holt–Winters and stepwise autoregressive combined}}$$
$$= 0.99$$

Thus, on the average, a slight improvement in accuracy is obtained when Box–Jenkins is combined with one or other of the fully automatic procedures. However, the most striking aspect of these calculations is the strong performance of the combined Holt–Winters and stepwise autoregressive forecast relative to that of Box–Jenkins. Clearly this combination produces a fully automatic forecast of considerable merit.

Now, the examples of combination given so far are not really designed to show the procedure in its best possible light. After all, it would be reasonable to expect combination to be most profitable when the individual forecasts are very dissimilar in nature. For example, combination of forecasts generated by an econometric model with those generated by an efficient time series analysis method ought to be potentially useful. Granger and Newbold [1975] give an example which uses 20 one-step ahead quarterly forecasts of real inventory investment generated by the Wharton–EFU econometric model. Forecasts of the same quantity were generated from a univariate Box–Jenkins procedure and combined with the model forecasts. The results are shown in Table 8.4. In this particular case, even though the Box–Jenkins forecast is on average considerably better than the econometric forecast, the latter apparently contains very useful information absent in the former. Thus, with the exception of formulas (8.2.4) and (8.2.5) with $v = 1$—extreme cases that will very rarely be appropriate in practice—all the procedures yield a combined forecast that considerably outperforms the better individual forecast. The examples given so far serve to emphasize the potential value of combination in situations where an optimal or near-optimal forecast is not available. It was seen that combination of Box–Jenkins with a univariate fully automatic forecast produced only a small improvement. This is to be expected since intelligent application of the principles of Box and Jenkins should very often lead to a forecast that is near-optimal, given the restricted information set consisting only of past values of the time series. However, examination of nonoptimal forecasts yields very different results, as was seen in the combination of Holt–Winters and stepwise autoregressive. In the wider context of macroeconomic forecasting, a much larger quantity of potentially valuable information is available. It is doubtful whether the concept of fully efficient use of this information is of much practical value. For example, the results of

Table 8.4 *Sums of squared errors of forecasts of real inventory investment, except farm sector (U.S.A)*

	Box–Jenkins 64.4				
S.S.E.	Wharton–EFU 120.0				
	Combined Forecasts				

Formula (8.2.4)		Formula (8.2.5)		Formula (8.2.6)		
SSE	v	SSE	v	α	SSE
73.0	1	58.4	1	0.5	45.1
49.7	3	49.4	1	0.7	41.7
46.6	6	46.8	1	0.9	39.0
43.6	9	43.5	3	0.5	44.2
45.1	12	45.0	3	0.7	41.0
Formula (8.2.7)		Formula (8.2.8)		3	0.9	38.8
SSE	W	SSE		6	0.5	42.2
44.1	1.00	44.0		6	0.7	40.3
44.4	1.50	46.3		6	0.9	38.6
45.1	2.00	47.5		9	0.5	40.2
47.2	2.50	48.4		9	0.7	39.4
				9	0.9	38.1
				12	0.5	41.6
				12	0.7	40.4
				12	0.9	38.2

Table 8.4 indicate that, in forming the econometric forecast, information on past values of real inventory investment was almost certainly not efficiently employed. It is in this wider context that combination of forecasts is likely to prove most fruitful.

Many of the results discussed earlier in this section can be extended to the combination of more than two forecasts, as discussed by Reid [1969] and Newbold and Granger [1974]. Let $f_n^{(j)}$, $j = 1, 2, \ldots, M$, be a set of one-step ahead forecasts of X_n. Assume, as before, that these forecasts are unbiased. Now write $\mathbf{f}_n' = (f_n^{(1)}, f_n^{(2)}, \ldots, f_n^{(M)})$, and the individual forecast errors as $\mathbf{e}_n = X_n \mathbf{1} - \mathbf{f}_n$ with $E(\mathbf{e}_n \mathbf{e}_n') = \Sigma$ and where $\mathbf{1}' = (1, 1, \ldots, 1)$. Then the combined forecast can again be written as a weighted average of individual forecasts, so that

$$C_n = \mathbf{k}_n' \mathbf{f}_n, \qquad \mathbf{k}_n' \mathbf{1} = 1, \qquad 0 \leqslant k_n^{(j)} \leqslant 1 \qquad \text{for all } j$$

where $\mathbf{k}_n' = (k_n^{(1)}, k_n^{(2)}, \ldots, k_n^{(M)})$. It is straightforward to show that the variance of the combined forecast error is minimized by taking the value of $\mathbf{k}_n$ given by

$$\mathbf{k}_0 = (\Sigma^{-1} \mathbf{1})/(\mathbf{1}' \Sigma^{-1} \mathbf{1}),$$

an expression that simplifies to (8.2.2) when $M = 2$. Since the elements of Σ will not be known in practice, this expression cannot be used as it stands. However, formulas (8.2.4)–(8.2.8) can be generalized as follows

$$\hat{\mathbf{k}}_n = (\hat{\Sigma}^{-1} \mathbf{1})/(\mathbf{1}' \hat{\Sigma}^{-1} \mathbf{1}) \quad \text{where} \quad (\hat{\Sigma})_{i,j} = v^{-1} \sum_{t=n-v}^{n-1} e_t^{(i)} e_t^{(j)} \qquad (8.2.9)$$

$$\hat{k}_n^{(i)} = \left(\sum_{t=n-v}^{n-1} e_t^{(i)2} \right)^{-1} \Bigg/ \left(\sum_{j=1}^{M} \left(\sum_{t=n-v}^{n-1} e_t^{(j)2} \right)^{-1} \right) \qquad (8.2.10)$$

$$\hat{k}_n^{(i)} = \alpha \hat{k}_{n-1}^{(i)} + (1 - \alpha) \left(\sum_{t=n-v}^{n-1} e_t^{(i)2} \right)^{-1} \Bigg/ \left(\sum_{j=1}^{M} \left(\sum_{t=n-v}^{n-1} e_t^{(j)2} \right)^{-1} \right), \quad 0 < \alpha < 1$$

$$(8.2.11)$$

$$\hat{\mathbf{k}}_n = (\hat{\Sigma}^{-1} \mathbf{1})/(\mathbf{1}' \hat{\Sigma}^{-1} \mathbf{1})$$

where

$$(\hat{\Sigma})_{i,j} = \left(\sum_{t=1}^{n-1} W^t e_t^{(i)} e_t^{(j)} \right) \Bigg/ \left(\sum_{t=1}^{n-1} W^t \right), \qquad W \geqslant 1 \qquad (8.2.12)$$

$$\hat{k}_n^{(i)} = \left(\sum_{t=1}^{n-1} W^t e_t^{(i)2} \right)^{-1} \Bigg/ \left(\sum_{j=1}^{M} \left(\sum_{t=1}^{n-1} W^t e_t^{(j)2} \right)^{-1} \right), \qquad W \geqslant 1 \quad (8.2.13)$$

Newbold and Granger [1974] considered the combination of Box–Jenkins, Holt–Winters, and stepwise autoregressive one-step ahead forecasts for 80

Table 8.5 *Percentage number of series for which the combined Box–Jenkins, Holt–Winters, and stepwise autoregressive forecast outperforms Box–Jenkins*

Formula (8.2.9)	v	Formula (8.2.10)	v	Formula (8.2.11) α	
50.00	1	50.00	1	0.5	60.00
52.50	3	65.00	1	0.7	61.25
47.50	6	60.00	1	0.9	60.00
52.50	9	63.75	3	0.5	63.75
51.25	12	62.50	3	0.7	61.25
Formula (8.2.12)		Formula (8.2.13)	3	0.9	61.25
	W		6	0.5	62.50
48.75	1.00	61.25	6	0.7	61.25
47.50	1.50	65.00	6	0.9	60.00
47.50	2.00	63.75	9	0.5	65.00
48.75	2.50	63.75	9	0.7	62.50
			9	0.9	62.50
			12	0.5	62.50
			12	0.7	62.50
			12	0.9	62.50

time series using the above formulas, restricting weights to lie between zero and one. The results are summarized in Table 8.5.

The geometric mean for the ratio of the average squared error of the combined forecast to that of Box–Jenkins for formula (8.2.10) with $v = 12$ was 0.92. Two conclusions can be drawn from these results. First, the formulas that ignore correlation between the individual forecast error series continue to do considerably better than those that attempt to take it into account. Second, whatever combining method is employed, a further slight improvement in forecast accuracy results from the addition of the third forecast.

8.3 The Evaluation of Forecasts

The problem of how to evaluate a set of economic forecasts or a forecasting model has received a good deal of attention in the past few years, and many extensive evaluation exercises have been carried out. In part this is a result of the multitude of large-scale macroeconomic models that have sprung up. All these models have their distinctive features, reflecting the multitude of fashions in which one can view macroeconomic behavior. Moreover, from the standpoint of economic theory, the specification of any one of these models is generally defensible–indeed, great care has been taken to ensure that it should be. However, from the same viewpoint, none is regarded as perfect–reality is far too complex, and in many areas theory too ill defined for this to be possible. Of course, practitioners can and do argue the merits of particular models on the basis of theoretical specification, but it is generally difficult to make out a case for one formulation over another, or even to argue the intrinsic worth of a complex model, on a priori grounds alone. Again, the inevitable suboptimality of any set of economic forecasts achieved through use of the methodology outlined in Section 6.3 argues strongly for their detailed evaluation.

An evaluation of forecast performance can, and probably should, be carried out at two levels. At the subjective level, one could look particularly closely at any large errors, or perhaps any failures to detect turning points, which have been made and try to determine the cause of these inadequacies. It may be, for example, that failure to allow for particular circumstances is responsible. If these circumstances could have been foreseen at the time forecasts were made, then the forecast-generating mechanism should be altered to preclude the possibility of making the same mistake again. The inherent danger in such an approach, and one of the reasons it is not of itself sufficient, lies in a natural tendency for the practitioner to explain away all his errors in terms of events (strikes, dramatic shifts in government policy, and so on) that could not possibly have been anticipated at the time forecasts were prepared. Such an attitude is hardly likely to lead to improved performance in the future. An objective evaluation of forecast performance is, in our opinion, of the utmost importance. This has been accepted for a number of years, and various criteria have been made available, principally by Theil [1958, 1966] and by Mincer and Zarnowitz [1969]. More recently, Granger and Newbold [1973] have argued that many of the evaluation criteria previously employed were insufficiently demanding, and have proposed a more stringent alternative.

An objective evaluation of a set of forecasts might seek to answer one or more of the following three questions:

(a) Is one set of forecasts better than its competitors?
(b) How "good," in some sense, is a particular set of forecasts?
(c) Can the forecast-generating mechanism be modified in some way so as to yield improved forecast performance?

Each of these points will be discussed in turn, but it is first necessary to have a criterion for forecast accuracy. The relevant concept in this connection is the cost of error function introduced in Chapter 4. It is assumed that for every forecast error e there is an associated cost $C(e)$. It is reasonable to assume that this function is such that

$$C(0) = 0$$

and

$$C(e_1) > C(e_2) \qquad \text{if either} \quad e_1 > e_2 > 0 \quad \text{or} \quad e_1 < e_2 < 0$$

Furthermore, in many applications it may be quite reasonable to assume a symmetric cost function, but as noted in Section 4.2 this will not always be the case. Given these assumptions, there remains, of course, an infinity of functions from which to choose. In many instances—for example, macroeconomic forecasting by some independent agency—the costs of making a forecast error are typically notional rather than real since decisions are not generally made as a result of these forecasts. In situations where forecasts lead to decisions an accountant might be able to quantify the costs of particular errors. However, he may not be able to do so with any great precision since these costs themselves will not occur until sometime in the future, and hence must be predicted. The problem is therefore circular in nature. Given these difficulties, the practitioner is generally forced to choose a specific cost function, the most popular by far being

$$C(e) = ae^2, \qquad a > 0 \tag{8.3.1}$$

Three factors argue for the choice of the quadratic cost of error function. It is often a priori not an unreasonable assumption, it is mathematically more tractable than any alternative, and it bears an obvious relationship to the least squares criterion generally used in estimating forecasting models. Thus forecasts will be judged here in terms of expected squared error.

In the remainder of this chapter only the cost function (8.3.1) will be employed, though, as noted in Section 4.2 and in Granger [1969a], some progress can be made with alternative functions. In fact, if one's primary interest is in comparing two forecasts, the choice of cost function within a very wide range may not be too critical. Granger and Newbold [1973] consider the very general case of two forecasting procedures producing zero-mean white noise errors $e^{(j)}$, $j = 1, 2$, which are identically distributed apart from a scaling constant, i.e., $e^{(j)}$ has cumulative density function $F(x/S_j)$, so that the larger is S_j the more spread is the distribution of $e^{(j)}$. Consider now a general cost function satisfying $C(0) = 0$ and that monotonically increases from $e = 0$, so that $C(e_1) > C(e_2)$ if either $e_1 > e_2 > 0$ or $e_1 < e_2 < 0$. It can be shown that for *any* such cost function, the predictor producing errors distributed with the lower scale factor S_j has smaller expected cost. Of course, in practice, expected cost must be estimated from sample values, and while this result holds for the population, it will not necessarily do so in a finite sample.

Let X_t, $t = 1, 2, \ldots, N$, be an observed time series and let f_t, $t = 1, 2, \ldots, N$, be a series of forecasts available at time $t - h$; that is, f_t is a forecast of X_t made h time periods previously. The forecast errors are then simply

$$e_t = X_t - f_t, \qquad t = 1, 2, \ldots, N$$

and the criterion following from (8.3.1), expected squared forecast error, is estimated by

$$D_N^2 = \frac{1}{N} \sum_{t=1}^{N} e_t^2 \tag{8.3.2}$$

Consider now the first potential requirement of an evaluation exercise: the comparison of two or more sets of forecasts of the same quantity. Suppose two competing forecasting procedures produce errors $e_t^{(1)}$ and $e_t^{(2)}$, $t = 1, 2, \ldots, N$. Then, if expected squared error is to be the criterion, the procedure yielding the lower average squared error (8.3.2) over the sample period will be judged superior. One would like to determine, where possible, whether one procedure performed significantly better than the other under the usual criteria of statistical significance. It is tempting in this context to employ the usual variance ratio or F test. However, this would be inappropriate for two reasons. First, it is not reasonable to assume in general that the errors produced by one procedure will be uncorrelated with those produced by another. Second, for forecasts of more than one step ahead, the error series are not typically white noise even for optimal forecasts. In fact, for optimal forecasts, the h-step ahead forecast errors constitute in general a moving average process of order $h - 1$. A valid test of one-step ahead forecast errors, under assumptions that frequently may not be too unreasonable, can easily be derived. Suppose that $(e_t^{(1)}, e_t^{(2)})$, $t = 1, 2, \ldots, N$, constitutes a random sample from a bivariate normal distribution with means zero, variances σ_1^2 and σ_2^2, and correlation coefficient ρ. In particular, then, it is assumed that the individual forecasts are unbiased and the forecast errors not autocorrelated—properties that are certainly desirable, though not always attained, in one-step ahead forecasts. Consider, now, the pair of random variables $e^{(1)} + e^{(2)}$ and $e^{(1)} - e^{(2)}$. Now

$$E[(e^{(1)} + e^{(2)})(e^{(1)} - e^{(2)})] = \sigma_1^2 - \sigma_2^2$$

and so the two error variances, and hence, given the assumption of unbiasedness, the two expected squared errors, will be equal if and only if this pair of random variables is uncorrelated. Thus the usual test for zero correlation, based on the sample correlation coefficient

$$r = \frac{\sum\limits_{t=1}^{N} (e_t^{(1)} + e_t^{(2)})(e_t^{(1)} - e_t^{(2)})}{\left[\sum\limits_{t=1}^{N} (e_t^{(1)} + e_t^{(2)})^2 \sum\limits_{t=1}^{N} (e_t^{(1)} - e_t^{(2)})^2 \right]^{1/2}}$$

can be employed, under the stated assumptions, to test equality of expected squared forecast errors. In fact (see, e.g., Lehmann [1959]) this test is uniformly most powerful unbiased.

Attempts to compare forecasts on the basis of criteria that are not monotonic functions of D_N^2 of (8.3.2) can lead to inconsistencies. For example, consider the "inequality coefficient"

$$U_1 = D_N \left/ \left[\left(\frac{1}{N} \sum f_t^2 \right)^{1/2} + \left(\frac{1}{N} \sum X_t^2 \right)^{1/2} \right] \right. \tag{8.3.3}$$

The numerator of this expression and the second quantity in the denominator (which will in any case be fixed in any comparisons) are harmless enough. Introduction of the first term in the denominator, simply an estimate of the standard deviation of the predictor series, adds into the expression a term that is irrelevant in judging forecast quality. It is this factor that gives the inequality coefficient the potential to produce misleading results, as the following simple example demonstrates. Suppose that a time series X_t is generated by the first-order autoregressive process $X_t = aX_{t-1} + \epsilon_t$. Let a be a fixed number with $0 \leqslant a < 1$, and consider the set of one-step ahead predictors of X_t given by

$$f_t = \beta X_{t-1}, \qquad 0 \leqslant \beta \leqslant 1$$

Let us examine the limiting case as sample size N tends to infinity, so that population values replace the corresponding sample quantities. Now,

$$\lim_{N \to \infty} D_N^2 = \left[(1 - a^2) + (\beta - a)^2\right] \operatorname{var}(X)$$

and

$$\lim_{N \to \infty} \left(\frac{1}{N} \sum f_t^2 \right) = \operatorname{var}(f) = \beta^2 \operatorname{var}(X)$$

It then follows that

$$\lim_{N \to \infty} U_1^2 = 1 - \left[2\beta(1 + a)/(1 + \beta)^2\right]$$

But U_1^2, and hence U_1, is minimized, not by taking $\beta = a$, which is the optimal forecast, but by $\beta = 1$, whatever the value of a. The reason is that (8.3.3) gives great weight to the variance of the predictor series, which in this particular case is highest at $\beta = 1$. Needless to say, any evaluation criterion capable of yielding such misleading results is dangerous and its use should be avoided.

The second goal of an evaluation exercise—assessment of the worth, in some absolute sense, of a set of forecasts—is by far the most difficult of the three. Numerous possibilities, of varying degrees of utility, have been proposed and the following brief review is intended to point out which lines of attack are in our opinion most fruitful. The first and most obvious policy, when no competitor against which to judge a set of forecasts is available, is to construct one. It is becoming increasingly common for forecasters to compare their product with so-called "naïve" forecasts—that is, forecasts derived without the benefit of economic theory, and generally consisting of some function of past values of the series of interest. For example, the coefficient

$$U_2^2 = N^{-1} \sum (X_t - f_t)^2 / N^{-1} \sum X_t^2 \tag{8.3.4}$$

proposed by Theil [1966] compares mean squared error of a forecast with that of the "no change rule" (future values forecast as last available observed value) if X is taken to denote actual change and f predicted change. More recently, the naïve competitors contemplated have become rather more sophisticated. For example, Cooper [1972] employs autoregressive models and Dhrymes et al. [1972] propose univariate Box–Jenkins models, which given the results of Section 5.6 would appear to concur with the suggestion of Mincer and Zarnowitz [1969] that the best available extrapolative univariate procedure be employed. Although, as will be seen in the next section, this kind of approach is frequently used, its value is by no means obvious. For example, why consider only univariate time series technques as competitors when it may well be possible to produce superior forecasts through use of the multivariate time series procedures discussed in the previous chapter? If a sophisticated model fails to outperform a naïve extrapolation rule, then clearly something is seriously wrong; but suppose it does succeed in such an undemanding task; ought one really then to be wholly satisfied? We think not. A more demanding alternative is to consider the combination of a forecast with its competitor. If the variance of the combined forecast error is not significantly less than that of the forecast of interest, then the competing forecast would appear to possess no additional useful information. Granger and Newbold [1973] term this concept "conditional efficiency" and urge that model builders be dissatisfied with their products if they are not conditionally efficient with respect to univariate Box–Jenkins forecasts.

A similar idea is considered by Nelson [1972b]. Let $f_t^{(1)}$ denote a set of one-step ahead forecasts derived from a model and $f_t^{(2)}$ a corresponding set of univariate Box–Jenkins forecasts. Consider the regression

$$X_t = kf_t^{(1)} + (1 - k)f_t^{(2)} + \epsilon_t \qquad or \qquad (X_t - f_t^{(2)}) = k(f_t^{(1)} - f_t^{(2)}) + \epsilon_t \quad (8.3.5)$$

The weight k can be estimated by ordinary least squares and its difference from unity tested for significance. (Nelson also considers the generalization of (8.3.5) in which the weights are not constrained to sum to one and Hatanaka [1974a] further generalizes to an equation with a constant term and several alternative forecasts.) Nelson's procedure is certainly a practical approach to computing conditional efficiency, although in some circumstances we may wish to allow for the possibility of changing weights through time. In fact, Nelson applied combination only to sample period fitted values, and employed the technique as a means of improving model forecasts, using weights derived over the period of fit to generate postsample forecasts. This may well be useful, but we would also regard the conditional efficiency criterion as being of great potential value in actual forecast evaluation.

An alternative line of attack is to seek some measure of forecast quality not dependent on the construction of an artificial set of competing forecasts. It was for this purpose that the inequality coefficient U_1 of (8.3.3) was designed. It satisfies $0 \leqslant U_1 \leqslant 1$, taking the value zero only if the forecasts are perfect, that is, $f_t = X_t$ for $t = 1, 2, \ldots, N$. It takes the value one when $f_t = -bX_t$, $t = 1, 2, \ldots, N$, for any $b \geqslant 0$. However, the objections we have raised concerning this coefficient rule out its profitable use in any evaluation exercise.

Mincer and Zarnowitz [1969] attempt a definition of forecast efficiency based on the regression

$$X_t = \alpha + \beta f_t + e_t \tag{8.3.6}$$

A forecast is deemed "efficient" if $\alpha = 0$ and $\beta = 1$ in this expression, and efficiency is tested by the application of ordinary least squares to the available sample. One can raise two practical objections to such an approach. First, in order to obtain consistent estimates it is necessary to assume f_t is uncorrelated with e_t—a requirement which will certainly hold for optimal forecasts, but which need not necessarily do so in practical applications of forecast evaluation. Second, one must assume, for the usual test procedures to be valid, that the error series e_t is white noise. This will not necessarily be so for suboptimal one-step ahead forecasts, and is not generally so even for optimal forecasts of more than one step ahead. However, a far more fundamental objection can be raised to this concept of efficiency, as can be seen from the following simple example. Suppose that X_t is generated by the random walk model

$$X_t = X_{t-1} + \epsilon_t$$

and consider the set of one-step ahead predictors

$$f_t^{(j)} = X_{t-j}, \qquad j = 1, 2, 3, \ldots$$

Now, for any j, (8.3.6) will in theory have $\alpha = 0$ and $\beta = 1$, and hence all these forecasts will be deemed "efficient" by the criterion of Mincer and Zarnowitz. Their criterion is certainly a desirable one for a forecast and hence constitutes a necessary condition for forecast efficiency, but according to any acceptable interpretation of that word it cannot be regarded as a *sufficient* condition. The best one can say of the Mincer–Zarnowitz condition is that any forecast f satisfying $\alpha = 0$ and $\beta = 1$ in (8.3.6) is, as noted by Hatanaka [1974a], optimal in the class $c + df$ where c and d are any constants. It would seem to us that any acceptable definition of forecast efficiency must contemplate the existence of an optimal forecast based on the given information set. The efficiency of a particular forecast would then be the ratio of expected squared error (or some other criterion if a different cost function is employed) of the optimal forecast to that of the given forecast. Unfortunately, this concept is in general of very little operational worth since the optimal forecast error variance is typically unknown. An exception is the case where the information set is restricted to include only past values of the variable to be forecast and in addition only linear predictors are considered. If X_t has known spectrum $f(\omega)$, then Kolmogorov [1941b] shows that the minimum attainable forecast error variance is given by

$$I = \exp\left[\frac{1}{2\pi} \int_{-\pi}^{\pi} \log 2\pi f(\omega) \, d\omega \right]$$

The possibility of estimating this quantity from a given set of data is considered by Janacek [1975]. For general purposes, however, we feel that the only concept of efficiency of any practical value is that of conditional efficiency discussed earlier.

In the absence of a competing set of forecasts, attempts at evaluation have centered on comparisons of the forecast and actual time series. There are three difficulties with such an approach. First, the properties of an optimal predictor series typically differ from those of the actual series. For, let f_t be the optimal predictor of X_t, based on a particular information set, and let e_t be the forecast error so that

$$X_t = f_t + e_t \qquad (8.3.7)$$

Now, for the forecast to be optimal in the expected squared error sense, the error series must have mean zero and be uncorrelated with the predictor series. The first of these requirements implies that $E(X_t) = E(f_t)$ and the second that $\mathrm{var}(X_t) = \mathrm{var}(f_t) + \mathrm{var}(e_t)$. Thus, unless the error series takes on the value zero with probability one, the predictor series will have smaller variance than the real series. Second, many of the measures of forecast quality one might propose are not invariant to linear transformations of the variable to be forecasted. In particular, one gets different results when comparing predicted and actual values than when comparing predicted change and actual change. For example, perhaps the simplest form of presentation of a set of forecasts is to graph predicted and actual values simultaneously. For the levels of economic time series, such pictures almost invariably look extremely impressive. This is simply a result of the fact that most such series follow integrated processes, whose smooth graphs are rather easy to duplicate with a predictor series. Indeed, on this criterion, Box and Newbold [1971] demonstrated that a random walk can appear to give reasonable predictions of another *independent* random walk. However, if one graphs predicted change and actual change, the forecasts typically appear in a much less favorable light since changes in economic series are far less smooth than levels. Since the object of an evaluation exercise ought to be to examine forecast quality as critically as possible, we shall for the remainder of this section assume that X_t is a series of changes and f_t a corresponding series of predicted changes. The third problem is much more difficult to surmount and arises because in most practical situations one has no idea of the minimum attainable forecast error variance. This would not be too serious a problem if all the series one encountered were inherently equally difficult to forecast. However, this is by no means the case. For example, it is not too difficult to predict reasonably well (at least over relatively short horizons) changes in consumption. On the other hand, changes in stock market prices, as many can confirm from bitter experience, are considerably more difficult to anticipate. In the latter case, then, the forecaster ought to be satisfied with results that appear considerably less impressive than those obtained in the former. However, provided these difficulties are kept in mind, it is possible to find simple measures that do convey some idea of the quality of a set of forecasts.

To begin, assume that the series to be forecast has fixed mean and variance, given by

$$E(X_t) = \mu_x \quad \text{and} \quad E\left[(X_t - \mu_x)^2\right] = \sigma_x^2$$

One can consider three attributes of a predictor series f_t; its mean μ_f, its variance σ_f^2, and the degree of its correlation ρ with X_t. Now, expected squared forecast error can be written as

$$E\left[(X_t - f_t)^2\right] = (\mu_f - \mu_x)^2 + (\sigma_f - \rho\sigma_x)^2 + (1 - \rho^2)\sigma_x^2 \qquad (8.3.8)$$

Taking μ_x and σ_x to be fixed numbers, it is clear that expected squared error is minimized by finding as large a ρ as possible—that is, a predictor series as strongly correlated with the actual series as possible—and simultaneously

$$\mu_f = \mu_x, \qquad \sigma_f = \rho\sigma_x \qquad (8.3.9)$$

The second of these conditions confirms the already stated conclusion that, for optimal forecasts, the variance of the predictor series is less than that of the actual. In fact, these requirements have already been met in different form, for provided f_t and e_t in (8.3.6) are uncorrelated, it is straightforward to verify that $\alpha = 0$ and $\beta = 1$ in that expression if and only if (8.3.9) holds. Thus, for one-step ahead forecasts, the conditions (8.3.9) can be tested in actual samples by the usual linear regression procedures. Further, provided (8.3.9) does hold for different forecasts of the same quantity, the best forecast will be the one most correlated with the actual values. Thus the population correlation coefficient, or in practice its sample estimate, can be used as a measure of forecast quality. If such a procedure is to be adopted, however, the conditions (8.3.9) should also be checked. We emphasize again that the correlation recommended is that between predicted and actual changes. It is trivially easy to obtain a predictor series that is apparently highly correlated with the level of any economic time series, making the correlation coefficient between the two series virtually meaningless.

An alternative measure, which is in fact equivalent if the forecast and error series are uncorrelated, follows naturally from (8.3.7). Define PM to be the ratio of error variance to variance of the series to be forecast, so that

$$PM = \sigma_e^2 / \sigma_x^2 \qquad (8.3.10)$$

It is straightforward to verify that, if f_t and e_t are uncorrelated, $PM = 1 - \rho^2$. The square on the correlation coefficient in this expression is at first sight worrying, but in fact if ρ is negative f_t and e_t must be correlated. Now, given our assumption, PM lies between zero and one, taking the value zero if f is a perfect forecast—that is, the series X is purely deterministic— and the value one if f_t is the mean of X for all t. The measure (8.3.10), or its sample estimate, can be viewed as a criterion for judging forecast performance or, if f is an optimal forecast based on a particular information set, as a measure of the predictability of a time series. In the latter context, it is clear that PM

Table 8.6 *Probability (p) that forecast and actual values have the same signs*

PM	0	0.1	0.3	0.5	0.7	0.9	1
ρ	1	0.949	0.837	0.707	0.548	0.316	0
p	1	0.968	0.903	0.833	0.753	0.644	0.5

increases as the forecast horizon is extended, and that it is nonincreasing as the information set considered is widened.

An interesting application of this measure involves the probability that the forecast and actual series will have the same sign if they have zero mean, or alternatively if their deviations from their common mean will have the same sign. On the assumption that these series are distributed as bivariate normal with correlation ρ, it follows from Kendall and Stuart [1963, p. 351] that this probability is given by

$$p = \frac{1}{2} + \frac{1}{\pi} \text{ arc sin } \rho = \frac{1}{2} + \frac{1}{\pi} \text{ arc cos } PM$$

Some specific values are shown in Table 8.6.

Finally, turning to the third requirement of an evaluation exercise, one can propose diagnostic checks on forecast performance. The object here is not simply to assess the quality of a set of forecasts, but rather to suggest possible strategies for improving the forecast generating mechanism. In this context, the measure of conditional efficiency with respect to univariate Box–Jenkins forecasts is relevant, for, should a set of forecasts be inefficient by this criterion, the implication is that the forecaster has not optimally taken account of information provided by past values of the time series. An examination of the specification of lag and error structure in the relevant equations of the model may well suggest remedies for this defect.

Theil [1958] has observed two decompositions of average squared forecast error (8.3.2); these were thought to provide insight into the causes of forecast error and are frequently calculated in actual evaluation exercise. They are given by

$$D_N^2 = \frac{1}{N} \sum (X_t - f_t)^2 = (\bar{f} - \bar{X})^2 + (s_f - s_x)^2 + 2(1 - r)s_f s_x \quad (8.3.11)$$

and

$$D_N^2 = (\bar{f} - \bar{X})^2 + (s_f - rs_x)^2 + (1 - r^2)s_x^2 \quad (8.3.12)$$

where $\bar{f}$ and $\bar{X}$ are the sample means of the predictor and predicted series, s_f and s_x are the sample standard deviations, and r is the sample correlation between the two series. The equality (8.3.11) leads to the definition of the quantities

$$U^M = (\bar{f} - \bar{X})^2 / D_N^2, \qquad U^s = (s_f - s_x)^2 / D_N^2, \qquad U^c = 2(1 - r)s_f s_x / D_N^2$$

Clearly $U^M + U^s + U^c = 1$ and Theil suggests that these quantities, which are often calculated in actual evaluation exercises, have useful interpretations. We doubt that this is so, for consider the following simple example. Let X_t be generated by the first-order autoregressive process

$$X_t = aX_{t-1} + \epsilon_t, \qquad 0 \leqslant a \leqslant 1$$

The optimal forecast of X_t, based on the information set consisting of all past values of X, is given by $f_t = aX_{t-1}$ and, for this predictor, as sample size tends to infinity,

$$U^M = 0, \qquad U^s = \frac{1-a}{1+a}, \qquad U^c = \frac{2a}{1+a}$$

As a varies from 0 to 1, U^s and U^c can take on any values, subject only to the restrictions $0 \leqslant U^s$, $U^c \leqslant 1$ and $U^s + U^c = 1$. Thus interpretation of these quantities is impossible. In the more general forecasting context, the difficulty is that some series are inherently not easy to forecast. Thus, however wide the information set employed, a high correlation between predictor and predicted will not be achieved. In such situations, as is seen from (8.3.9), the standard deviation of the forecast series will be markedly less than that of the actual series for optimal forecasts, and consequently U^s can be expected to differ substantially from zero. On the other hand, for more predictable series, the value taken by U^s can be expected to be lower for optimal forecasts.

The decomposition (8.3.12) is the sample analogue of (8.3.8) and so as such is easier to interpret. It leads to the definition of the quantities

$$U^M = (\bar{f} - \bar{X})^2 / D_N^2, \qquad U^R = (s_f - rs_x)^2 / D_N^2, \qquad U^D = (1 - r^2)s_x^2 / D_N^2$$

As already noted, for optimal forecasts U^M and U^R should not differ significantly from zero, and hence U^D should be close to unity. In fact these requirements are equivalent to those that the sample estimates of α and β in (8.3.6) not differ significantly from zero and unity, respectively. Such a test is worth performing in conjunction with the plotting of Theil's "prediction realization diagram"—a plot of predicted against actual values, yielding a spread of points around the line of perfect forecasts $f_t = X_t$, which could give useful information concerning inadequacies in forecast performance, provided it is handled carefully. A potential danger, however, is to leap to the conclusion that forecast changes tend to underestimate the magnitude of large absolute changes, and that some simple remedy is available for correcting this defect. This phenomenon, however, can simply result from the fact that optimally the standard deviation of the predictor series is less than that of the actual, and hence ought only to be corrected for if a predictor series more strongly correlated with the true series of interest can be found—a task that may be by no means simple.

It should now be clear that examination of the relationship between the forecast and actual series is fraught with danger. However, more clear-cut and frequently more useful, conclusions can often be drawn by looking at the error series e_t. This can, of course, be tested directly for zero mean, but more important is an examination of its time series properties. Optimal h-step ahead forecasts have errors with autocorrelations of order h and higher equal to zero, for otherwise the forecast error will be correlated with something that is known at the time the forecast is made, and hence the forecast could have been improved upon. Ideally, one would like to do a full correlogram or spectral analysis of the error series, but rarely will a sufficiently long set of data be available to make such an exercise worthwhile. As a minimum, one-step ahead forecast errors can be tested for randomness against the alternative of first-order autocorrelation by comparing the von Neumann ratio

$$Q = \frac{(N-1)^{-1} \sum_{t=2}^{N} (e_t - e_{t-1})^2}{N^{-1} \sum_{t=1}^{N} (e_t - \overline{e})^2}$$

with tabulated values given by Hart [1942]. In fact, as will be seen in the following section, in many evaluation exercises involving large-scale econometric models, one-step ahead forecasts are "corrected" for autocorrelated errors. However, the point here is that if a procedure produces such errors, then the model itself should be corrected, not simply the forecasts it generates.

8.4 A Survey of the Performance of Macroeconomic Forecasts

There now exist several large scale macroeconomic models that have been, or are capable of being, employed in the production of forecasts. Such models are generally also used for purposes other than forecasting. For example, they may be thought of as vehicles for the explanation of or for testing theories about macroeconomic behavior. Again, models are frequently used to assess the likely consequences of various policy alternatives. However, our only concern here is with forecasting, although it must be added that we would be extremely dubious about the use of a model for any of these other purposes if it were incapable of forecasting well. In the last few years a number of studies of the forecast performance of such models have been published, and although on the surface some of the evidence appears contradictory and its interpretation subject to great controversy, we feel that taken together these exercises do in fact show a good deal of consistency and allow a number of definite conclusions to be drawn.

At the outset, it is necessary to distinguish between the ability of a model to forecast well and the ability of a team of skilled economists, aided by a model, to forecast well. It seems to us to be both valid and worthwhile to examine both these questions. In fact, these extremes are not the only possibilities, and it is useful to examine intermediate stages between the two. Nevertheless the distinction is worth making, and in fact plays an important part in our conclusions. Two kinds of exercises that are frequently carried out, but whose results will not be discussed here, should be mentioned. First, it is common to see in the literature reports of the "forecasting ability" of a model over the period for which it is fit to data. These seem to us to be of rather limited value since, given sufficient effort, one can almost always obtain a reasonably good fit to a set of sample data. The worth of such a fit however ought to be verified outside the sample period before placing any great confidence in a model. Second, following Adelman and Adelman [1959], a number of studies (several of which are contained in Hickman [1972]) solve econometric models (with and without stochastic error structures) for the time path of endogenous variables given an assumed, generally deterministic, path for the exogenous variables. The objective is to assess, often through spectral analysis, whether or not the resulting simulated series exhibit the kind of fluctuations typically found in macroeconomic time series. Such exercises cannot truly be regarded as forecasting, and moreover, it is difficult to see how they could convey any useful information about the forecasting ability of a model. Hence they are outside the scope of the present discussion, and their results will not be considered here.

As a starting point to our discussion, we shall consider the results of a massive study carried out by Cooper [1972]. This author examines seven previously specified quarterly models of the United States economy, designated Friend–Taubman, Fromm, Klein, Liu, OBE, Wharton–EFU and Goldfeld. (Appropriate references to sources of these models are given in Cooper's paper.) Each model is fitted to a sample of 48 quarterly observations, beginning in the first quarter of 1949. In addition, for the endogenous variables of interest, a univariate autoregressive model is estimated over the same sample period, the order of the autoregressive scheme being determined in each case by the scheme with smallest residual variance, but with an arbitrary cut-off point of eight quarters. The "naïve" forecasts obtained from the autoregressive models were compared with forecasts derived from the seven econometric models over a postsample period of twenty quarters, beginning in the first quarter of 1961. Only one-quarter ahead forecasts were considered. The econometric "forecasts" were obtained by solving the models after insertion of actual future values of the exogenous variables. No attempt was made to modify the econometric forecasts by either mechanical or judgmental means. Thus Cooper's comparisons might be regarded as an attempt at evaluation of the forecasting ability of models as they stand–not as evaluations of the forecasting ability of econometricians working with the help of these models. Table 8.7 presents a broad summary of Cooper's comparisons over 33 macroeconomic time series. It is seen that, over this group of series, the naïve univariate time series procedure outperforms *all* the econometric models on the majority of occasions. (In fact, because of

Table 8.7   *Number of times, in 33*
series, a particular forecasting method
performed best, using mean squared error
as criterion, in Cooper's study

Method	Number of times best
Fromm	7
Liu	1
Klein	1
OBE	2
Wharton–EFU	4
Friend–Taubman	0
Goldfeld	0
Naïve	18

differences in their specifications, some models did not produce forecasts of particular series considered in this table.) These results are startling enough, but more detailed examination of Cooper's findings reveals, if anything, a situation even less complimentary to the econometric models. Every model examined is outperformed, using mean squared forecast error as the criterion for comparison, on a substantial majority of occasions by the simple autoregressive predictor. Of even greater concern is the frequency with which the model forecasts are beaten by very substantial margins indeed by their supposedly "naïve" competitor.

Needless to say, Cooper's paper has not exactly met with uncritical acclaim among the fraternity of econometric forecasters. Criticism, involving both the methodology employed and the conclusion drawn, has come from Goldfeld [1972], McCarthy [1972], Green et al. [1972b] and Howrey et al. [1974]. The debate raises a number of substantial points which ought to be considered before drawing conclusions from Cooper's study, or indeed from the other studies to be described in this section. Accordingly, it is necessary to devote a little space to the arguments involved, before attempting to put Cooper's results into perspective. As a first step, objections to Cooper's methodology and conclusions are summarized:

(1)   The various models considered in the study were not designed by their proprietors to cover the same data sets. Of course, to put these models on comparable footing, it was necessary for Cooper to estimate them all over the same time period. However, it can be argued that the structure of the economy shifts over time, and hence that a model appropriate to one time period may not be adequate for a different period. A particular difficulty is that some of the models examined were originally fitted by their authors to data outside the Korean war period. In refitting these models, it is argued,

some allowance, such as the introduction of dummy variables, should have been made for this.

(2)   A related, but rather deeper, point concerns the whole process of econometric model construction. A model, it is argued, is not a mere timeless prescription handed down from the gods on stone tablets. As was noted in Chapter 6, a postulated specification is typically modified several times according to departures from prior expectations in estimated coefficients. This experimentation in model building is charmingly termed tender loving care (TLC) by Howrey, Klein, and McCarthy. Cooper simply reestimated given sets of equations over a particular sample period, taking the models as fixed. The absence of TLC in such an exercise is seen as a serious departure from model building methodology and hence, it is implied, the forecasting ability of the estimated models will generally be a good deal poorer than that of models derived by practicing econometricans.

(3)   Along similar lines, it is further argued that continual examination of model specification is desirable, and that in the forecast period considered by Cooper several respecifications may well have been thought desirable. Hence it is inappropriate to examine the forecast performance of a fixed model over such a long time period.

(4)   Because the models built in the study contained nonlinearities in the variables, the estimation procedure employed by Cooper does not in general produce consistent parameter estimates.

(5)   No account was taken by Cooper of autocorrelation, either in the errors of individual equations in the models, or in the errors of forecasts of the endogenous variables. Thus the forecasts produced from the models are unnecessarily inefficient.

(6)   Rarely in economic forecasting is the primary interest in making predictions just one quarter ahead, yet this is the only lead time considered by Cooper. Further, because their design is such that they capture well short-run variations in a time series, autoregressive models might be expected to do relatively better in this comparison than in forecasting over longer horizons.

(7)   The mechanical extrapolation of fitted models in the production of forecasts does not faithfully depict econometric forecasting methodology. In particular, it ignores the crucial element of judgment which is almost inevitably present in the production of any economic forecast. This view sees the econometric model as a useful tool in the production of forecasts, but certainly not as some mechanical agency that can safely be allowed to do the job without human intervention.

The objections to Cooper's study are, of course, serious. That many of the points raised are valid does not however imply that nothing of value can be inferred from his results. The conclusion rather is that, as in any piece of applied statistical work, great care is needed in interpretation. It is doubtful whether any research of this kind could stand up unscathed under the minute dissection to which Cooper's article has been subjected. It is almost inevitable that in such circumstances objections of more or less validity would be raised. Typically, however, one ought not to conclude in such circumstances that the research can safely be ignored, but ought rather to examine in turn the arguments made and then attempt to assess what of value remains. Cooper, in a reply to the discussion following his paper, has commented on several of the above points, and we supplement his remarks with our own, point by point.

(1) The estimation of all the econometric models over the same time period if they were to be fairly compared was, of course, necessary. The relevant question is just how much modification is necessary to render a model, deemed appropriate over one time span, appropriate over another. (Typically there was, in this case, a good deal of overlap between the fitting period used by Cooper and that employed in building the original models.) If the modification required is only slight, then it might be reasonable to expect that the numerical conclusions derived would not be too seriously affected. If, on the other hand, more severe modification is thought necessary, one is forced to ask how useful as forecasting tools the models could be! For if instability of this kind extends beyond the period of fit, any forecasts derived from a model may be seriously misleading. The specific question of the Korean War period is answered by Cooper, who notes that, in those models where dummy shift variables were employed, these quantities explained only a very small proportion of variation in the dependent variables.

(2) Part of the above discussion applies also to the second point, for if the equations of a model are assumed to be so specific to the fitting period, how can one expect it to forecast well? The concept of tender loving care, however, requires further consideration, for two distinct points are involved. First, it is of course true that in attempting to explain macroeconomic behavior the investigator may have in mind a number of alternative model structures. In such circumstances it is desirable to test among these alternatives, and one such test may well involve comparing estimated coefficients with prior beliefs as to their probable values. However, there is inherent in this kind of approach the danger of data mining—that is, if one searches sufficiently diligently, one is very likely to find eventually a model that accords well with any reasonable prejudices the econometrician might hold. If nothing else, such experimentation certainly invalidates the usual significance tests, rendering some kind of validation of the model outside the sample period essential before one can have any great confidence in it. Further, if practitioners possess such strong prior beliefs about model coefficients, it is difficult to see why no attempt is made to incorporate them formally in the estimation procedure via Bayes' theorem.

(3) In some sense, given Cooper's results, the assertion here is almost a tautology, for surely if models performed so poorly, some modification would be attempted. Again, it is reasonable in practice to view forecasting as a learning situation in which procedures would be modified in light of unsatisfactory results. However, the question here is a rather different one, and

concerns the forecasting performance of given models. It appears self-evident that, in order to evaluate this performance, it is necessary to record results achieved by the models over a reasonably long time span. The concept of sudden and regular shifts from one model to another over time is rather disturbing, for if the underlying economic process is thought to be so highly nonstationary, one might ask why it is thought reasonable that a fixed coefficients model fitted to a span of 40 or 50 quarters of data would produce adequate forecasts. Surely in such circumstances it would be necessary to incorporate the possibility of nonstationarity of coefficients into the estimation procedure.

(4)   This point is certainly valid, although its force is hard to assess. In practice, what is really required is information about finite sample properties of estimators employed in fitting large-scale econometric models, rather than asymptotic properties. Unfortunately no such general results are available. Although it cannot be proved, our impression is that only a relatively small proportion of the often substantial differences between the forecasting abilities of econometric models and naïve time series procedures found by Cooper can reasonably be credited to any defects in the estimation procedure used in his study.

(5)   We firmly believe that, had Cooper fitted models with an appropriately specified time series error structure, substantial improvement in forecasting performance would have resulted. However, in not allowing for autocorrelated errors, Cooper was simply imitating the admittedly suboptimal behavior of many of the model builders themselves. There is a good deal of evidence to suggest that forecasts may be greatly improved if mechanical adjustments, compensating for autocorrelation in their errors, are made. However, the existence of severe autocorrelation in forecast errors strongly suggests the presence of autocorrelated error problems in the model itself. Unless these are treated at source, by taking account of them in the estimation of the model, the resulting forecasts, whether mechanically adjusted or not, will be unnecessarily inefficient.

(6)   While it is true that one is typically interested in forecasting further than one quarter ahead, it is difficult to see how a model that cannot forecast even one step ahead satisfactorily can be applied to the more ambitious task with any great confidence. In this type of exercise, future values of exogenous variables are assumed to be known, and hence, if the model has any value at all, it would be reasonable to expect its performance to improve, relative to that of the naïve model, the further ahead forecasts are made. However, in forecasting h steps ahead, one is attempting to estimate the change between the current value of a series and its value h steps into the future. This is simply the sum of h step changes, the first of which is the difference between the current value and the next—that is, the one-step ahead forecast less something that is already known. But if this particular element in the h-step ahead forecast is better predicted by the naïve method than by the model, it

follows that, whether or not the naïve predictor is outperformed *h* steps ahead, model forecasts should be judged deficient.

(7)   While generally true, the assertion here is of little relevance to questions concerning the predictive ability of models. The performance of econometric forecasters is a different matter entirely, and will be discussed later in this section. It should be added that a bias in favor of the models is introduced by calculating forecasts given *actual* future values of exogenous variables. Moreover, given the larger information set they take into account, one would certainly expect the models to produce better forecasts than univariate time series procedures.

The debate just summarized is by no means a simple one and, if nothing else, serves to illustrate the practical difficulties involved. What, if anything, can be salvaged from this mass of arguments and counterarguments? We feel that the following conclusion is not unreasonable. It appears that large-scale macroeconomic models, fitted without consideration of appropriate time series error structures, when extrapolated for the production of one-step ahead forecasts over a moderately long span of time, frequently perform on the average rather poorly in comparison with simple univariate time series models. Additional evidence in support of this conclusion is provided by Naylor et al. [1972] and by Nelson [1972b]. The latter paper is of some interest since of all published evaluations of model forecasts, it comes closest to employing the concept of conditional efficiency advocated in the previous section. Nelson compares forecasts from the FRB–MIT–PENN quarterly model of the United States economy with those derived from univariate Box–Jenkins models. Only one-quarter ahead forecasts are considered, and for the econometric model these are computed by setting exogenous variables equal to their actual future values. The forecast period covered only nine quarters and, over this time span with average squared error as criterion, Box–Jenkins outperformed the model for 9 of the 14 series considered, often substantially so. Nelson introduces the idea of combining forecasts only over the sample period. As is typical, the econometric model performs relatively better over the sample period, outperforming Box–Jenkins on 12 of the 14 series. However, in forming optimal weighted averages of the two forecasts, Nelson finds that the regression estimators for the weight given to Box–Jenkins differ significantly from zero at the 5% level for nine of the series. Hence, even the fit of the econometric model can hardly be regarded as satisfactory. Further results along these lines are given by Cooper and Nelson [1975].

That the qualification concerning time series error specification in our conclusion about econometric model performance is important is illustrated by results of Fair [1970, 1974]. Fair's model differs in three respects from those considered earlier. First, it is relatively small, containing just 14 stochastic equations and five identities. Second, it uses anticipations variables, and finally, in estimating the model, allowance is made for first-order autocorrelation in the individual equation error terms. When this model is projected forward for forecasting using, as in previous studies, actual future values of exogenous variables, forecasts of very high quality are found. In our opinion the chief reason that the Fair model is so successful compared to many other models in this respect is that it takes some account of time series error structure. Of course, the use of anticipations variables may also be a factor since, as noted by Adams and Duggal [1974], their inclusion in an econometric model can lead to an improvement in forecast performance. However, judging by the magnitudes involved, we would regard this as an insufficient explanation in itself of the very striking results obtained by Fair. Of course, one might speculate that the relative smallness of the Fair model helps explain its success. Such an explanation would certainly not appeal to traditional macroeconomic model builders!

A recent paper by Christ [1975] appears to show econometric model forecasts in far more favorable light, relative to univariate time series predictions, than do earlier studies. It may be that the quality of models has improved substantially over the last two or three years, or that the particular forecast period used was, for one reason or another, favorable to the models. However, comparison of Christ's results with those of earlier studies is made difficult by a number of factors. First, forecasts were computed for only three variables, one of which is simply the ratio of the other two. Furthermore, forecasts from some models contained judgmental adjustments, for some models the estimation and forecast period overlapped, while yet other models were continually reestimated through the forecast period.

Given the relatively poor predictive performance frequently observed in econometric models, the reader may be tempted to wonder just how econometric forecasters manage to stay in business. Of course, there is at the present time a market for forecasts of whatever quality in a good many fields, as a casual glance through the popular press will indicate. Frequently the appetite of the public does not seem to diminish in the face of past poor performance — an example of this observation is provided by the proliferation of forecasts of election results in the United Kingdom over the past few years. Nevertheless, the results achieved in particular studies by some econometric models are on occasion so poor as to lead to profound skepticism about the value of further forecasts obtained from such sources. Indeed, one can argue that studies such as Cooper's could not possibly reflect reality since any forecaster would very quickly modify his methods on observing such disastrous results. The fact of the matter is that the actual forecasts produced by econometricians are considerably better than the "forecasts" obtained by straightforward extrapolation of econometric models. These actual, or "ex ante," forecasts generally (but by no means invariably) outperform naïve univariate predictions. Their quality is, by any objective criterion, quite high, though it certainly cannot be said that the stage has been reached where the policy maker is always presented with forecasts that are sufficiently good for his purposes. Perhaps economic forecasting is intrinsically too difficult for this to be the case, or perhaps a sufficient improvement might be obtained through use of improved forecasting techniques. The additional factor introduced in ex ante forecasting is, of course, the judgment of the econometrician, and there can be little doubt that in the past this factor has been of crucial importance in the preparation of economic forecasts of reasonable quality. To review the discussion in Chapter 6 of this point, judgment is introduced in three places. First, of necessity, the forecaster must predict future values of exogenous variables before solving the model equations. In principle this ought to add to inaccuracy of forecasts compared to studies in which models are evaluated by solving the models given actual future values of exogenous variables. Second, the individual equations themselves are modified through their constant terms to take account of factors of importance whose effects are not formally incorporated in the model structure. Finally, the forecasts obtained through solution of the model can themselves be modified judgmentally, either directly or through reconsideration of the previously made constant term adjustments. The process by which the econometrician proceeds from model to forecast is by no means a simple one. However, since the transition appears to produce reasonably good forecasts from what might seem to be an unpromising starting point, it warrants closer examination. A number of studies of forecast performance have attempted to dissect the methodology employed in the generation of ex ante forecasts in an attempt to assess the contributions of various elements. This represents an extremely ambitious program of research, and the results obtained should be treated as tentative. Nevertheless it is worth enquiring as to what has been learned from such studies.

Evans et al. [1972] present a very detailed analysis of forecasts based on two large econometric models: Wharton–EFU and OBE. They analyze the effects of both mechanical and judgmental modifications to forecasts. As a first step, they simply solve the two models and calculate "forecasts" given actual future values of the exogenous variables. This exercise is similar to that of Cooper, although here forecasts up to six quarters ahead were calculated, and led to similar conclusions. In the authors' own words: "Similar results have led many economists to conclude that econometric models do not forecast very well. On the basis of the results brought forth in the previous section we are not in a position to contradict them." It was found that the inclusion of mechanical adjustments, to take account of serial correlation in the forecast errors, produced significant improvement in forecast quality one and two quarters ahead. However, such mechanical adjustments failed to lead to satisfactory forecast performance. On the other hand the true ex ante forecasts were of much higher quality, implying that the injection of judgment into the forecasting procedure was necessary for the achievement of satisfactory performance. An interesting footnote concerns the problem of forecasting future values of exogenous variables before computing ex ante forecasts. In this context, Evans, Haitovsky, and Treyz found a peculiar result. Duplication of the procedure by which ex ante forecasts were calculated, but using actual rather than predicted future exogenous variables, produced "forecasts" that showed no noticeable improvement for one model and were actually worse for the other. One might suspect faulty model specification as the cause of this strange phenomenon. To oversimplify, the results presented so far seem to indicate that econometric models frequently forecast rather poorly, but that econometricians do much better. This leads to an obvious question, posed rather succinctly by Evans, Haitovsky, and Treyz: "This study has shown that econometricians have had a better forecasting record to date than an analysis of the econometric models that they used would have led us to predict. Our results offer no substantive evidence that the same econometricians, forecasting without the benefit of an econometric model, would have done any better or any worse in their prediction." Unfortunately, no evidence exists on this point, for econometricians do not produce purely judgmental forecasts to compare with their "model-aided" ex ante predictions. However, Haitovsky and Treyz [1972] produce some examples in which econometric forecasts outperform judgmental forecasts from other sources. A second question would be to ask whether econometricians might not produce even better ex ante forecasts given a more promising starting point than the model they employ! It has already been noted that the Fair model, which at least takes some account of autocorrelated errors in the individual equations, yields reasonably good forecasts without judgmental modification. Fair [1974] has also found that, contrary to the results just presented for other models, these forecasts are considerably improved when actual values of exogenous variables are substituted for predicted.

A recent study by Hirsch et al. [1974], presents econometric models in rather more favorable light as forecasting tools. These authors examine just one model, the BEA (Bureau of Economic Analysis) quarterly model of the United States. Postsample forecasts up to six quarters ahead were made for eight variables, using mean squared error as evaluation criterion. The study was designed to examine in detail the impact of various stages in the forecast generating cycle and, as such, contains valuable information about the effect of judgmental adjustments. For comparison, univariate ARIMA forecasts were calculated to provide a naïve standard. The various forecasts derived were as follows:

(a) ARIMA Forecasts obtained using univariate Box–Jenkins procedure.

(b) PSS Extrapolation of an essentially a priori specified model structure, using actual future values of exogenous variables.

(c) EAF(1) Ex ante forecasts, but with only mechanical constant term adjustments, usually simply to take account of serial correlation in the errors, but occasionally also of special features such as strikes and changes in tax laws. Projections of exogenous variables are employed. These and subsequent forecasts are not based on a fixed model structure, but rather on one that was modified from time to time.

(d) EAF(2) As EAF(1), except that judgmental adjustment of constant terms is used instead of or as well as mechanical adjustment.

(e) EAF(3) These are the actual ex ante forecasts, and differ from EAF(2) in that final forecasts are modified judgmentally.

(f) EPF These are the same as EAF(3), except that actual rather than projected future exogenous variables are employed.

A crude summary of the results of this study is shown in Table 8.8. The first striking result from this table is that the PSS forecast performs rather better than would be expected on the basis of Cooper's findings. Perhaps the quality of the model used here is higher, but it is tempting to look for an alternative explanation in an attempt to reconcile these results with earlier findings. The estimation period employed for the model generating PSS forecasts was 1954 I–1971 IV, while the forecast period in all cases was 1970 III–1973 II. Hence there is an overlap between fitting and forecast periods, so that for example six of the 12 one-quarter ahead PSS "forecasts" are in fact not forecasts at all but fitted values. It is difficult to assess the extent to which this biases the comparison in favor of the model; but certainly some bias is present, and the results should accordingly be treated with caution.

The forecast period used in this study was very short, making it difficult to generalize from the results. For ARIMA and PSS there were 12 one-quarter ahead forecast, 11 two-quarter ahead forecasts, and so on. However, for the other forecasting procedures: "The forecast made during 1971 III was omitted since, because of an extreme uncertainty regarding the implication of the New Economic Policy at the time it was not possible to produce meaningfully distinct EAF(1) and EAF(2) forecasts." Hence the reader is invited to compare ARIMA forecasts over the whole period with ex ante forecasts that exclude predictions made at a time of "extreme uncertainty," when presumably forecasting was correspondingly difficult! Again there is no way to assess the effect of the bias introduced into the comparison.

It may also be that, had a different short span of time been employed for forecast evaluation, the model would have performed relatively less well. As a

Table 8.8 *Number of times Box–Jenkins is outperformed over eight time series*

Quarters ahead	PSS	EAF(1)	EAF(2)	EAF(3)	EPF
1	5	6	6	7	4
2	6	6	6	7	6
3	6	7	6	7	$6\frac{1}{2}$ [a]
4	6	7	7	7	7
5	5	7	7	7	7
6	5	8	8	8	7

[a]Note: $\frac{1}{2}$ indicates a tie in one comparison.

matter of fact, in an earlier study of the same model, Hirsch [1973] found, using a different forecast period, a relatively poorer performance for the model in comparison to a univariate time series procedure.

Taking into account the points just made, we feel that this study produces very little evidence to contradict three previously made conclusions about large-scale econometric models constructed with little or no regard for the time series structure of residuals. First, unaided by judgment, these models tend to forecast rather poorly compared to univariate time series procedures. Second, when judgment is injected, forecast performance improves markedly, improvements being noted at each stage of judgmental modification. Finally, in computing ex ante forecasts it appears that knowledge of actual future values of exogenous variables would have been of little or no use to these forecasters.

8.5 Econometric Forecasting and Time Series Analysis

At this stage we shall attempt to pull together some material from previous chapters, and from earlier sections of this chapter, in order to assess the current state of the art of economic forecasting and to anticipate possible future developments. The most common approach to economic forecasting is through the construction of an econometric model, which together with the forecaster's judgmental modifications is then used to produce forecasts. The degree of judgmental content in any set of forecasts varies from one econometrician to another and also depends on the state of the economy at the time forecasts are made. Nevertheless, the vast majority of econometricians accept judgmental modification as essential, and few produce unaided model forecasts for public consumption. (An exception is Fair [1974].) Purely time series methods are employed relatively infrequently, except in micro sales forecasting.

It is interesting to compare the recent development of time series and econometric model building methodology. Time series analysts have in the past few years been greatly concerned with the problem of how best to relate just two or three time series. A practical procedure for relating a pair of series when causality is unidirectional was given by Box and Jenkins [1970], and more recent research, as described in Chapter 7, has been concerned with modeling the relationship between two series when there is feedback. The appropriate methodology is by no means simple, even in the two-variable case, and a practical general multivariate time series procedure of proven value still awaits development. At the same time as time series analysts were engaged in such modest tasks, econometricians were building models of several hundred equations, using data that time series analysts would not regard either as plentiful or of particularly high quality. This dichotomy suggests two possibilities. Either, in their anxiety to produce large systems that faithfully depict the many relationships in a complex economy, econometricians have failed to touch some extremely important bases on the way, or time series analysts have, for practical purposes, been concerning themselves with problems of minor importance, at least as regards economic forecasting. The reader may already have guessed that we prefer the first explanation. We feel, in fact, that there is an abundance of evidence to support such a preference.

Of course, the econometrician has going for him one thing that the time series analysts lack. He can appeal to economic theory to suggest a priori a plausible model structure. There is no doubt that this is of enormous importance, for we simply could not conceive of the possibility of building a general model linking, say, a hundred series, using only the principles of time series analysis, particularly given the relatively short series typically available. If one wishes to build such a large system, it is essential to have some theory to suggest an appropriate structure, and hence greatly reduce the range of possible model formulations that need to be considered. Is this, then, the complete answer? Given a reasonable theory of economic behavior, is it legitimate for econometricians to proceed to build models, paying little or no regard to time series analysis? We think not, for a number of reasons. For one thing there are a number of areas in which economic theory is not terribly well developed. For example, no completely satisfactory explanation of inflation exists. Again, there are often competing theories of economic activity, so that, for example, the role of money supply is an area in which there is much disagreement among economists. More important though, even when a model is built, it will fit the data imperfectly, producing residuals from each equation of the system, and unless these residuals are negligibly small (which typically they are not) their time series properties (about which economic theory has little or nothing to say) will be of great relevance to forecasting. The econometrician ignores this factor at his peril. It has already been shown in Chapter 6 how failure to attack the problem of autocorrelated residuals can lead to nonsense regressions and consequently inefficient forecasts. The poor forecast performance of models that ignore time series error structure, as described in the previous section, is also suggestive, particularly since Fair [1970, 1974], with a model that allows the individual error processes to be first-order autoregressive, produces forecasts of reasonable quality. This may also explain why knowledge of actual future values of exogenous variables would be of more use to Fair than it appears to be to other model builders. Is Fair's procedure all that is required? Clearly it is of some help, but there is no reason to expect it to be wholly adequate. We doubt whether first-order autoregressive will always be appropriate in this situation, and in cases where it is not, the possibility of the production of spurious regressions, far too frequently, has been demonstrated in Chapter 6. Moreover, one ought also to consider the relationship between error series from different equations, so that, even after a model has been built, there may still be a multivariate time series problem. Finally, practically any econometric model will include, quite rightly, some lag specifications in individual equations. From a forecasting point of view, the desirability of appropriate time lag specification is obvious. However, here again economic theory is of little help. Theory may postulate the existence of lags, but is rarely completely specific about their structure. One might, of course, experiment with various possibilities, but it would seem far more promising to employ the time series analysis techniques of Chapter 7 to suggest a possible specification.

In summary, then, we feel that if economic forecasting models are to be built, it is necessary and desirable that they should incorporate as far as possible the benefits of economic theory. (At the same time they can be used as tools for the empirical testing of elements of that theory.) However, it is equally important that due consideration be paid to problems of time series behavior, if one entertains the hope of producing reasonably accurate forecasts. We would expect that econometric models built in the next few years will indeed take greater account of the principles of time series analysis, and consequently will be superior forecasting tools. We doubt whether any model built in this way would produce results like those found by Cooper. In fact, the process has already begun, and one or two small models recently built (see, e.g., Hendry [1974] and Wall et al. [1975]) do indeed take a good deal more account of time structure than any of their predecessors.

One is left with the element of judgment. Even now econometricians, using models with poor forecasting ability, succeed through judgmental adjustments in frequently producing reasonably good forecasts. However, this recalls the man who, when asked for directions, replied: "Well, you shouldn't start from here." It seems to us entirely reasonable to suppose that it is possible to build quantitative economic models that are considerably better forecasting tools than many of those in current use. Judgmental adjustments, however, would probably still be desirable to account for the influence of unquantifiable factors, or indeed any factors that for one reason or another were not formally incorporated in the model structure. It would appear likely, in such circumstances, that, given a superior starting point, the model builder could produce, on the average, better forecasts.

CHAPTER NINE

NONLINEARITY, NONSTATIONARITY, AND OTHER TOPICS

One fortune teller to another:
No doubt about it, Zelda, the future isn't what it used to be.

CARTOON, SUN NEWSPAPER
London, 29 February 1972

9.1 Instantaneous Data Transformations

In previous chapters a number of important assumptions have been made and utilized. It is the objective of this chapter to briefly investigate the effect of the relaxation of assumptions such as linearity, stationarity, and normality. It will be seen that in certain ways the problems that arise when relaxing these assumptions individually are not unrelated.

There is no completely convincing reason to analyze time series data in precisely the form in which they are provided by a company, government office, or whatever. It has already been seen that it is frequently advisable to consider the first difference of a series rather than the raw series. One possible change worth considering is the analysis of an instantaneously transformed series $Y_t = T(X_t)$ rather than the given series X_t, where $T(\ )$ is some well-behaved function. There are two main reasons that have been proposed for making such a transformation. The first is that Y_t may be a Gaussian series, with several consequent advantages, and the second is that the error from a linear model fitted to the transformed series may have a constant (homogeneous) variance, whereas a linear model fitted to the original series X_t may not have this property.

The main advantage of having a Gaussian series is that it is then known that the optimal single series, least-squares forecast will be a linear forecast, comprising a linear combination of past observed values. For non-Gaussian series it is possible that a nonlinear forecast will be superior, in a least-squares sense, to a linear one. That this possibility is not a necessity can be seen by considering the series generated by

$$X_t = aX_{t-1} + \epsilon_t \qquad (9.1.1)$$

where ϵ_t is a zero mean white noise series having a nonnormal distribution, such as a rectangular or double exponential distribution. The optimal h-step forecast will be $f_{n,h} = a^h X_n$ and so is linear despite X_t not being a Gaussian process.

An example of a series with nonhomogeneous residual variances is that generated by

$$X_t = A(t)Z_t \qquad (9.1.2)$$

where Z_t is a stationary series with mean μ and $A(t)$ is a positive smooth function of time. Differencing the X_t may produce a series that is stationary in mean but not in variance. If $Z_t > 0$, all t, the obvious transformation to consider is the logarithmic, so that

$$Y_t = \log X_t = \log A(t) + log\ Z_t \qquad (9.1.3)$$

and Y_t now has a trend only in mean, which could possibly be removed by differencing. In economics it is often observed that series have similar trends in mean and standard deviation, but for the logarithmic transformation to be appropriate, these quantities have to be proportional and there is no reason to suppose that actual series have such a property. To illustrate a different possibility, consider the following rather contrived model for X_t

$$X_t = a^2 X_{t-1} + \eta_t \qquad (9.1.4)$$

where η_t is white noise but whose standard deviation is linearly related to X_{t-1} and so is not constant through time. This model can be derived from an AR(1) series Y_t given by

$$Y_t = aY_{t-1} + \epsilon_t \qquad (9.1.5)$$

where ϵ_t is a zero-mean, Gaussian white noise series and then considering the model for $X_t = Y_t^2$, which is (9.1.4) with $\eta_t = \epsilon_t^2 + 2a\epsilon_t Y_{t-1}$. If one is given the series X_t to analyze, the proper transformation to apply is naturally $Y_t = X_t^{1/2}$ since Y_t is both Gaussian and has a constant variance error term.

In practice, the most popular instantaneous transformation is the logarithmic, although there has recently been an increased interest in the class of transformations introduced by Box and Cox [1964] (see Box and Jenkins [1973])

$$Y_t = \frac{\left[(X_t + m)^\theta - 1\right]}{\theta} \qquad (9.1.6)$$

which involves two parameters, m and θ, and which, for $\theta = 0$, corresponds to the logarithmic. The question of how to choose the parameters m and θ will be discussed later.

For whatever reason, instantaneous transformations are often used in practice, and it is important to study the time series properties of the resulting series and also the question of how to forecast X_{n+h} given that one only has available a model for $Y_t = T(X_t)$, where $T(\ )$ is the transformation used. These questions have been considered in some detail by Granger and Newbold [1976], and so only the main results will be presented here, largely without proof.

Let X_t be a stationary, Gaussian series with mean μ, variance σ^2 and autocorrelation sequence corr $(X_t, X_{t-\tau}) = \rho_\tau$. Set $Z_t = (X_t - \mu)/\sigma$ so that Z_t, $Z_{t-\tau}$ will be jointly distributed as bivariate normal with zero means, unit variances and correlation ρ_τ. Consider an instantaneous transformation of the form $Y_t = T(Z_t)$ where $T(\ )$ can be expanded in terms of Hermite polynomials in the form

$$T(Z) = \sum_{j=0}^{m} \alpha_j H_j(Z) \tag{9.1.7}$$

and where m can be infinite. The jth Hermite polynomial $H_j(Z)$ is a polynomial in Z of order j with, for example, $H_0(Z) = 1$, $H_1(Z) = Z$, $H_2(Z) = Z^2 - 1$, $H_3(Z) = Z^3 - 3Z$, and so forth. If X and Y are normally distributed random variables with zero means, unit variances, and correlation ρ, these polynomials have the important orthogonal properties

$$E[H_n(X)H_k(X)] = 0, \qquad n \neq k$$
$$= n!, \qquad n = k \tag{9.1.8}$$

and

$$E[H_n(X)H_k(Y)] = 0, \qquad n \neq k$$
$$= \rho^n n!, \qquad n = k \tag{9.1.9}$$

Using these properties, it is easy to show that $E(Y_t) = \alpha_0$ and

$$\operatorname{cov}(Y_t, Y_{t-\tau}) = \sum_{j=1}^{m} \alpha_j^2 j! \rho_\tau^j$$

Thus, the linear properties of the transformed series can be determined. It follows, for example, that if

$$Y_t = a + bX_t + cX_t^2 \tag{9.1.10}$$

then

$$\operatorname{cov}(Y_t, Y_{t-\tau}) = (b + 2c\mu)^2 \sigma^2 \rho_\tau + 2c^2 \sigma^2 \rho_\tau^2$$

and if $Y_t = \exp(X_t)$, then

$$\operatorname{cov}(Y_t, Y_{t-\tau}) = \exp(2\mu + \sigma^2)(\exp(\sigma^2 \rho_\tau) - 1)$$

Further, if X_t is MA(q), Y_t has autocovariances appropriate for MA(q), but if X_t is AR(p) and Y_t is given by (9.1.10), then its autocovariances correspond to a mixed ARMA($\frac{1}{2}p(p + 3)$, $\frac{1}{2}p(p + 1)$) process.

It may also be shown that if X_t is an integrated ARIMA(p, 1, q) process, then Y_t will also be such that its first differences can be modeled as an ARMA process.

If X_t is Gaussian, or is at least assumed to have this property, and a model has been formed for it, then standing at time n an optimal forecast, $f_{n,\,h}$ of X_{n+h} can be easily formed by the methods discussed in Chapters 4 and 5, and it will be linear in $X_{n-j}, j \geq 0$. Suppose, however, that X_{n+h} is not the quantity for which a forecast is required, but rather

$$Y_{n+h} = T\left(\frac{X_{n+h} - \mu}{\sigma}\right)$$

needs to be forecast. An example would be an economist who has built a model for log price but actually wants to forecast price. Two forecasts of Y_{n+h} can be considered easily, the optimal forecast under a least squares criterion given by

$$g_{n,\,h}^{(1)} = E[Y_{n+h}|I_n] \tag{9.1.11}$$

where I_n is the information set $X_{n-j}, j \geq 0$, and the naïve forecast

$$g_{n,\,h}^{(2)} = T\left[\frac{f_{n,\,h} - \mu}{\sigma}\right] \tag{9.1.12}$$

Write $X_{n+h} = f_{n,\,h} + e_{n,\,h}^{(x)}$ so that $e_{n,\,h}^{(x)}$ is the h-step forecast error of X_{n+h} when the optimal forecast $f_{n,\,h}$ is used at time n, and denote

$$S^2(h) = \mathrm{var}(e_{n,\,h}^{(x)}) \tag{9.1.13}$$

Since Y_{n+h} is a function of a Gaussian variable, it can be shown that the optimal, generally nonlinear, forecast is given by

$$g_{n,\,h}^{(1)} = \sum_{j=0}^{m} \alpha_j A^j H_j(P) \tag{9.1.14}$$

where

$$A = \left(1 - S^2(h)/\sigma^2\right)^{1/2}, \qquad P = (f_{n,\,h} - \mu)/(\sigma^2 - S^2(h))^{1/2} \tag{9.1.15}$$

A little algebra then yields the unconditional expected squared forecast error as

$$\mathrm{var}(e_{n,\,h}^{(y)}) = \mathrm{var}\left(Y_{n+h} - g_{n,\,h}^{(1)}\right) = \sum_{j=1}^{m} \alpha_j^2 j! (1 - A^{2j}) \tag{9.1.16}$$

In deriving this result it is necessary to use the facts that, under the conditions stated, $f_{n,\,h}$ is normally distributed and that

$$E[g_{n,\,h}^{(1)}] = E[Y_{n+h}] = \alpha_0 \tag{9.1.17}$$

Define as a measure of forecastability of a series Y_t, having finite variance, using a given information set, the quantity

$$R_{h,y}^2 = \frac{\text{var}(f_{n,h}^{(y)})}{\text{var}(Y_{n+h})} \qquad (9.1.18)$$

where $f_{n,h}^{(y)}$ is the optimal forecast of Y_{n+h}, possibly a nonlinear forecast, based on the information set and using a least-squares criterion. Clearly, the nearer this quantity is to unity, the more forecastable is the series. If the series one is interested in does not have finite variance, an integrated process being an example, then the definition should be applied after the series has been differenced sufficiently often to produce a finite variance series, if this is possible. Since

$$\text{var}(Y_{n+h}) = \sum_{j=1}^{m} \alpha_j^2 j! \qquad (9.1.19)$$

it follows from this and (9.1.16) that

$$R_{h,y}^2 = \frac{\sum_{j=1}^{m} \alpha_j^2 j! A^{2j}}{\sum_{j=1}^{m} \alpha_j^2 j!}$$

Assuming that X_t is nondeterministic, so that $S^2(h) > 0$, then

$$R_{h,x}^2 = A^2 \qquad \text{and} \qquad 0 \leqslant A^2 < 1$$

It thus follows that

$$R_{h,y}^2 < R_{h,x}^2, \qquad m > 1 \quad \text{and} \quad A > 0.$$

This implies that any nonlinear instantaneous transformation of a Gaussian process X_t is always less forecastable than X_t, provided X_t is not simply a white noise series. We call this the *forecastability theorem*.

As an example, suppose that X_t has $\mu = 0$ and $\sigma = 1$ and is Gaussian, but that a forecast is required of $Y_{n+h} = \exp(X_{n+h})$. This would correspond to the example given earlier where X_t is log price and Y_t is price. Let, as before, $f_{n,h}^{(x)}$ be the optimal forecast of X_{n+h} using I_n: $X_{n-j}, j \geqslant 0$, and $S^2(h)$ be the variance of the h-step forecast error of X_{n+h}. The optimal forecast of Y_{n+h} is then given by

$$g_{n,h}^{(1)} = \exp(f_{n,h} + \tfrac{1}{2} S^2(h))$$

and the naïve forecast is

$$g_{n,h}^{(2)} = \exp(f_{n,h})$$

which is seen to be biased and, naturally, to lead to a higher expected squared error. In this case it is fairly easily shown that

$$R_{h,y}^2 = \frac{\exp(1 - S^2(h)) - 1}{e - 1}$$

and of course

$$R^2_{h, x} = 1 - S^2(h)$$

Some representative values for these quantities are:

$S^2(h)$	0.2	0.5	0.8
$R^2_{h, x}$	0.8	0.5	0.2
$R^2_{h, y}$	0.713	0.378	0.129

The results presented here are very general, although the normality assumption is rather a strong one and the formulas involved can be very complicated for some transformations, such as that in (9.1.6) with fractional θ. The first use of Hermite polynomials in tackling the problems just discussed was by Barrett and Lampard [1955].

9.2 Forming Nonlinear Forecasts

In the previous section, a Gaussian series was transformed into a non-normal one and then the optimal nonlinear forecast found for the transformed series. If this sequence can be reversed, so that an instantaneous transformation can be found for the given series X_t of the form $Y_t = T(X_t)$ such that now Y_t is Gaussian, then the theory of the previous section can be used to find a nonlinear forecast for X_t that should be superior to the best linear forecast, at least in theory. The obviously difficult part of this procedure is to find the proper transformation. Suppose that a specific class of transformations are to be considered, involving a vector of parameters $\boldsymbol{\theta}$, so that

$$Y_t(\boldsymbol{\theta}) = T(X_t, \boldsymbol{\theta}) \tag{9.2.1}$$

and that there exists a particular value for $\boldsymbol{\theta}$, denoted $\boldsymbol{\theta}_0$, such that $Y_t(\boldsymbol{\theta}_0)$ is a Gaussian process, in the sense that an ARIMA model can be fitted to $Y_t(\boldsymbol{\theta}_0)$ of the usual form

$$a(B)(1 - B)^d Y_t(\boldsymbol{\theta}_0) = b(B)\epsilon_t \tag{9.2.2}$$

ignoring seasonal possibilities, and where ϵ_t is Gaussian white noise. The only question that now remains is how to find $\boldsymbol{\theta}_0$. There are at least two ways to approach this question and these do not necessarily give identical answers and could be thought of as having somewhat different objectives. The first approach is to write down the likelihood function for $Y_t(\boldsymbol{\theta})$. If it is supposed that an ARIMA model is fitted to $Y_t(\boldsymbol{\theta})$ for each $\boldsymbol{\theta}$, giving residuals $\epsilon_t(\boldsymbol{\theta})$ and the parameters of the model are chosen so that the estimate of the variance of $\epsilon_t(\boldsymbol{\theta})$ is minimized, giving $\hat{\sigma}^2_\epsilon(\boldsymbol{\theta})$, the log likelihood becomes

$$L(\boldsymbol{\theta}) = -\tfrac{1}{2} n \log \hat{\sigma}^2_\epsilon(\boldsymbol{\theta}) + \log J(X, \boldsymbol{\theta}) \tag{9.2.3}$$

where J is the Jacobian of the transformation. $\boldsymbol{\theta}_0$ is then chosen as the value of θ that maximizes $L(\theta)$. An alternative approach can be based on the forecastability theorem introduced in the previous section. This can be phrased as saying that, in terms of population values,

$$R^2(\theta) = 1 - \frac{\hat{\sigma}_\epsilon^2(\theta)}{\text{var}(Y_t(\theta))} \qquad (9.2.4)$$

will have a maximum at $\theta = \theta_0$.

Both $L(\theta)$ and $R^2(\theta)$ depend on $\hat{\sigma}_\epsilon^2(\theta)$, and at first sight this is a difficult quantity to obtain as θ varies since the results of the previous section show that the ARIMA model may have to be reidentified for every θ value used. However, $\sigma_\epsilon^2(\theta)$ can be estimated without going through an actual model-building process by using result (4.5.7), which states that

$$\log \sigma_\epsilon^2(\theta) = \frac{1}{2\pi} \int_{-\pi}^{\pi} \log 2\pi s(\omega, \theta) \, d\omega \qquad (9.2.5)$$

where $s(\omega, \theta)$ is the spectrum of $Y_t(\theta)$. The use of this result to obtain a "nonparametric" estimate of $\log \sigma_\epsilon^2(\theta)$ was suggested by Brubacher and Wilson [1975]. The properties of this estimate have been studied by Janacek [1975], who found that the log of the periodogram, introduced in Section 1.4, provided an adequate estimate of the log-spectrum in these circumstances. Expressions for the bias, variance and distribution of the estimate of $\log \sigma_\epsilon^2(\theta)$ were also obtained. Thus, the suggested procedure is to form $Y_t(\theta)$ for various values of θ, to form estimates of $\log \sigma_\epsilon^2(\theta)$ from (9.2.5), and to use these estimates to find θ_0 using either (9.2.3) or (9.2.4).

9.3 General Nonlinear Models

The models considered in the previous two sections are of the form

$$a(B)T(X_t) = b(B)\epsilon_t \qquad (9.3.1)$$

and so constitute a particular class of nonlinear models. More general models might take the form

$$T_0(X_t) = \sum_{j=1}^{p} a_j T_j(X_{t-j}) + b(B)\epsilon_t \qquad (9.3.2)$$

or, in terms of $Y_t = T_0(X_t)$,

$$Y_t = \sum_{j=1}^{p} a_j S_j(Y_{t-j}) + b(B)\epsilon_t \qquad (9.3.3)$$

where $S_j(\ )$, $j = 1, \ldots, p$, is a sequence of transformations that are not necessarily identical, so that Y_t is modeled as the sum of transformed values of previous Y's plus a moving average term. To further generalize, transformed pairs of previous values, such as $S(Y_{t-1}, Y_{t-2})$, could be added to the right-hand side of the equation. However, it seems very unlikely that identification of the transformations to use would be possible without either extremely large amounts of data, a very specific underlying nonlinear theory, or limitation of the transformations used to a very particular class, such as polynomials. A class of models that might be considered, for example, would be of the form

$$X_t = \sum_{j=1}^{p} a_j X_{t-j} + \sum_{j=1}^{r} \sum_{k=1}^{r'} c_{jk} X_{t-j} X_{t-k} + b(B)\epsilon_t \qquad (9.3.4)$$

although higher order terms could also be included. A problem with such models is that they are unlikely to be nonexplosive over long time spans. Consider the very simple model

$$X_t = aX_{t-1}^2 + \epsilon_t \qquad (9.3.5)$$

where ϵ_t is zero-mean white noise with a symmetric distribution. If $a > 0$ and $X_{t-1} > 1/a$, the probability that $X_t > X_{t-1}$ is greater than $\frac{1}{2}$ and so the series is inclined to explode. If the series begins with a small value and ϵ_t has small variance, X_t may well appear stationary for some time, but will almost certainly eventually become unstable. In practice, a series generated by (9.3.5) will have three modes: a stable one that may last only for a very short time, an explosive one, and an intermediate mode from which the series will either return to the stable mode or become explosive. Once in the explosive mode, the series will almost certainly remain in it. Since highly nonstable models may be thought unrealistic and are also very difficult to analyze statistically, it seems doubtful that polynomial, autoregressive models such as (9.3.5) need extensive consideration at this time. Even if such models were fitted, it is very difficult to use them to forecast more than one step ahead since to go further ahead a forecast of X_{n+1}^2, say, is required. But since X_t will almost certainly not be Gaussian, the methods discussed in Section 9.1 cannot be used.

A related class of models, but with fewer problems, has been considered by Nelson and Van Ness [1973a, b]. In these models, a series X_t is generated from a second series Y_t in a nonlinear fashion. If just quadratic terms are used, the model takes the form

$$X_t = m + \sum_{j=1}^{p} a_j Y_{t-j} + \sum_{j=1}^{q} \sum_{k=1}^{q'} c_{jk} Y_{t-j} Y_{t-k} + \epsilon_t \qquad (9.3.6)$$

By defining

$$Z_{j,t} = Y_t Y_{t-j} - \lambda_j^{(y)} \qquad \text{where} \quad \lambda_j^{(y)} = \text{cov}(Y_t, Y_{t-j})$$

then (9.3.6) can be written

$$X_t = m' + \sum_{j=1}^{p} a_j Y_{t-j} + \sum_{j=1}^{q} \sum_{k=1}^{q'} c'_{jk} Z_{j,t-k} + \epsilon_t \qquad (9.3.7)$$

and the coefficients can be estimated as for any linear multivariate model, provided X_t, Y_t are jointly stationary. Nelson and Van Ness also consider an estimation procedure based on spectral estimates, but in a simulation study they find that if p, q, q' are not large, the regression approach gives better estimates. The class of models considered in (9.3.6) appears unnecessarily restricted, and the addition of lagged, linear terms in X_t to the right-hand side of the equation produces a clearly superior class. It should be emphasized that such models are relevant only when there is one-way causality from Y_t to X_t and that consideration of a feedback situation is very much more difficult. The possibility of improving forecasts by including nonlinear Y_t terms can be tested using stepwise regression methods, but the usual worries when considering a large number of alternative models, sometimes called "data-mining," should be borne in mind. The model should not be evaluated in terms of its goodness-of-fit over some sample period but rather by its forecasting ability, compared to a linear model, over some later time period.

9.4 Forecasting Nonstationary Processes

In the previous chapters the only series considered in any detail were either assumed to be stationary or could be made stationary by suitable differencing or by removal of a trend in mean. There are many reasons for believing that the series generated by the real world belong to broader classes of non-stationary processes. The main problem of relaxing the assumption of stationarity is that the number of possible nonstationary models is very large indeed and that many cannot be analyzed given just a single realization over a limited time period. The analyst is therefore forced to consider those models that he has a chance of correctly analyzing from the available data, except on those rare occasions where a sound economic theory provides a specific nonstationary model.

An important class of nonstationary series is that generated by linear models with time changing parameters such as

$$X_t = m_t + \sum_{j=1}^{p} a_j(t)X_{t-j} + \sum_{j=0}^{q} b_j(t)\epsilon_{t-j} \qquad (9.4.1)$$

where ϵ_t is a zero-mean, white noise series with constant variance. X_t will have mean μ_t that can be found by solving the difference equation

$$\mu_t - \sum_{j=1}^{p} a_j(t)\mu_{t-j} = m_t$$

In the completely theoretical, but very convenient, situation where m_t, $a_j(t)$, and $b_j(t)$ are assumed to be known perfectly for all t, the μ_t become known and can be subtracted from the data, so that nothing is lost in this case by taking μ_t to be identically zero for all t. Under the same assumptions, it can be shown that the optimal prediction of X_{n+h} based on the information set I_n: $X_{n-j}, j \geqslant 0$, is given by

$$f_{n,h} = \sum_{j=1}^{p} a_j(n+h)f_{n,h-j} + \sum_{j=0}^{q} b_j(n+h)\hat{\epsilon}_{n+h-j} \qquad (9.4.2)$$

where

$$f_{n,k} = X_{n+k}, \quad k \leqslant 0$$

$$\hat{\epsilon}_{n+k} = \frac{X_{n+k} - f_{n+k-1,1}}{b_0(n+k)}, \quad k \leqslant 0$$

$$\hat{\epsilon}_{n+k} = 0, \quad k > 0$$

This has been proved by Whittle [1965] and by Abdrabbo and Priestley [1967] and represents the natural generalization of the equivalent prediction formula for a stationary ARMA process, as considered in Section 4.4. As an example, consider the nonstationary AR(1) process $X_t = a(t)X_{t-1} + \epsilon_t$ for which the optimal forecasts obey the difference equation $f_{n,h} = a(n+h)f_{n,h-1}$ so that

$$f_{n,h} = \left(\prod_{j=1}^{h} a(n+j) \right) X_n \qquad (9.4.3)$$

In the more realistic situation, when the parameters $a_j(t)$ and $b_j(t)$ are not known, then with the information set $X_{n-j}, j \geqslant 0$, it is conceivable that they could be estimated for t ranging up to n but as (9.4.2) and (9.4.3) show, to form optimal forecasts, the future values of these parameters are also required. This means that $a(n+h)$, $h > 0$, needs to be somewhat predictable from $a(n-k)$, $k \geqslant 0$, and this requirement places an important, but natural, restriction on the class of nonstationary processes that are worth consideration in practice. If the $a_j(t)$ and $b_j(t)$ are trending smoothly over time, the processes could be called slowly changing. Such processes have been considered from the spectral viewpoint by Priestley [1965].

It is not clear how best to identify an ARMA model with time-changing coefficients, but a possible starting point is to take p and q in (9.4.1) to be time invariant and to identify their values by using the methods discussed in Chapter 3, treating the series as though it were stationary. Given a single realization of the series, it is clearly not possible to estimate $a_j(t)$, $b_j(t)$ for all t, but it is possible to obtain an estimate of the "average" values of these parameters over some time interval. It is necessary to assume that the roots of $a_t(z)$ lie outside the unit circle for all t, where

$$a_t(z) = 1 - \sum_{j=1}^{p} a_j(t)z^j$$

and also that the parameters change slowly with time. Dividing the data into possibly overlapping segments of length k, a stationary model can be fitted to each of these segments and the estimates obtained considered as average values of the actual parameter values over the time segment. As an example, suppose the model to be fitted is

$$X_t = a(t)X_{t-1} + \epsilon_t$$

Considering the time period $T - K/2 \leqslant t \leqslant T + K/2$, fit a model of the form

$$X_t = aX_{t-1} + \eta_t$$

where η_t is taken to be white noise. Then the estimate $\hat{a}$ of a obtained can be thought of as an estimate of

$$\frac{1}{K+1} \sum_{t=T-K/2}^{T+K/2} a(t)$$

which might be taken to be an approximation for $a(T)$ if $a(t)$ changes slowly with time. Repeating this procedure with different values of T will give a sequence of rather crude estimates of $a(T)$ up to $T = n - K/2$. It should then be possible to extrapolate this sequence to obtain estimates of $a(t)$ for $t > n - K/2$. Since the data segments will probably overlap considerably, it is better to try to predict the change in $\hat{a}(T)$ from one segment to the next as the series of estimates $\hat{a}(T)$ will be highly autocorrelated. There is no difficulty in principle in extending this method to deal with any time-changing ARMA process, although whether there is any advantage in forecasting terms over using a time-invariant model is not clear since these estimated errors may accumulate to considerable importance for the time-change model. An alternative approach would be to estimate the auto-covariance sequence for each time segment, extrapolate these values, and then deduce the corresponding parameter values for $t > n - K/2$. We know of no results comparing these alternatives or evaluating the usefulness for forecasting economic series of ARMA models with time-changing coefficients.

Using segments of data might be thought to be wasteful since relevant data are being ignored, i.e., that occurring earlier than the starting time of the segment. To overcome this, a further method of estimating time-changing parameters could be used, based on discounted least-squares, as suggested by Gilchrist [1967]. Considering a simple AR(1) model with parameter a, this parameter could be estimated by minimizing

$$I_T(\theta) = \sum_{t=1}^{T} (X_t - aX_{t-1})^2 \theta^{2(T-t)} \tag{9.4.4}$$

where θ is a parameter, of modulus less than one, chosen to give greater weight to the most recent data. Note that (9.4.4) can be considered as the minimization of

$$I_T(\theta) = \sum_{t=1}^{T} (Y_t - (a\theta^{-1})Y_{t-1})^2 \qquad \text{where} \quad Y_t = X_t\theta^{T-t} \tag{9.4.5}$$

and so is equivalent to fitting an AR(1) model with parameter $a\theta^{-1}$ to the nonstationary series Y_t. There is no difficulty in generalizing this procedure to more complicated models, and by allowing T to increase, estimates of time-changing coefficients result. A problem with this approach is in deciding exactly what is being estimated. If the parameters are $a_j(t)$, $b_j(t)$, the estimates obtained relate to a weighted average over time of these quantities. The "average lag" concept developed from the theory of distributed lags (see Dhrymes [1971]) suggests that if the parameters are slowly changing with time, the estimate can be taken to be of $\hat{a}(T - m(\theta))$, where $m(\theta)$ is the average lag of the weights used and is given by $m(\theta) = \theta/(1 - \theta)^2$. Thus, for example, if $\theta = 0.5$, $m(\theta) = 2$, if $\theta = 0.75$, $m(\theta) = 12$ and if $\theta = 0.9$, $m(\theta) = 90$. The figures for $m(\theta)$ also indicate the equivalent degrees of freedom, or data length, involved for different values of θ. The choice of θ must be left to the analyst and will depend on his assessment of the rate of change through time of the parameters of the model. Again, once estimates of $a_j(t)$, $b_j(t)$ have been obtained, they can be extrapolated and used to forecast X_{n+h}.

One further class of nonstationary processes can easily be optimally predicted, being those generated by explosive ARMA models with time-invariant parameters. Suppose, for example, that X_t is generated by

$$X_t = aX_{t-1} + \epsilon_t \qquad (9.4.6)$$

where ϵ_t is zero-mean white noise but $|a| > 1$, so that X_t is not stationary. The optimal forecast of X_{n+h} based on X_{n-j}, $j \geq 0$, is easily shown to be $f_{n,h} = a^h X_n$, and so $|f_{n,h}|$ will be monotonically increasing. There is usually no problem in realizing that an explosive model is an appropriate one for data generated by such processes, although proper identification is much more difficult. For example, if in (9.4.6), $a = 1.1$, then $a^8 > 2$, so that X_{t+8} will be roughly twice the magnitude of X_t. Only if a is very near one, such as $a = 1.02$, will the series be almost indistinguishable from an integrated process. If X_t is generated by the ARMA model $a(B)X_t = b(B)\epsilon_t$, then if $a(z)$ has only a single, and hence real, root inside the unit circle, a sensible procedure is to form $Y_t(\rho) = X_t - \rho X_{t-1}$ for various values of $\rho > 1$ until a series is produced with correlogram appropriate for a stationary series. Once an appropriate value of ρ has been found, the identification and estimation of a model for the derived stationary series can proceed as usual, and forecasts of X_{n+h} are easily derived. If $a(z)$ has several roots inside the unit circle, this approach will not be appropriate.

9.5 Testing for Normality

Most of the tests of significance used in time series analysis assume that the series is Gaussian, that is, every subset of the series $(X_{t_1}, X_{t_2}, X_{t_3}, \ldots, X_{t_k})$ has a multivariate normal distribution. Given only a single realization, it is clearly impossible to test such an assumption, but it is possible to test whether or not the marginal distribution is normal. There is evidence that some series that occur in economics are not normally distributed, and it has been proposed that the actual distribution has infinite variance. For a discussion of this possibility and its consequences, see Granger and Orr [1972].

A test for normality of the marginal distribution of a series X_t,

$t = 1, \ldots, n$, has been suggested by Lomnicki [1961]. Define

$$m_j = \frac{1}{n} \sum_{t=1}^{n} \left(X_t - \overline{X} \right)^j, \qquad j = 2, 3, 4$$

$$G_1 = m_3 m_2^{-3/2}, \qquad G_2 = m_4 m_2^{-2} - 3$$

Then Lomnicki proved that if X_t was both Gaussian and stationary and if n is sufficiently large, then both G_1 and G_2 are normally distributed with zero means and variances:

$$\text{var}(G_1) = 6 \sum_{\tau = -\infty}^{\infty} \rho_\tau^3 \qquad \text{and} \qquad \text{var}(G_2) = 24 \sum_{\tau = -\infty}^{\infty} \rho_\tau^4$$

In practice, these variances would have to be estimated from the data by replacing the infinite sums by finite sums of powers of estimates of the ρ_τ. The test statistics then become

$$\text{statistic 1} = \frac{G_1}{\left(\widehat{\text{var}}\, G_1 \right)^{1/2}} \qquad \text{and} \qquad \text{statistic 2} = \frac{G_2}{\left(\widehat{\text{var}}\, G_2 \right)^{1/2}}$$

Although these statistics are asymptotically independent, it is better to construct an approximate joint test by asking whether either falls outside the interval $\pm \lambda_{\alpha/2}$, where

$$\text{Prob}(|X| \leqslant \lambda_{\alpha/2}) = 100(1 - \alpha) \qquad \text{with} \quad X \sim N(0, 1).$$

In an unpublished thesis, Webb [1972] has investigated the power of this test by simulation methods, using samples ranging in size from $n = 20$ to $n = 240$ and white noise series from normal, rectangular, exponential, and Cauchy distributions. The test was found to behave fairly well for the longer series but to be less satisfactory for the short series, when statistic 2 was not found to be normally distributed even when the data used were normal. When the test was applied to MA and AR series generated from the previous white noises, the null hypothesis of normality was rejected less frequently, except when Cauchy white noise was used. The reason for this is easily found since it can be shown that weighted sums of nonnormal variates become more nearly normal, in a relevant sense. This is connected with the well-known result that applying a low-band pass filter to a nonnormal series produces a series that is approximately normal; see Mallows [1967] for details. The implication of these results is that the normality test should be applied to prewhitened data, such as the residuals of a single-series Box–Jenkins model, rather than to the raw series.

9.6 Forecasting Several Steps Ahead from an Estimated Model

The forecasting theory developed in Chapters 4 and 7 was based on the assumption that the coefficients in a model were given. In practice, of course, these quantities must be estimated from sample data and coefficient estimates substituted into the corresponding formulas. To see the effect of this on forecasting more than one step ahead, consider the first-order autoregressive process

$$X_t = aX_{t-1} + \epsilon_t \tag{9.6.1}$$

Standing at time n, the optimal linear forecast of X_{n+h} is

$$f_{n,h} = a^h X_n \tag{9.6.2}$$

The usual procedure is to substitute $\hat{a}$ for a in (9.6.2), where $\hat{a}$ is the least-squares estimate of a in (9.6.1). Unfortunately, $\hat{a}^h$ is typically a biased estimator of a^h, for $h > 1$. For the case $h = 2$, an unbiased estimator is given by

$$\widehat{(a^2)} = \hat{a}^2 + \hat{\sigma}_{\hat{a}}^2 \tag{9.6.3}$$

where $\hat{\sigma}_{\hat{a}}^2$ is the estimated variance of $\hat{a}$. If it is assumed that $\hat{a}$ is normally distributed, similar modifications, based on the moments of the normal distribution, can be obtained for general h.

In practice, for moderately large samples, the correction factors involved in expressions like (9.6.3) will not be terribly large and, as noted by Box and Jenkins [1970], confidence intervals for forecasts will be but little affected.

A fairly general treatment of the problem of forecasting several steps ahead, with coefficient estimation and prediction treated simultaneously, is given in a Bayesian formulation by Chow [1973].

9.7 Other Forecasting Problems and Techniques

There are a number of forecasting situations in economics and proposed techniques that have not been explicitly covered in this book. Some of these fit well into the structure that has been presented, others do not. As an example of the former, the anticipations or expectations data collected by government agencies or public opinion firms are potentially important, but can be dealt with either within the multiseries information sets considered in Chapter 7 or within the combination of forecasts method discussed in Chapter 8. An example of an important problem that does not easily fit in with the methods presented earlier is the forecasting of turning points in the economy. Single series models are probably not very helpful in predicting turning points, but multiseries and econometric models could be. The most popular method currently employed involves leading indicators. It is tempting to try to evaluate possible indicators by building multiseries models involving them and then using the evaluation methods of Chapter 8. However, it could be argued that the leading indicators do not generally lead *except* at turning points. This involves a very complicated cost function to be used in the evaluation stage. A further technique that could be used is to consider turning points as a sample from a point process and to build single or multiple models to forecast such a process. However, methods for doing this are not yet well developed, to the best of our knowledge.

The methods discussed in this book are generally most suitable only for short or medium term forecasting. For longer term forecasting one has to turn to trend extrapolation and technological forecasting, using techniques such as Delphi, scenario writing, or the construction of global models. Some appropriate references are Martino [1972], Kahn and Wiener [1967], and Nordhaus [1973]. In many ways these techniques are less scientific than the shorter term forecasting methods, necessarily being based on less specific assumptions, and their usefulness is still being evaluated.

Some of the other forecasting techniques that have not been discussed range from the sublime to the ridiculous. They have been left out because it is thought that they are either too complicated and costly to use or because it is doubted that they are superior to the methods that have been presented, or both. It is nevertheless important that new methods continue to be developed, evaluated, and refined since it is by no means clear that the presently available methods extract all of the information available in the present about development in the future.

REFERENCES

ABDRABBO, N. A., and M. B. PRIESTLEY [1967], On the prediction of nonstationary processes, *J. Roy. Stat. Soc. B* **29**, 570–585.

ADAMS, F. G., and V. G. DUGGAL [1974], Anticipations variables in an econometric model: performance of the anticipations version of Wharton Mark III, *Int. Econ. Rev.* **15**, 267–284.

ADELMAN, I., and F. L. ADELMAN [1959], The dynamic properties of the Klein-Goldberger model, *Econometrica* **27**, 596–625.

ANDERSEN, L. C., and K. M. CARLSON [1974], St. Louis model revisited, *Int. Econ. Rev.* **15**, 305–327.

ANDERSON, R. L. [1942], Distribution of the serial correlation coefficient, *Ann. Math. Stat.* **13**, 1–13.

ANDERSON, T. W. [1971], "The Statistical Analysis of Time Series," New York: Wiley.

ANDERSON, T. W., and T. SAWA [1973], Distributions of estimates of coefficients of a single equation in a simultaneous system and their asymptotic expansions, *Econometrica* **41**, 683–714.

BACHELIER, L. [1900], Théorie de la spéculation, *Ann. Sci. Ecole. Norm. Sup., Paris, Ser. 3* **17**, 21–86.

BALL, R. J. (ed.) [1973], "The International Linkage of National Economic Models," Amsterdam: North Holland Publ. Co.

BARNARD, G. A. [1963], New methods of quality control, *J. Roy. Stat. Soc. A* **126**, 255–259.

BARNARD, G. A., G. M. JENKINS, and C. B. WINSTEN [1962], Likelihood inference and time series, *J. Roy. Stat. Soc., A*, **125**, 321–352.

BARRETT, J. F., and D. G. LAMPARD [1955], An expansion for some second order probability distributions and its application to noise problems, *I.R. E. Trans. PGIT*, **IT–1**, 10–15.

BARTLETT, M. S. [1946], On the theoretical specification of sampling properties of autocorrelated time series, *J. Roy. Stat. Soc. B* **8**, 27–41.

BATES, J. M., and C. W. J. GRANGER [1969], The combination of forecasts, *Oper. Res. Q.* **20**, 451–468.

BATTY, M. [1969], Monitoring an exponential smoothing forecasting system, *Oper. Res. Q.* **20**, 319–325.

BHANSALI, R. J. [1973], A monte-carlo comparison of the regression method and the spectral method of prediction, *J. Am. Stat. Assoc.* **68**, 621–625.

BHANSALI, R. J. [1974], Asymptotic properties of the Wiener-Kolmogorov predictor I, *J. Roy. Stat. Soc. B* **36**, 61–73.

BHATTACHARYYA, M. N., and A. P. ANDERSEN [1974], A post-sample diagnostic test for a time series model, Working paper, Dep. of Economics, Univ. of Queensland, Australia.

BOX, G. E. P., and D. R. COX [1964], An analysis of transformations, *J. Roy. Stat. Soc. B* **26**, 211–243.

BOX, G. E. P., and G. M. JENKINS [1962], Some statistical aspects of adaptive optimization and control, *J. Roy. Stat. Soc. B* **24**, 297–343.

BOX, G. E. P., and G. M. JENKINS [1970], "Time Series Analysis, Forecasting and Control," San Francisco: Holden Day.

BOX, G. E. P., and G. M. JENKINS [1973], Some comments on a paper by Chatfield and Prothero and on a review by Kendall, *J. Roy. Stat. Soc. A*, **136**, 337–345.

BOX, G. E. P., G. M. JENKINS, and D. W. BACON [1967], Models for forecasting seasonal and non-seasonal time series. *In* "Advanced Seminar on Spectral Analysis of Time Series" (B. Harris, ed.). New York: Wiley.

Box, G. E. P., and P. Newbold [1971], Some comments on a paper of Coen, Gomme, and Kendall, *J. Roy. Stat. Soc. A* **134**, 229–240.

Box, G. E. P., and D. A. Pierce [1970], Distribution of residual autocorrelations in autoregressive integrated moving average time series models, *J. Am. Stat. Assoc.* **65**, 1509–1526.

Brown, R. G. [1962], "Smoothing, Forecasting and Prediction of Discrete Time Series," Englewood Cliffs, New Jersey: Prentice Hall.

Brubacher, S. R., and G. T. Wilson [1975], Selecting data transformations for time series, Dept. of Mathematics, Univ. of Lancaster.

Chatfield, C., and D. L. Prothero [1973], Box–Jenkins seasonal forecasting: problems in a case study, *J. Roy. Stat. Soc. A* **136**, 295–336.

Chow, G. C. [1973], Multiperiod predictions from stochastic difference equations by Bayesian methods, *Econometrica* **41**, 109–118.

Christ, C. F. [1975], Judging the performance of econometric models of the U.S. economy, *Int. Econ. Rev.* **16**, 54–74.

Cleveland, W. S. [1972], The inverse autocorrelations of a time series and their applications, *Technometrics* **14**, 277–293.

Cogger, K. O. [1974], The optimality of general order exponential smoothing, *Oper. Res.* **22**, 858–867.

Cooper, J. P., and C. R. Nelson [1975], The ex ante prediction performance of the St. Louis and F.R.B.–M.I.T.–Penn. econometric models and some results on composite predictors, *J. Money, Credit, Banking* **7**, 1–32.

Cooper, R. L. [1972], The predictive performance of quarterly econometric models of the United States. *In* "Econometric Models of Cyclical Behavior" (B. G. Hickman, ed.). New York: Columbia Univ. Press.

Cramer, H. [1961], On some classes of non-stationary stochastic processes. *In* "Proceedings of 4th Berkeley Symposium on Mathematical Statistics and Probability: Vol. 2, Contributions to Probability Theory" (J. Neyman, ed.). Berkeley: Univ. of California Press.

Croxton, F. E., and D. J. Cowden [1955], "Applied General Statistics," 2nd ed. Englewood Cliffs, New Jersey: Prentice Hall.

Davies, N., M. B. Pate, and M. G. Frost [1974], Maximum autocorrelations for moving average processes, *Biometrika* **61**, 199–200.

Davis, H. T. [1941], "The Analysis of Economic Time Series," Bloomington, Indiana: Principia Press.

Dhrymes, P. J. [1970], "Econometrics," New York: Harper and Row.

Dhrymes, P. J. [1971], "Distributed Lags: Problems of Formulation and Estimation," San Francisco: Holden Day.

Dhrymes, P. J., et al. [1972], Criteria for evaluation of econometric models, *Ann. Econ. Soc. Meas.* **1**, 291–324.

Draper, N. R., and H. Smith [1966], "Applied Regression Analysis," New York: Wiley.

Durbin, J. [1960], The fitting of time series models, *Rev. Inst. Int. Stat.* **28**, 233–244.

Durbin, J. [1970], Testing for serial correlation in least squares regression when some of the regressors are lagged dependent variables, *Econometrica* **38**, 410–421.

Durbin, J., and G. S. Watson [1950], Testing for serial correlation in least squares regression I, *Biometrika* **37**, 409–428.

Durbin, J., and G. S. Watson [1951], Testing for serial correlation in least squares regression II, *Biometrika* **38**, 159–178.

Durbin, J., and G. S. Watson [1971], Testing for serial correlation in least squares regression III, *Biometrika* **58**, 1–19.

Eckstein, O., E. W. Green, and A. Sinai [1974], The Data Resources model: uses, structure and analysis of the U.S. economy, *Int. Econ. Rev.* **15**, 595–615.

Evans, M. K., Y. Haitovsky, and G. I. Treyz [1972], An analysis of the forecasting properties of U.S. econometric models. *In* "Econometric Models of Cyclical Behavior" (B. G. Hickman, ed.). New York: Columbia Univ. Press.

EWAN, W. D., and K. W. KEMP [1960], Sampling inspection of continuous processes with no autocorrelation between successive results, *Biometrika* **47**, 239–271.

FAIR, R. C. [1970],"A Short-Run Forecasting Model of the United States Economy," Lexington, Massachusetts: D. C. Heath.

FAIR, R. C. [1974], An evaluation of a short-run forecasting model, *Int. Econ. Rev.* **15**, 285–303.

FISHER, F. M. [1966], "The Identification Problem," New York: McGraw-Hill.

FISHMAN, G. S. [1969], "Spectral Methods in Econometrics," Cambridge, Massachusetts: Harvard Univ. Press.

FROMM, G., L. R. KLEIN, and G. R. SCHINK [1972], Short- and long-term simulations with the Brookings model. *In* "Econometric Models of Cyclical Behavior" (B. G. Hickman, ed.). New York: Columbia Univ. Press.

GILCHRIST, W. G. [1967], Methods of estimation involving discounting, *J. Roy. Stat. Soc. B* **29**, 355–369.

GODFREY, M. D. and H. KARREMAN [1967], A spectrum analysis of seasonal adjustment, *In* "Essays in Mathematical Economics in Honor of Oskar Morgenstern" (M. Shubik, ed.). New Jersey: Princeton Univ. Press.

GOLDBERGER, A. S. [1962], Best linear unbiased prediction in the generalized linear regression model, *J. Am. Stat. Assoc.* **57**, 369–375.

GOLDBERGER, A. S. [1964], "Econometric Theory," New York: Wiley.

GOLDFELD, S. M. [1972], Discussion of paper by R. L. Cooper. *In* "Econometric Models of Cyclical Behavior" (B. G. Hickman, ed.). New York: Columbia Univ. Press.

GRANGER, C. W. J. [1966], The typical spectral shape of an economic variable, *Econometrica* **34**, 150–161.

GRANGER, C. W. J. [1969a], Prediction with a generalized cost of error function, *Oper. Res. Q.* **20**, 199–207.

GRANGER, C. W. J. [1969b], Investigating casual relations by econometric models and cross-spectral methods, *Econometrica* **37**, 424–438.

GRANGER, C. W. J. [1975], Some consequences of the valuation model when expectations are taken to be optimum forecasts, *J. Finance* **30**, 135–145.

GRANGER, C. W. J., and M. HATANAKA [1964], "Spectral Analysis of Economic Time Series," New Jersey: Princeton Univ. Press.

GRANGER, C.W. J., and A. O. HUGHES [1968], Spectral analysis of short series — a simulation study, *J. Roy. Stat. Soc. A* **131**, 83–99.

GRANGER, C. W. J., and A. O. HUGHES [1971], A new look at some old data: the Beveridge wheat price series, *J. Roy. Stat. Soc. A* **134**, 413–428.

GRANGER, C. W. J., and O. MORGENSTERN [1970], "Predictability of Stock Market Prices." Lexington, Massachusetts: D. C. Heath.

GRANGER, C. W. J., and M. MORRIS [1976], Time series modeling and interpretation, *J. Roy. Stat. Soc. A* **38**, 246–257.

GRANGER, C. W. J., and P. NEWBOLD [1973], Some comments on the evaluation of economic forecasts, *Appl. Econ.* **5**, 35–47.

GRANGER, C. W. J., and P. NEWBOLD [1974], Spurious regressions in econometrics, *J. Econometrics* **2**, 111–120.

GRANGER, C. W. J., and P. NEWBOLD [1975], Economic forecasting: the atheist's viewpoint, *In* "Modelling the Economy" (G. A. Renton, ed.). London: Heinemann Educational Books.

GRANGER, C. W. J., and P. NEWBOLD [1976], Forecasting transformed series, *J. Roy. Stat. Soc. B* **38**, 189–203.

GRANGER, C. W. J., and D. ORR [1972], "Infinite variance" and research strategy in time series analysis, *J. Am. Stat. Assoc.* **67**, 275–285.

GREEN, G. R., M. LIEBENBERG, and A. A. HIRSCH [1972a], Short- and long-term simulations with the O.B.E. econometric model, *In* "Econometric Models of Cyclical Behavior" (B. G. Hickman, ed.). New York: Columbia Univ. Press.

GREEN, G. R., M. LIEBENBERG, and A. A. HIRSCH [1972b], Comment on paper by R. L. Cooper,

In "Econometric Models of Cyclical Behavior" (B. G. Hickman, ed.). New York: Columbia Univ. Press.

GRENANDER, U., and M. ROSENBLATT [1957], "Statistical Analysis of Stationary Time Series," New York: Wiley.

HAITOVSKY, Y., and G. I. TREYZ [1972], Forecasts with quarterly macro-econometric models, equation adjustments and benchmark predictions: the U.S. experience, *Rev. Econ. Stat.* **54**, 317–325.

HANNAN, E. J. [1960], "Time Series Analysis," London: Methuen.

HANNAN, E. J. [1969], The identification of vector mixed autoregressive moving average systems, *Biometrika* **56**, 223–225.

HANNAN, E. J. [1970], "Multiple Time Series," New York: Wiley.

HANNAN, E. J. [1971], The identification problem for multiple equation systems with moving average errors, *Econometrica* **39**, 751–765.

HARRISON, P. J. [1965], Short-term sales forecasting, *Appl. Stat.*, **14**, 102–139.

HARRISON, P. J. [1967], Exponential smoothing and short-term sales forecasting, *Manage. Sci.* **13**, 821–842.

HARRISON, P. J., and O. L. DAVIES [1964], The use of cumulative sum (Cusum) techniques for the control of routine forecasts of product demand, *Oper. Res.* **12**, 325–333.

HARRISON, P. J., and C. F. STEVENS [1971], A Bayesian approach to short term forecasting, *Oper. Res. Q.* **22**, 341–362.

HART, B. L. [1942], Significance levels for the ratio of the mean square successive difference to the variance, *Ann. Math. Stat.* **13**, 445–447.

HATANAKA, M. [1974a], A simple suggestion to improve the Mincer-Zarnowitz criterion for the evaluation of forecasts, *Ann. Econ. Soc. Meas.* **3**, 521–524.

HATANAKA, M. [1974b], On the identification and estimation of the dynamic simultaneous equations model with autoregressive disturbances, Discussion Paper 83, Institute of Social and Economic Research, Osaka Univ., Japan.

HAUGH, L. D. [1972], "The identification of time series interrelationships with special reference to dynamic regression," Ph.D. Thesis, Dept. of Statistics, Univ. of Wisconsin, Madison.

HAUGH, L. D., and G. E. P. BOX [1974], Identification of dynamic regression (distributed lag) models connecting two time series, Technical Report 370, Dept. of Statistics, Univ. of Wisconsin, Madison.

HENDRY, D. F. [1974], Stochastic specification in an aggregate demand model of the United Kingdom, *Econometrica* **42**, 559–578.

HICKMAN, B. G. (ed.) [1972], "Econometric Models of Cyclical Behavior," New York: Columbia Univ. Press.

HICKMAN, B. G. [1975], Project LINK in 1972: retrospect and prospect, *In* "Modelling the Economy" (G. A. Renton, ed.). London: Heinemann Educational Books.

HIRSCH, A. A. [1973], The B.E.A. quarterly model as a forecasting instrument, *Survey of Current Business* **53** (August), 24–38.

HIRSCH, A. A., B. T. GRIMM, and G. V. L. NARASIMHAM [1974], Some multiplier and error characteristics of the B.E.A. quarterly model, *Int. Econ. Rev.* **16**, 616–631.

HOLT, C. C. [1957], "Forecasting seasonals and trends by exponentially weighted moving averages," Carnegie Institute of Technology, Pittsburgh, Pennsylvania.

HOUTHAKKER, H. S. and L. D. TAYLOR [1966], "Consumer Demand in the United States," Cambridge: Harvard Univ. Press.

HOWREY, E. P. [1968], A spectrum analysis of the long-swing hypothesis, *Int. Econ. Rev.* **9**, 228–252.

HOWREY, E. P., L. R. KLEIN, and M. D. MCCARTHY [1974], Notes on testing the predictive performance of econometric models, *Int. Econ. Rev.* **15**, 366–383.

JANACEK, G. [1975], Estimation of the minimum mean squared error of prediction, *Biometrika* **62**, 175–180.

JENKINS, G. M. [1974], Personal communication.

JENKINS, G. M., and D. G. WATTS [1968], "Spectral Analysis and its Applications," San Francisco: Holden Day.

JOHNSTON, J. [1972], "Econometric Methods," 2nd ed. New York: McGraw-Hill.

KAHN, H., and A. J. WIENER [1967], "The Year 2000," New York: MacMillan.

KALMAN, R. E. [1960], A new approach to linear filtering and prediction problems, *J. Basic Eng.* **82**, 35–45.

KALMAN, R. E. [1963], New methods in Wiener filtering theory, *In* "Proceeding of First Symposium on Engineering Application of Random Function Theory and Probability" (J. L. Bogdanoff and F. Kozin, eds.). New York: Wiley.

KENDALL, M. G. [1954], "Exercises in Theoretical Statistics," London: Griffin.

KENDALL, M. G. [1973], "Time Series Analysis," London:Griffin.

KENDALL, M. G., and A. STUART [1963], "The Advanced Theory of Statistics," Vol. I, London: Griffin.

KLEIN, L. R. [1960], The efficiency of estimation in econometric models, Cowles Foundation Paper 157, Yale Univ., New Haven, Connecticut.

KLEIN, L. R. [1971a], "An Essay on the Theory of Economic Prediction," Chicago: Markham.

KLEIN, L. R. [1971b], Forecasting and policy evaluation using large scale econometric models: the state of the art, *In* "Frontiers of Quantitative Economics" (M. D. Intriligator, ed.). Amsterdam: North Holland Publ. Co.

KLEIN, L. R., and A. S. GOLDBERGER [1955], "An Econometric Model of the United States 1929–1952," Amsterdam: North Holland Publ. Co.

KOLMOGOROV, A. [1939], Sur l'interpolation et l'extrapolation des suites stationnaires, *C. R. Acad. Sci. Paris* **208**, 2043–2045.

KOLMOGOROV, A. [1941a], Interpolation und extrapolation von stationären zufälligen folgen, *Bull. Acad. Sci. (Nauk) U.R.S.S. Ser. Math.* **5**, 3–14.

KOLMOGOROV, A. [1941b], Stationary sequences in Hilbert space (in Russian), *Bull. Math. Univ. Moscow* **2** (6), 1–40.

KOOPMANS, L. H. [1974], "The Spectral Analysis of Time Series," New York: Academic Press.

LABYS, W. C., and C. W. J. GRANGER [1970], "Speculation, Hedging and Forecasts of Commodity Prices," Lexington, Massachusetts: D. C. Heath.

LEHMANN, E. L. [1959], "Testing Statistical Hypotheses," New York: Wiley.

L'ESPERANCE, W. D., and D. TAYLOR [1975], The power of four tests of autocorrelation in the linear regression model, *J. Econometrics* **3**, 1–21.

LEWIS, P. A W. (ed.) [1972], "Stochastic Point Processes," New York: Wiley.

LOMNICKI, Z. A. [1961], Tests for departure from normality in the case of linear stochastic processes, *Metrika* **4**, 37–62.

MALINVAUD, E. [1966], "Statistical Methods of Econometrics," Amsterdam: North Holland Publ. Co.

MALLOWS, C. L. [1967], Linear processes are nearly Gaussian, *J. Appl. Prob.* **4**, 313–329.

MANN, H. B., and A. WALD [1943], On the statistical treatment of linear stochastic difference equations, *Econometrica* **11**, 173–220.

MARQUARDT, D. W. [1963], An algorithm for least squares estimation of nonlinear parameters, *J. Soc. Ind. Appl. Math.* **2**, 431–441.

MARTINO, J. P. [1972], "Technological Forecasting for Decision Making," New York: American Elsevier.

MCCARTHY, M. D. [1972], Discussion of paper by R. L. Cooper, *In* "Econometric Models of Cyclical Behavior" (B. G. Hickman, ed.). New York: Columbia Univ. Press.

MEISELMAN, D. [1962], "The Term Structure of Interest Rates," Englewood Cliffs, New Jersey: Prentice Hall.

MINCER, J., and V. ZARNOWITZ [1969], The evaluation of economic forecasts, *In* "Economic Forecasts and Expectations" (J. Mincer, ed.). New York: National Bureau of Economic Research.

MUTH, J. F. [1960], Optimal properties of exponentially weighted forecasts, *J. Am. Stat. Assoc.* **55**, 299–306.

NAYLOR, T. H., T. G. SEAKS, and D. W. WICHERN [1972], Box–Jenkins methods: an alternative to econometric models, *Rev. Inst. Int. Stat.* **40**, 123–137.

NEAVE, H. R. [1970], An improved formula for the asymptotic variance of spectrum estimates, *Ann. Math. Stat.* **41**, 70–77.

NEAVE, H. R. [1972], Observations on "Spectral analysis of short series—a simulation study" by Granger and Hughes, *J. Roy. Stat. Soc. A* **135**, 393–405.

NELSON, C. R. [1972a], "The Term Structure of Interest Rates," New York: Basic Books.

NELSON, C. R. [1972b], The prediction performance of the F.R.B. – M.I.T. – PENN model of the U.S. economy, *Am. Econ. Rev.* **62**, 902–917.

NELSON, C. R. [1973], "Applied Time Series Analysis for Managerial Forecasting," San Francisco: Holden Day.

NELSON, J. Z., and J. W. VAN NESS [1973a], Formulation of a nonlinear predictor, *Technometrics* **15**, 1–12.

NELSON, J. Z., and J. W. VAN NESS [1973b], Choosing a nonlinear predictor, *Technometrics* **15**, 219–231.

NERLOVE, M. [1964], Spectral analysis of seasonal adjustment procedures, *Econometrica* **32**, 241–286.

NERLOVE, M., and S. WAGE [1964], On the optimality of adaptive forecasting, *Manage. Sci.* **10**, 207–229.

NERLOVE, M., and K. F. WALLIS [1966], The use of the Durbin–Watson statistic in inappropriate situations, *Econometrica* **34**, 235–238.

NEWBOLD, P. [1973a], Forecasting Methods, Civil Service College Occasional Papers 18, London: H.M.S.O.

NEWBOLD, P. [1973b], Bayesian estimation of Box–Jenkins transfer function–noise models, *J. Roy. Stat. Soc. B* **35**, 323–336.

NEWBOLD, P. [1974], The exact likelihood function for a mixed autoregressive–moving average process, *Biometrika* **61**, 423–426.

NEWBOLD, P., and N. DAVIES [1975], Error mis-specification and spurious regressions, Dept. of Mathematics, Univ. of Nottingham.

NEWBOLD, P., and C. W. J. GRANGER, [1974], Experience with forecasting univariate time series and the combination of forecasts, *J. Roy. Stat. Soc. A* **137**, 131–146.

NORDHAUS, W. D. [1973], World dynamics: measurement without data, *Econ. J.* **83**, 1156–1183.

PARZEN, E. [1961], Mathematical considerations in the estimation of spectra, *Technometrics* **3**, 167–190.

PARZEN, E. [1967], "Time Series Analysis Papers," San Francisco: Holden Day.

PAYNE, D. J. [1973], "The determination of regression relationships using stepwise regression techniques," Ph.D. Thesis, Dept. of Mathematics, Univ. of Nottingham.

PIERCE, D. A. [1974], Relationships—and the lack thereof – between economic time series, with special reference to money, reserves and interest rates, Special Studies Paper 55, Federal Reserve Board, Washington, D. C.

PIERCE, D. A., and L. D. HAUGH [1975], The assessment and detection of causality in temporal systems, Technical Report 83, Dept. of Statistics, Univ. of Florida, Gainesville.

PRESTON, A. J., and K. D. WALL [1973], An extended identification problem for state space representations of econometric models, Discussion Paper 6, Programme of Research into Econometric Methods, Dept. of Economics, Queen Mary College, London.

PRIESTLEY, M. B. [1965] Evolutionary spectra and non-stationary processes, *J. Roy. Stat. Soc. B* **27**, 204–237.

QUENOUILLE, M. H. [1949], Approximate tests of correlation in time series, *J. Roy. Stat. Soc. B* **11**, 68–84.

QUENOUILLE, M. H. [1957], "The Analysis of Multiple Time Series," London: Griffin.

REID, D. J. [1969], "A comparative study of time series prediction techniques on economic data," Ph.D. Thesis, Dept. of Mathematics, Univ. of Nottingham.

RENTON, G. A. (ed.) [1975], "Modelling the Economy," London: Heinemann Educational Books.

RICHARDSON, D. H. [1968], The exact distribution of a structural coefficient, *J. Am. Stat. Assoc.* **63**, 1214–1226.

ROSENBLATT, M. [1957], Some purely deterministic processes, *J. Math. Mech.* **6**, 801–810.

SAWA, T. [1969], The exact sampling distribution of ordinary least squares and two-stage least squares estimates, *J. Am. Stat. Assoc.* **64**, 923–937.

SCHUSTER, A. [1898], On the investigation of hidden periodicities, *Terr. Magn. Atmos. Electr.* **3**, 13–41.

SHEPPARD, D. K. [1971], "The Growth and Role of U.K. Financial Institutions 1880–1962," London: Methuen.

SIMS, C. A. [1972], Money, income and causality, *Am. Econ. Rev.* **62**, 540–552.

SUITS, D. B. [1962], Forecasting and analysis with an econometric model, *Am. Econ. Rev.* **52**, 104–132.

SUPPES, P. [1970], "A Probabilistic Theory of Causality," Amsterdam: North Holland Publ. Co.

THEIL, H. [1958], "Economic Forecasts and Policy," Amsterdam: North Holland Publ. Co.

THEIL, H. [1966], "Applied Economic Forecasting," Amsterdam: North Holland Publ. Co.

THEIL, H., and S. WAGE [1964], Some observations on adaptive forecasting, *Manage. Sci.* **10**, 198–206.

TINBERGEN, J. [1939], "Statistical Testing of Business Cycle Theories," Vol. II, Business Cycles in the United States of America 1919–1932. Geneva: League of Nations.

TRIGG, D. W. [1964], Monitoring a forecasting system, *Oper. Res. Q.* **15**, 271–274.

TRIGG, D. W., and A. D. LEACH [1967], Exponential smoothing with an adaptive response rate, *Oper. Res. Q.* **18**, 53–59.

VAN HORNE, J. C. [1971], "Financial Management and Policy," 2nd ed. Englewood Cliffs, New Jersey: Prentice Hall.

WAGLE, B., J. Q. G. H. RAPPOPORT, and V. A. DOWNES [1968], A program for short-term sales forecasting, *The Statistician* **18**, 141–147.

WALKER, G. [1931], On periodicity in series of related terms, *Proc. Roy. Soc. London A* **131**, 518–532.

WALL, K. D., A. J. PRESTON, J. W. BRAY, and M. H. PESTON [1975], Estimates of a simple control model of the U.K. economy, *In* "Modelling the Economy" (G. A. Renton, ed.). London: Heinemann Educational Books.

WALLIS, K. F. [1972], Testing for fourth order autocorrelation in quarterly regression equations, *Econometrica* **40**, 617–636.

WEBB, R. A. J. [1972], A simulation study of Lomnicki's test for departure from Normality in the case of linear stochastic processes, M.Sc. Thesis, Dept. of Mathematics, Univ. of Nottingham.

WHITTLE, P. [1963], "Prediction and Regulation," London: English Universities Press.

WHITTLE, P. [1965], Recursive relations for predictors of non-stationary processes, *J. Roy. Stat. Soc. B* **27**, 523–532.

WICHERN, D. W. [1973], The behavior of the sample autocorrelation function for an integrated moving average process, *Biometrika* **60**, 235–239.

WIENER, N. [1949], "The Extrapolation, Interpolation and Smoothing of Stationary Time Series," Boston: M.I.T. Press.

WILKS, S. S. [1962], "Mathematical Statistics," New York: Wiley.

WILSON, G. T. [1969], Factorization of the generating function of a pure moving average process, *SIAM J. Num. Anal.* **6**, 1–7.

WOLD, H. [1954a], "A Study in the Analysis of Stationary Time Series," 2nd ed. Uppsala: Almquist and Wicksell.

WOLD, H. [1954b], Causality and econometrics, *Econometrica* **22**, 162–177.

WOLD, H. (ed.) [1964], "Econometric Model Building: Essays on the Causal Chain Approach," Amsterdam: North Holland Publ. Co.

WINTERS, P. R. [1960], Forecasting sales by exponentially weighted moving averages, *Manage. Sci.* **6**, 324–342.

YULE, G. U. [1926], Why do we sometimes get nonsense correlations between time series?–a study in sampling and the nature of time series, *J. Roy. Stat. Soc.* **89**, 1–64.

YULE, G. U. [1927], On a method for investigating periodicities in disturbed series with special reference to Wolfer's sunspot numbers, *Phil. Trans. Roy. Soc. London A* **226**, 267–298.

ZELLNER, A., and F. PALM [1974], Time series analysis and simultaneous equation econometric models, *J. Econometrics* **2**, 17–54.

AUTHOR INDEX

SUBJECT INDEX

A

Accumulated series, 2
Aggregation, 30
Airline passenger miles, 269
Aitken's generalized least squares estimator, 189
Aliasing, 55
Amplitude, 10
Anticipations variables, 202
ARIMA process, 41, 64, 71
ARMA model-spectrum of, 54
ARMA, optimum forecast, 124
ARMA process, 25, 64, 150
Autocorrelated errors, 199
Autocorrelation function, 26, 72
Autocorrelation generating functions, 82
Autocorrelations, 5
Autocovariance generating function, 7, 19, 22, 24, 43
Autocovariances, 5, 13, 73
Automobile demand, 197
Autoregressive process, 13
Average lag, 314

B

Backward ϵ-memory, 115
Backward-looking filter, 31
Backward operator, 5, 6
Bayesian formulation, 185, 201, 316
Bayes' theorem, 175, 198, 293
Bivariate model building, 232, 244
Bivariate series, 230
Bivariate series, forecasting, 262

Black box, 32
Box–Cox transformation, 304
Box–Jenkins forecasts, 149, 184, 272, 273, 283
Box–Jenkins model, 269, 295, 315
Box–Jenkins procedure, 162, 176, 268
Box–Pierce test, 93, 106
Brown's generalized exponential smoothing, 176, 180, 183
Bureau of Economic Analysis model, 298
Bureau of Labor Statistics, 66

C

Causality, 224
Causality, definition, 225
Coefficient of multiple correlation, 187
Coherence function, 57
Coincidental situations, 30
Combination of forecasts, 179, 269, 270, 316
Commodity market, 23
Comparison of forecasts, 179
Conditional efficiency, 283
Conditional estimation, 84
Conditional forecasting, 189, 197
Conditional variables, 112
Confidence intervals, 201
Construction begun in England and Wales, 102, 105, 154, 158, 160
Consumption, 197
Consumption–investment equations, 262
Continuous time series, 2
Convolution, 7, 22
Corrected coefficient of multiple correlation, 188
Correlogram, 5, 42, 66
Cospectrum, 57

ECONOMIC THEORY, ECONOMETRICS, AND MATHEMATICAL ECONOMICS

Consulting Editor: Karl Shell

UNIVERSITY OF PENNSYLVANIA
PHILADELPHIA, PENNSYLVANIA

Edmund S. Phelps. Studies in Macroeconomic Theory, Volume 1: *Employ-ment and Inflation.* Volume 2: *Redistribution and Growth*

Marc Nerlove, David M. Grether, and José L. Carvalho. Analysis of Economic Time Series: *A Synthesis*

Thomas J. Sargent. Macroeconomic Theory

Jerry Green and José Alexander Scheinkman (Eds.). General Equilibrium, Growth and Trade: *Essays in Honor of Lionel McKenzie*

Michael J. Boskin (Ed.). Economics and Human Welfare: *Essays in Honor of Tibor Scitovsky*

Carlos Daganzo. Multinomial Probit: *The Theory and Its Application to Demand Forecasting*

L. R. Klein, M. Nerlove, and S. C. Tsiang (Eds.). Quantitative Economics and Development: *Essays in Memory of Ta-Chung Liu*

Giorgio P. Szegö. Portfolio Theory: *With Application to Bank Asset Management*

M June Flanders and Assaf Razin (Eds.). Development in an Inflationary World

Thomas G. Cowing and Rodney E. Stevenson (Eds.). Productivity Measurement in Regulated Industries

Robert J. Barro (Ed.). Money, Expectations, and Business Cycles: *Essays in Macroeconomics*

Ryuzo Sato. Theory of Technical Change and Economic Invariance: *Application of Lie Groups*

Iosif A. Krass and Shawkat M. Hammoudeh. The Theory of Positional Games: *With Applications in Economics*

Giorgio Szegö (Ed.). New Quantitative Techniques for Economic Analysis

John M. Letiche (Ed.). International Economic Policies and Their Theoretical Foundations: A Source Book

Murray C. Kemp (Ed.). Production Sets

Andreu Mas-Colell (Ed.). Noncooperative Approaches to the Theory of Perfect Competition

Jean-Pascal Benassy. The Economics of Market Disequilibrium

Tatsuro Ichiishi. Game Theory for Economic Analysis

David P. Baron. The Export-Import Bank: *An Economic Analysis*

Réal P. Lavergne. The Political Economy of U. S. Tariffs: *An Empirical Analysis*

Halbert White. Asymptotic Theory for Econometricians

In preparation

Thomas G. Cowing and Daniel L. McFadden. **Microeconomic Modeling and Policy Analysis:** *Studies in Residential Energy Demand*